Pelican Books
Political Leaders of the Twentieth Century

Mao Tse-tung

Stuart R. Schram was born in Minnesota, U.S.A., in 1924.
He took his B.A. at the University of Minnesota in 1944,
followed by a Ph.D. in political science at Columbia
University in 1954. From 1954 to 1967 he carried out research
at the Centre d'Étude des Relations Internationales of the
Fondation Nationale des Sciences Politiques in Paris, where,
with Hélène Carrère d'Encausse, he was responsible for the
Soviet and Chinese Section. He is now Professor of Politics (with
reference to China) in the University of London and Head of the
Contemporary China Institute of the School of Oriental and African
Studies. In his research he is chiefly interested in the role of ideology
in politics, especially in Communist countries, and in the
history of Leninist theories together with their application by
the Communist movement in Asia.

Professor Schram has also published *Le Marxisme et l'Asie, 1853–
1964* (with Hélène Carrère d'Encausse), and
The Political Thought of Mao Tse-tung.

Apart from his research, Professor Schram is interested in music,
poetry, the theatre and nature.

Political Leaders of the
Twentieth Century

Mao Tse-tung

Stuart Schram

With 29 plates

Penguin Books

Penguin Books Ltd, Harmondsworth, Middlesex, England
Penguin Books Inc., 7110 Ambassador Road,
Baltimore, Maryland 21207, U.S.A.
Penguin Books Australia Ltd, Ringwood, Victoria, Australia

First published 1966
Reprinted (with revisions) 1967
Reprinted 1968, 1969, 1970

Made and printed in Great Britain by
Hazell Watson & Viney Ltd
Aylesbury, Bucks
Set in Monotype Times

Contents

List of Plates

Acknowledgements

Every book of history or biography owes something to the criticisms and suggestions of persons who have made a special study of some aspect of the period treated, but by the nature of the subject this one owes more to the kindness of my colleagues than most. Much has been done since the publication of such pioneering works as Brandt, Schwartz, and Fairbank's *A Documentary History of Chinese Communism* and Benjamin Schwartz's *Chinese Communism and the Rise of Mao* a decade and a half ago, but much remains to be done. Several important monographs, abundantly utilized in the present work, have appeared in recent years, but an even larger number are in preparation or in the press. The authors of several of these have been kind enough to comment on my text, to direct my attention to important materials, and in some cases even to allow me to read portions of their own manuscripts. I should like to thank especially Walter Gourlay and Roy Hofheinz, now engaged in finishing monographic studies at Harvard University (the first on the Kuomintang in 1921–7, the second on Chinese Communist rural politics in 1927–45); Donald L. Klein – currently working, at the Harvard East Asian Research Center, on a biographical dictionary of the Chinese Communist leaders – who not only pointed out a number of minor errors, but supplied me with the substance of most of the capsule biographies of figures secondary to Mao's own story which appear in this book; Sidney Liu, a specialist on the military history of Chinese Communism, who commented at length on the passages dealing with the Kiangsi Republic; and Ezra Vogel, likewise of the Harvard East Asian Research Center, who made a number of useful observations, especially regarding the portion of the book dealing with the period since 1949. My colleagues in Paris, Claude Cadart and Jean Chesneaux, also made suggestions and provided information. Finally, I am most grateful to Professor Pai Yü (in the usual transcription, or Yu Beh, as he prefers to write it himself) of National Chengchi University in Taipei, who has allowed me to make liberal use of his reminiscences of encounters with Mao from the time when they were classmates in Changsha in 1916–18 to the years of Kuomintang-Communist collaboration in 1925–6.

Thanks are due to Edgar Snow and to his publisher, Victor Gollancz, for permission to utilize a number of brief quotations from his irreplaceable work *Red Star Over China*. Mao's own story as told to Mr Snow in 1936 remains a fundamental source for his life down to that point. Acknowledgement is made to Stanford University Press for permission to reproduce a quotation from *Soviet Russia and the East*, by X. J. Eudin and R. C. North, and to the University of Washington Press for permission to use a passage from *Mao's China*, edited by Boyd Compton.

I should also like to record here my debt to Jerome Ch'ên's volume *Mao and the Chinese Revolution* (Oxford University Press, 1965), which was the first serious and properly documented biography of Mao. I disagree with Mr Ch'ên about a considerable number of points of fact and interpretation. (With a few exceptions, it has seemed to me to be superfluous to point these out; the reader who is sufficiently interested in the subject will easily discover them.) But there is no doubt that if I had not been able to compare my own versions of events with his, and benefit from his extensive knowledge of certain aspects of the Chinese background of Mao's life, the present book would have suffered even more from the lack of monographic studies than is actually the case.

A word is perhaps in order regarding two aspects of the choice of sources for citing Mao's works. The current official canon has been, as I shall frequently have occasion to point out in these pages, so extensively rewritten that it is a very unreliable guide to what Mao actually wrote or thought at any given moment of his life. It is for this reason that, whenever the precise wording or nuance of the text is important, and whenever the passage in question is included in my study *The Political Thought of Mao Tse-tung*, I have cited from the versions given there, which are based insofar as possible on the contemporary Chinese texts. Secondly, for the numerous writings not included in the above anthology, I have cited systematically the English translation published in Peking (4 volumes, Foreign Languages Press, 1960–

65), and not that in 5 volumes published earlier in London by Lawrence and Wishart and in New York by International Publishers. The latter edition has now been repudiated by the Chinese, and it therefore seems preferable to cite the authorized translation. Volumes II and III of the Peking edition became available only when this book was in the galley stage. It seemed hardly necessary to complicate the printer's task by replacing the text of the London translation by that of the Peking version in the body of the work. I have therefore limited myself to changing the references in the notes. The reader who wishes to turn to the *Selected Works* may consequently find there a translation which is not altogether identical with the one I have cited, but he will have no difficulty in locating the passage in question.

The past year has seen the publication of a number of translations of poems by Mao Tse-tung: Jerome Ch'ên's version of the thirty-seven now extant, in his biography already mentioned; a translation and analysis of ten poems, by Professor Paul Demiéville (*Mercure de France*, April 1965); an article by C. N. Tay entitled 'From *Snow* to *Plum Blossoms*' (*Journal of Asian Studies*, February 1966); and the official translation of the ten poems published in January 1964 (*Chinese Literature*, No. 5, 1966). These have incited me to re-examine some of the versions which I had previously made or adapted, and to introduce a certain number of changes. I should like to thank Professor Donald Holzman of Paris for his kindness in aiding me in this effort with his profound knowledge of the Chinese language and of Chinese literature.

Thanks to those, mentioned above, who have so kindly aided me or whose published work I have exploited, I hope that this book is reasonably accurate. I have tried to write without bias, if not without convictions. For the errors which undoubtedly remain, and for such interpretations and hypotheses as may be found unconvincing, I must accept the sole responsibility.

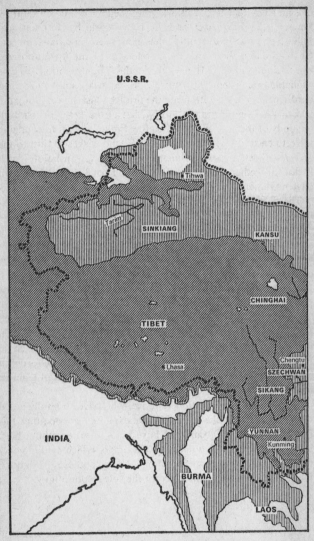

China

Today

Introduction

On the evening of 22 October 1911 Mao Tse-tung, then not quite eighteen years old, stood on a hill in Changsha, the capital of his native Hunan province. He had lived through a day of feverish excitement. The anti-Manchu revolution begun in Wuhan twelve days earlier had broken out in Changsha that morning, and in a few hours had swept away the old imperial administration. Now, as Mao watched, a forest of white banners sprung up about the city, bearing the inscription: '*Ta-Han Min-kuo Wan Sui!*' 'Long Live the Great Han Republic!'

In a sense, these words epitomize both the situation of China at that moment, and the direction in which, over a period of half a century, Mao Tse-tung has striven to transform his country. *Min-kuo*, the term used then to translate the Western concept 'republic', means literally 'people's country'. This was at the time a very revolutionary idea indeed. According to the traditional Confucian outlook, the sole political function of the people was to obey their legitimate ruler. Only in the last decade of the nineteenth century had voices been raised to affirm that the conscious participation of every citizen in the life of society could release the energies necessary to make China rich and powerful once more. This notion had been developed first by Yen Fu, the pioneer in the introduction of Western ideas, whose translations of Mill, Spencer, and Montesquieu Mao was soon to devour avidly. It had been taken up by the reformers of 1898, and then, in more radical form, by Sun Yat-sen and his revolutionary comrades of the T'ung Meng Hui, who had already made a series of abortive attempts at a rising to overthrow the alien Manchu dynasty and establish a republic. Now at last the 'people's country' was being proclaimed.

But this 'people's country' was to be at the same time the 'Great Han People's Country'. Just *how* great, His Britannic Majesty's Consul in Changsha discovered a few days later, when he found himself obliged to return to the revolutionary authorities in the city a document announcing the new régime, in which the characters for 'Great Han Republic' had been elevated two spaces higher on the page than those for 'Great English Kingdom'.

Throughout his political career, Mao Tse-tung has taken as his goal the radical transformation of Chinese society in order to liberate the energies of its citizens, and he has been guided in this enterprise by ideas of Western origin, first nineteenth-century liberalism and then Leninist collectivism. But at the same time he has been determined that this 'new China', this 'people's China', should resume her rightful place among the nations – the first place.

The high value which the Chinese place on their own culture has been a subject of exasperation and puzzlement to Europeans ever since they began their efforts, a century or so ago, to 'civilize' the Middle Kingdom by a judicious combination of gunboats, trade, and missionaries. The simple fact that China has long been regarded by its inhabitants as the 'central country' (to translate *Chung-kuo* into language slightly more modern than 'Middle Kingdom') is not in itself surprising. Most great and powerful nations have been inclined, at least in the days of their ascendancy, to view themselves as the centre of the world. More significant, and more distinctive, is the traditional Chinese attitude according to which theirs has been the only genuine civilization, and the Chinese empire co-extensive with the civilized world.

These conceptions were the natural product of a three-thousand-year history during which China as we know it today had been sometimes unified and sometimes divided, occasionally conquered from the outside, but almost never brought into contact with a civilization equally advanced, still less superior. The Mongol conquerors of the thirteenth century, like the Manchu conquerors of the seventeenth, had soon adopted Chinese ways. The only occasion, prior to the Western impact, on which China had been seriously influenced by ideas coming from abroad lay more than a thousand years in the past, when Buddhism had been imported from India. And though this alien faith survived as the religion of a minority and exercised a certain influence on the development of Confucian thought itself, it was soon engulfed and profoundly modified by the indigenous tradition.

Apart from the fact that it had developed largely in isolation, Chinese tradition possessed two factors which set it radically apart from the traditions of South and South-East Asia: a sense of history, and a concern for politics as one of the essential dimensions of human activity. To be sure, the traditional Chinese view of history envisaged a cyclical process of dynastic construction and decay, not a unilinear advance toward progress and/or redemption, while the Confucian view of politics was, as already suggested, hierarchical and largely undemocratic. China's encounter with the modern world was therefore fated to shake the tradition to its foundations. But at the same time, by directing the attention of the Chinese to the historical and political dimensions of existence, this tradition prepared them exceptionally well to survive in the modern world. And that same sense of history and of political destiny was to make it both easy and inevitable for Mao and his colleagues to reaffirm the continuity of today's China with that of the empire of the Hans, despite the transformations occasioned by one of the most radical social upheavals in world history.

But if today's China, as it emerges from half a century of revolution which Mao Tse-tung himself has materially shaped both as theorist and as man of action, appears to be moving toward a certain fusion of Communist and traditional elements, this tendency toward synthesis follows a long period in which Chinese and Western patterns and ideas struggled with one another in utter confusion. By the end of the nineteenth century, it had become obvious that China must learn from the Europeans in order to acquire the strength to resist their incursions; but at the same time the feeling of humiliation occasioned by this necessity was in direct proportion to the massive sentiment of superiority which the Chinese had always entertained towards the barbarians. It is into this world of anguish over China's future, and uncertainty as to what could or should be salvaged of her past, that Mao Tse-tung was born.

Chapter 1
The World of Mao's Boyhood

Mao Tse-tung was born on 26 December 1893 in the village of Shaoshan, Hsiangt'an *hsien*,* Hunan province. His father had begun life as a poor peasant, but later raised himself to a somewhat more comfortable status. Mao himself affirmed in his autobiography, as told to Edgar Snow in 1936, that when he was ten his father had already become a middle peasant, and further supplemented his income by small trading. Later still, according to Mao, his father became a rich peasant and grain merchant. He defined these categories by indicating that as middle peasants his family owned 15 *mou*† of land, on which they raised 60 piculs‡ of rice a year; as rich peasants they owned 22 *mou* on which they raised 84 piculs.§

While the poverty of the Chinese countryside was such that a very modest degree of well-being might suffice to qualify a man as a 'middle' or even 'rich' peasant, Mao may have exaggerated slightly the status achieved by his family. Perhaps he was moved by some lingering filial piety towards a father with whom he had often been in violent conflict; perhaps, too, he may have wished, in 1936, to convince the sons of rich peasants, numerous among

---

* Territorial division (sometimes translated 'county') which formerly constituted the lowest level in the imperial administration, beneath which the gentry were left to deal with matters themselves. The term and the geographic unit survived in both Republican and Communist China.

† Traditional Chinese unit of area, about 0·15 acre.

‡ One picul or *tan* = 133⅓ lb.

§ Edgar Snow, *Red Star Over China* (Gollancz, 1938), pp. 126–7. Despite many errors of detail, some of which I shall have occasion to point out in the course of my book, Mao's own story as told to Mr Snow remains by far the most important single source regarding his life. Even when it is not altogether accurate, it is extremely interesting as the expression of Mao's own vision of his past. In order to avoid the unnecessary multiplication of footnotes, I shall henceforth give a precise reference to this work only when I wish to compare it with another source, or when I cite a particularly long or significant passage; but it should be clearly understood that, unless otherwise stated, all references to what Mao himself has said regarding his own life down to 1936 are from *Red Star Over China*.

the students, that their class origin was not necessarily a barrier to revolutionary activity. In any case, whatever the degree of wealth attained by Mao's father, it is certain that his poor peasant origins had left their mark on the family in a lack of social graces and of a concern either for comfort or appearances. Above all, his father was totally devoid of any interest in learning except for purely practical purposes.

Mao's father had had two years of schooling, and knew enough characters to keep books. His mother, to whom he was deeply attached, was wholly illiterate.[1] She was a devout Buddhist, and Mao remained one too, until his adolescence. The significance of this fact for his future development is hard to evaluate. There are dialectical elements in Buddhist thought which might explain Mao's tendency, from his student days, to reason in terms of the unity of opposites. On the other hand, popular Buddhism, such as that of Mao's mother must have been, is a matter of piety rather than philosophy.

With his father, Mao often clashed violently. Indeed, the key to his revolutionary vocation has sometimes been sought in his relations with his parents. (Authors writing on Taiwan commonly treat his 'unfilial' attitude toward his father as *prima facie* evidence that he was inherently perverse and fated to end as a bandit or rebel of some kind.) In fact, such patterns are common enough, especially in a culture where the father ruled with as heavy a hand as he did in China at the time. Moreover it is clear, even from Mao's autobiography as told to Edgar Snow, and still clearer on the basis of such supplementary testimony as is available, that his attitude towards his father was not simply one of unremitting hatred. This is not to say that an observer interested in personality structure could not draw useful conclusions from what we know of Mao's relations with his family. And he may indeed have learned, from some of the instances of revolt against paternal authority described in his autobiography, that in general superiors are more inclined to respect those who dare to stand up to them. But the key to the formation of his *political* personality can better be sought in his contact with the larger world about him.

From the age of eight Mao attended a primary school of the traditional type, in which the students learned to recite the Confucian classics without understanding them. To supplement this dry fare, Mao, like all Chinese schoolboys of his generation, read the popular novels compiled a few centuries earlier from oral traditions, covering them up with a classic when his teacher walked by. There is no doubt that these novels influenced him profoundly, especially the historical *Romance of the Three Kingdoms*, and *Water Margin* (translated by Pearl Buck under the title *All Men Are Brothers*), with its tales of heroic bandits in revolt against a corrupt court and bureaucracy. Nevertheless, if Mao 'disliked' the classics, as he tells us, he learned to know them well, as his subsequent writings with their frequent classical references abundantly show. Moreover, it is worth remembering that the *Romance of the Three Kingdoms* is utterly impregnated with Confucian principles, its main theme being the contrast between Liu Pei, the founder of the Minor Han dynasty, portrayed as the ideal Confucian ruler, and the ruthless realist Ts'ao Ts'ao (who was not really so bad as he is made out to be in the novel). Even *Water Margin* itself, which is *the* great traditional novel of peasant revolt, does not really criticize the imperial system itself, but merely its abuses; the goal of the bandit heroes is to defend the true Confucian way in a time of troubles, when the emperor is not fulfilling his proper role. So these novels, however subversive they may have been considered by Mao's teachers, encouraged him at best to be a rebel *within* the framework of tradition.

At the age of thirteen Mao was obliged by his father to abandon his study of the classics at the Shaoshan Primary School, and devote himself full-time to working on the land and aiding with the accounts. (He had been helping part-time in the fields from the age of six.) He continued to read, not only his favourite novels, but political writings of the time, such as *Sheng-shih Wei-yen* (*Words of Warning to an Affluent Age*), by the comprador Cheng Kuan-ying. The latter work was above all a plea for technical and economic modernization and constitutional monarchy, but it also denounced the foreigners' treatment of the Chinese in Shanghai. According to Mao, this book further stimulated his desire to resume his education, and he ran away from home to study with an unemployed law student, subsequently studying more of the classics under an old scholar who lived in the village. But these piecemeal expedients did not satisfy him, and at last, when he was sixteen, he enrolled, despite his father's opposition, in a modern higher primary school, the Tungshan Primary School, in nearby Hsiang Hsiang *hsien*, his mother's native district.

It is from his arrival at this school and his exposure to the wider horizons of a busy market town that we can date the real beginnings of Mao's intellectual and political development, but the foundations had already been laid during his early adolescence in his native village. In 1904 his imagination had been struck by the reports which reached Shaoshan regarding the exploits of Huang Hsing, who had escaped from Changsha after the failure of an attempt to assassinate the principal imperial officials gathered there to celebrate the birthday of the Empress Dowager. Though Mao does not mention it in his autobiography, some echo probably reached him of the great anti-Manchu uprising on the border between Hunan and Hupei which took place in the winter of 1906–7, under the direction of the Ko Lao Hui, a secret society very influential among the peasantry of the area. Three years afterwards, in the spring of 1910, when famine led to popular revolt followed by repressions and executions, Mao and his fellow students discussed the event for many days. Mao says he felt the rebels were 'ordinary people' like his own family, and 'deeply resented the injustice of the treatment given to them'. Similarly, when members of the Ko Lao Hui revolted shortly afterwards in Shaoshan itself against an unjust landlord, establishing themselves in the mountains, Mao and his fellow students 'all sympathized with the revolt' and regarded its leader, who was ultimately executed, as a hero.*

While these incidents may indeed, as Mao claims, have left an indelible impression on his mind, his attitude toward social problems at that time was ambiguous. During the famine of 1910 the poor villagers of Shaoshan demanded grain from their more fortunate neighbours, but Mao's father continued to export it to the city as usual; the hungry villagers had then seized one of his shipments, and he had been furious. 'I did not sympathize with

---

* Snow, op. cit., pp. 132–3. Mao's chronology for these events in his autobiography is exceedingly vague. I have endeavoured to date them from the historical context.

him,' declared Mao. 'At the same time I thought the villagers' method was wrong also.'*

If Mao's views on social problems had not yet taken shape, however, his views on China's relations with the outside world were both clear and energetic. He himself dates his attainment of 'a certain amount of political consciousness' from the reading (shortly before his departure for Hsianghsiang) of a pamphlet deploring the loss to China of Korea, Taiwan, Indochina, Burma, and other territories and tributary states. In 1936 he still recalled clearly the opening sentence: 'Alas, China will be subjugated!' After reading this pamphlet, he tells us, he 'felt depressed' about the future of his country and 'began to realize that it was the duty of all the people to help save it'.[2]

In having a clearer vision of external problems than of internal ones, Mao was typical of the China of the time. The repeated and bitter humiliations, beginning with the Opium War of 1840 and continuing with the burning of the Summer Palace in 1860 by the Anglo-French expeditionary force, the defeat by the Japanese (who had always been regarded as a petty and barbarous people) in 1895, and the repression of the Boxer Uprising in 1900 – to mention only a few of the most important – had made it increasingly evident that drastic measures must be taken to save the country from utter dependency and helplessness. In the earliest period even the most audacious of the great servants of the empire had imagined that this could be done by borrowing from the barbarians merely techniques, leaving Chinese ideas and Chinese social and political patterns virtually untouched. By the time of Mao's boyhood it had become increasingly evident that the nations of the West drew their strength not only from machines, but from ideas and institutions which made it possible

---

* This incident offers a typical example of the methods of Mao Tse-tung's official biographer, Li Jui, who attributes to Mao the opinion that his father's behaviour was 'extremely inhumane' (an expression which does not occur in the Chinese translation of Edgar Snow's book, which is his only source), and says nothing of Mao's disapproval of the villagers' behaviour. Li Jui, *Mao Tse-tung t'ung-chih ti ch'u-ch'i ko-ming huo-tung (The Early Revolutionary Activity of Comrade Mao Tse-tung)*, Peking, 1957, pp. 7–8. Obviously it cannot be allowed to appear that Chairman Mao, even in his adolescence, was so benighted as to disapprove of revolutionary action by the masses.

for them to mobilize the energies of all their people. This was the idea which, for more than a decade, Yen Fu had been preaching in his essays and in the prefaces to his translations.* This lesson had been so far accepted that in 1906, two years before her death, even the reactionary Empress Dowager, who had beheaded T'an Ssu-t'ung in 1898 for his reformist ideas, agreed to certain timid steps towards a more constitutional type of monarchy. A few years later, influential voices were to be raised praising democracy and individual freedom for their own sake. But, for the moment, the arguments in China regarding what should be borrowed from the West tended to treat parliaments, factories, and ideas all alike as merely means to the end of strengthening and saving the nation. A few ultra-conservatives still regarded even the survival of the nation as less important than the preservation of the Chinese 'way', which they considered alone capable of giving meaning and dignity to life. But such 'culturalism', which had been the rule a few decades before, was rapidly giving place to a modern nationalism, for which the supreme goal was the preservation of the Chinese state.† There was wide disagreement over the effect of various internal measures on the fate of the nation. K'ang and Liang, the leaders of the Reform Movement of 1898, were partisans of constitutional monarchy; Sun Yat-sen and his collaborators Wang Ching-wei and Hu Han-min preached republicanism. Both groups read fragments of Marx and carried on polemics about 'socialism'.‡ But regarding the goal of national salvation, they were all unanimous.

* In his brilliant study of Yen Fu's career, *In Search of Wealth and Power: Yen Fu and the West* (Cambridge, Mass., Harvard University Press, 1964), Benjamin Schwartz has shown how Yen Fu interpreted the Western liberal thinkers in such a way as to serve his own goal of making China strong, and how in so doing he may well have revealed the hidden tendencies of the nineteenth-century Utopia, even though superficially he distorted the meaning of such classical texts of liberalism as the writings of Adam Smith, Herbert Spencer, or John Stuart Mill. For some reflections on Mao's debt to Yen Fu, see my review of Schwartz's book in the *Revue française de science politique*, No. 1, 1965.

† This transition has been brilliantly chronicled by Joseph Levenson in the first volume of his work *Confucian China and its Modern Fate* (London, Routledge & Kegan Paul, 1958).

‡ See R. Scalopino and H. Schiffrin, 'Early Socialist Currents in the

At the Tungshan Higher Primary School Mao came for the
first time in direct contact with these currents of thought. A
cousin of his had sent him a volume of Liang Ch'i-ch'ao's organ,
the *Hsin-min Ts'ung-pao*, and another book about the Reform
Movement of 1898. Mao read and re-read them, and came
greatly to admire K'ang Yu-wei and Liang Ch'i-ch'ao. He also
learned something of history, and some facts regarding the contemporary world. It was only there, in 1910, that he learned of the
death of the Emperor and the Empress Dowager, which had
occurred two years before. From a teacher who had studied in
Japan he heard tales of Japan's power, and of her pride in the
victory of 1905 over Russia. In a book called *Great Heroes of the
World* he read of Catherine and Peter the Great, Wellington and
Napoleon, Gladstone, Rousseau, and Lincoln. He was especially
struck by the chapter on George Washington, which he summed
up to a fellow student as follows: 'Washington won victory and
built up his nation only after eight long, bitter years of war.' His
general conclusion was that China needed great men such as
these, in her efforts to become wealthy and powerful and avoid
the fate of Annam, Korea, and India.*

Among the rulers of ancient China, he was particularly
fascinated by the legendary Emperors Yao and Shun, and also by
the exploits of Ch'in Shih Huang-ti and Han Wu-ti, and he read
many books about them. It will be seen that from Japan to
America, and from the modern West to ancient China, the figures
which struck Mao's imagination and fired his enthusiasm were at
once warriors and nationbuilders. (No doubt it was principally
for this that he admired Ch'in Shih Huang-ti, the creator of the
unified empire in the third century BC, and Han Wu-ti, the 'Martial Emperor' who extended it in battles against the Huns, and
not because the former burned books and the latter instituted
forced labour, though recent events leave some doubt about the
book burning.) This taste for the martial unquestionably cor-

Chinese Revolutionary Movement', *Journal of Asian Studies*, Vol. 18, No. 3,
May 1959, pp. 321–42.

* Snow, op. cit., p. 135; Hsiao San, *Mao Tse-tung t'ung-chih ti ch'ing-
shao nien shih-tai* (*The Childhood and Youth of Comrade Mao Tse-tung*),
p. 28. His admiration for Napoleon has survived down to the present day;
in 1964, Mao told a French parliamentary delegation that while Robespierre
was a great revolutionary, he was personally more impressed by Napoleon.

responds to a fundamental aspect of Mao's personality, which we shall encounter at every turn in his career. But it is also important to note that, at a time when China owed her humiliations above all to the military superiority of the West, and when, on the contrary, Japan had for the first time scored a victory for the nations of Asia over the Europeans, the problem of military strength was a constant preoccupation of all politically conscious Chinese.

The year which Mao spent at the Tungshan Higher Primary School brought him not only a broader contact with the problems of the world as a whole, but new personal experience of the stratification of Chinese society. Most of his fellow students were sons of landlords, expensively dressed and schooled in polite behaviour. Mao had only one decent suit, and generally went about in a frayed coat and trousers. Moreover, since he had been obliged to interrupt his education because of his father's opposition and had pursued it at best in relatively unfavourable conditions, he was six years older than the others, and towered above them all. This tall, ragged, and uncouth 'new boy' met with a mixture of ridicule and hostility from the vast majority of his classmates. Like Mao's relations with his father, the psychological consequences of his experiences at Tungshan should not be exaggerated as a factor in his development, but this year can hardly have made him more kindly disposed towards the landlord class.*

The following winter, early in 1911, Mao 'began to long to go to Changsha' which, he had heard, was 'a magnificent place altogether', and enter the middle school there. He secured a recommendation from a teacher at the higher primary school, and took a steamer to the provincial capital, hardly daring to hope that he could 'actually become a student in this great school'. In fact he was admitted without difficulty. His stay in

---

* Mao's schoolmate and friend at this time, and later at the Normal School in Changsha, Hsiao Hsü-tung (Siao Yü) has given a highly coloured account of Mao's tribulations at Tungshan Higher Primary School, in which he strives, as he does throughout his memoirs, to present Mao's character as a mixture of arrogance, brutality, and stubbornness. This is certainly a caricature though no doubt a caricature which contains an element of truth. On the other hand, regarding Mao's relations with the other students and his mentality at the time, he provides useful information which adds a further dimension to Mao's own story. See *Mao Tse-tung and I Were Beggars* (Syracuse University Press, 1959).

Changsha, though short, marked a new stage in his political and intellectual development.

There he read a newspaper for the first time – an organ of Sun Yat-sen, which told of the attack on the Canton *yamen* under the leadership of his fellow Hunanese Huang Hsing, the principal leader of the T'ung Meng Hui after Sun Yat-sen himself. This was the last of the unsuccessful attempts at revolution inspired by Sun and his comrades; the next attempt was to be the 1911 revolution itself, which broke out on 10 October at Wuhan. During the intervening months, the political fever continued to mount in Changsha, as it did throughout the country. Mao was so excited by all this, and by his reading of the newspaper, that he wrote his first article and pasted it on the school wall. This first expression of political opinion was, as he later admitted himself, 'somewhat muddled'. In an attempt to combine his old admiration for K'ang Yu-wei and Liang Ch'i-ch'ao and his new enthusiasm for Sun Yat-sen and the revolutionaries of the T'ung Meng Hui, he proposed that a republic be established with Sun as president, K'ang as premier, and Liang as minister of foreign affairs.[3]

From political writing Mao progressed to political action. Like many other young Chinese at the time, he and a friend cut off their pigtails (which had been imposed on the Chinese by the alien rulers) and then forcibly removed the pigtails of a dozen other students who had promised to do so as well but had failed to keep their word.

Shortly after the uprising at Wuhan an emissary of the revolutionaries appeared at Mao's school in Changsha and made an impassioned patriotic speech. Mao was so impressed that a few days later he decided to go to Wuhan to join the revolutionary army. But before he could make suitable preparations he was forestalled by the outbreak of the revolution in Changsha itself. And so, as we have already seen, he stood on a hill in the city and watched the triumph of the insurgents. He can hardly have imagined, as he gazed at the banners hailing the 'Great Han Republic', that thirty-eight years later, almost to the day, it would fall to him to stand on the Gate of Heavenly Peace in Peking and to proclaim the establishment of the Chinese People's Republic.

Notes

1. Snow, *Red Star Over China*, p. 129.
2. ibid., p. 133.
3. ibid., p. 136.

Chapter 2
Student Days in Changsha

The fact that the revolution broke out in Changsha less than two weeks after the uprising in Wuhan was no accident, but the result of a prior agreement among the leaders of the T'ung Meng Hui in the two provinces. To understand why Mao's native province was thus in the forefront of the revolution it is necessary to know more of the conditions in Hunan at the time, and of the political traditions of the province in the late nineteenth and early twentieth centuries.

Hunan belongs to the sub-tropical region of Southern China, and its high mountain ranges, cut through by four major rivers, made it from ancient times a favourite haunt of bandits. The Hunanese, great eaters of red peppers, are noted for their vigorous personalities, and also for their political talents. (Mao has even put forward a theory regarding the relation between red pepper and revolution.) 'China can only be conquered', says a proverb, 'when all the Hunanese are dead.'* And in fact during the past century the natives of this province have distinguished themselves in the most diverse causes. The Taiping rebellion was finally crushed in large part by the efforts of the 'Hunanese Army' of the great scholar and servant of the empire Tseng Kuo-fan, himself a native of the province. To balance this eminent defender of the established order, there are such figures as T'an Ssu-t'ung, the most famous of the martyrs in the Reform Movement of 1898, and Huang Hsing. Huang was not merely Sun Yat-sen's second in command and the chief military leader of the T'ung Meng Hui, but a major revolutionary figure in his own right, whose experience went back to 1902. He belonged to a category extremely important in the political life of China during the first decades of this century, that of the returned students from Japan. It was only

* In 1920, Ch'en Tu-hsiu, the future secretary general of the Chinese Communist Party, wrote that this was not a vain boast. See the extracts from his article 'Salute to the Spirit of the Hunanese' in H. Carrère d'Encausse and S. Schram, *Le Marxisme et l'Asie, 1853–1964* (Paris, Armand Colin, 1965), pp. 291–2. A translation of this work is to be published shortly by Allen Lane The Penguin Press.

after the First World War that Chinese students in really large numbers began to go to Europe and America. Before that, modern ideas were acquired largely via Japan, which was near at hand and whose language was relatively easy for the Chinese to master. The Hunanese were particularly numerous among the students who went to Japan at this time; among them were not only Huang Hsing but Ch'en T'ien-hua, who was to be the principal ideologist of the T'ung Meng Hui from its foundation in 1905 until his suicide in 1908.

The two principal instruments on which Sun Yat-sen and Huang Hsing relied in their efforts to organize a force capable of overthrowing the Manchu empire were the secret societies and the 'New Army', a military force on Western lines organized by the imperial government to cope with the conditions of modern warfare. Both of these groups were particularly important in Hunan. The secret societies especially were remarkably numerous and active. In his founding address to the Hua Hsing Hui (Society for the Revival of China), established in Changsha in December 1903, Huang Hsing declared:

Speaking of Hunan province, there has been a rapid growth of revolutionary ideas among the army and the students. . . . Furthermore, members of the secret societies who also harbour anti-Manchu ideas have long spread and consolidated their influence, but they dare not start first; they are like a bomb full of gunpowder ready to blow, waiting for us to light the fuse.*

A quarter of a century later, Mao himself was to light this fuse and explode the revolutionary potential of the secret societies. Meanwhile, in 1904, Huang Hsing concluded an alliance with Ma Fu-i, the principal leader of the secret societies in Hunan, who was to carry out uprisings in the countryside to coincide with the assassination by bomb of the leading provincial officials and the capture of the city by Huang and his partisans among the military cadets. Some hundred thousand members of the Ko Lao Hui and other secret societies were recruited for this enterprise. As a result, the plot leaked out, Ma Fu-i was captured

---

* Chün-tu Hsüeh, *Huang Hsing and the Chinese Revolution* (Stanford, Stanford University Press, 1961), p. 18. The Hua Hsing Hui was combined with Sun Yat-sen's Hsing Chung Hui in 1905 to create the T'ung Meng Hui.

and executed, and Huang Hsing was obliged to flee to Japan. Two years later, in December 1906, a great uprising animated by the secret societies took place on the border between Hunan and Kiangsi.

We have already seen the effect produced on Mao's imagination by the echoes of these events which reached Shaoshan. The 1906 rising had not been planned by the T'ung Meng Hui and took Sun Yat-sen and Huang Hsing by surprise. Emissaries were immediately dispatched from Tokyo, where the T'ung Meng Hui had its headquarters, but despite heroic resistance on the part of the insurgents, who held out for more than a month, the movement was suppressed by the imperial forces. The pitiless repression which followed these events led to a lull in the revolutionary activities in Hunan, but it was a lull of only two or three years' duration.

It was during this brief period of reorganization that the future leaders of the 1911 revolution at Changsha, Chiao Ta-feng and Ch'en Tso-hsin, completed their studies and began their revolutionary activity. The former worked primarily among the secret societies, while the second was a lieutenant in the New Army, and both of them joined the T'ung Meng Hui several years before the revolution. It is an interesting manifestation of Mao's interest in the traditional forces of protest and revolt that in 1936 he should have presented both these men not as seasoned revolutionaries, but as representatives of the Ko Lao Hui.[1] (It is true that he was a simple observer at the time and may not have known anything about their background or identity.*)

The efforts of these numerous revolutionary leaders from

* Inasmuch as many episodes in Mao's life remain obscure, I have endeavoured, as far as possible, to document all important statements of fact regarding his own career. Beginning with Chapter 3, the text is therefore rather heavily studded with footnotes. By way of compensation, I have virtually abstained from giving any references for the general background information on China and Hunan at the time of Mao's boyhood presented in the first two chapters. The principal sources consulted regarding Chiao Ta-feng, Ch'en Tso-hsin, and the 1911 revolution in Hunan were the materials in Volume 6 of the documentary collection *Hsin-hai ko-ming* (Shanghai, Shanghai Jen-min Ch'u-pan-she, 1957), pp. 132–71; Li Shih-yo, *Hsin-hai ko-ming shih-ch'i liang-Hu ti-ch'ü ti ko-ming yün-tung* (Peking, San-lien Shu-tien, 1957); the contemporary files of the *North China Herald*; and Chün-tu Hsüeh, op. cit.

Hunan were aided by the difficult economic conditions in the province at the time. As a result of natural catastrophes, rice became very scarce in 1910 and the price more than doubled, but the authorities stubbornly refused to furnish grain from the public stocks at a more reasonable price. On 13 April the governor, while giving the appearance of listening to complaints from a crowd of over ten thousand people which had collected, sent for troops who opened fire, killing and wounding many. Furious, the crowd forced its way unarmed into the palace and burned it, despite the bullets of the troops.

Conditions and incidents such as this carried the resentment engendered among the peasantry by a multitude of privations to a new height. In the spring of 1911 the unsuccessful attack of Huang Hsing on the Canton yamen produced in Hunan and throughout the country the feeling that decisive events were at hand. In this context, Chiao Ta-feng entered into contact with the leaders of the T'ung Meng Hui in neighbouring Hupei province, where Wuhan is situated, and concluded an agreement according to which whenever a revolt broke out either in Wuhan or in Changsha, the revolutionaries of the other province promised to organize in turn, within ten days, a second uprising to support the initiative of their comrades. This accord was in force when the revolution broke out on 10 October 1911, and Chiao Ta-feng had the intention of applying it. Hesitations on the part of the moderate republican party in Changsha, headed by T'an Yen-k'ai, the president of the provincial Consultative Assembly, led to a few days' delay, but Hunan came down effectively on the side of the revolution within twelve days of its outbreak.

Mao has given us, in the story of his life as told to Edgar Snow, a picturesque account of the events on 22 October 1911. He speaks of a great battle before the walls of Changsha, and of workers taking the gates by assault from the inside to support the troops favouring the revolution who were attacking the city.[2] In fact, there was nothing that reasonably resembled a 'great battle'. At one of the two gates attacked by the revolutionaries, there was no resistance whatever; at the other, where Chiao Ta-feng himself led the attack, the defenders were quickly persuaded by his nationalist eloquence and rallied to the revolution after a brief show of opposition. Once inside the city, the two leaders Chiao and Ch'en united their forces for an attack on the governor's palace, but the latter, having discovered that his guard was not prepared to fight, fled through a hole in the back wall which he had prepared in advance for just such an eventuality. Nor is there any mention in the contemporary sources of an uprising by workers or by any civilians; the whole brief and virtually bloodless episode was conducted by the New Army. Obviously Mao (like many older and more experienced observers of revolutionary events) retained only a confused impression of the exciting scenes he had witnessed, and reconstructed this image a quarter of a century later in the light of his idea of what the 1911 revolution ought to have been.

But if his description of the revolutionary uprising itself is largely fictional, and if he erred, as pointed out earlier, in describing Chiao and Ch'en as simply leaders of the secret societies, there is a great deal more accuracy in his description of the political events that followed. Chiao and Ch'en, he said, 'were not bad men, and had some revolutionary intentions, but they were poor and represented the interests of the oppressed. The landlords and merchants were dissatisfied with them.' As a result, Mao explains, T'an Yen-k'ai, who represented the propertied interests of the province, organized a revolt against them and succeeded in having them killed; Mao saw their bodies lying in the street one day as he passed. This is, indeed, very much what happened. The 'disorderly elements' among the soldiers who killed Chiao and Ch'en a week after they had taken power were most probably instigated by T'an Yen-k'ai, who certainly had the support of the landlords and merchants. At the same time, it must be recognized that 'well-intentioned young men' had little place in the China of 1911; even Sun Yat-sen was soon to find himself helpless in the face of the generals, who held the real power, and he did not succeed in playing a really effective role in Chinese politics until he began to build up an army of his own, with the aid of Soviet advisers and Chinese Communist organizers.

Now that the revolution had spread to Changsha, Mao Tse-tung maintained his original intention of joining the army, but instead of going to Wuhan to do so, he was able to join a unit in Hunan. He thus found himself under the ultimate authority of T'an Yen-k'ai, who took over as supreme civil and military leader following the assassination of his rivals, and under the command of Colonel Chao Heng-t'i. It is highly improbable that either of these men noticed the existence of the eighteen-year-old Private Mao, though both of them were to know him well in subsequent years, Chao as an adversary and T'an Yen-k'ai, strangely enough, as an ally, at a time when T'an counted among the 'left wing' of the Kuomintang.

Mao's aim in joining the army was to 'help complete the revolution'. He does not appear to have done any fighting; just as armed conflict seemed inevitable between the supporters of Sun Yat-sen and those of Yüan Shih-k'ai, a former high imperial official who had gone over to the side of the republicans, the two men came to an agreement, Sun dissolved his government at Nanking, and unity and harmony were theoretically re-established. In fact, Yüan gathered all power into his own hands, and the leaders of the genuine republican party soon found themselves in exile or condemned to silence – when they were not imprisoned or assassinated. But in the spring of 1912, Mao concluded that the revolution was over, and decided to leave the army and return to his books, after half a year as a soldier.

If Mao thus did not actually participate to any significant degree in the 1911 revolution, his six months in the army marked a further stage in his education. It is doubtful whether, as his friend Emi Hsiao claims, his brief experience of military life led him as early as this to the conclusion that in China the army was the key to political power. Such a conclusion could in fact grow only out of the prolonged anarchy and division of the war-lord era. But Mao did broaden his contacts, making friends in particular with a miner and an ironsmith. At the same time, he had so far absorbed the student mentality that he spent a significant part of his monthly wages of seven dollars in buying water from the water carriers. The soldiers were expected to bring it themselves from outside the city, but Mao considered it beneath his dignity as an intellectual to carry anything.[3]

The time that he saved in this way he employed in reading newspapers, on which he spent the remainder of his wages. One of these contained articles on socialism, in which he encountered the term for the first time. He also read several pamphlets by Chiang K'ang-hu, a returned student from Japan who had founded a 'Chinese Socialist Party' in November 1911. Inspired by all this, he discussed socialism and social reformism with other soldiers, and corresponded on the subject with some of his student friends, but did not meet with any great response.

One can certainly not conclude from what Mao tells us of his 'enthusiasm' for socialism in 1912 that he was converted to the support of social revolution at this time. Five years later, as we shall see, his first published article contained scarcely a trace of radical ideas. But the thoughts planted by this first encounter with the idea of socialism may not have been completely without influence on his later development.

On leaving the army, Mao was resolved to return to school, but beyond that he was uncertain what type of education or career he should choose. In this he was altogether typical of China at the time. The old learning was to a considerable extent discredited; though the historical and philosophical writings of the past were still studied even in the most modern schools, only a handful of ultra-conservatives thought them a useful equipment for the modern world. Moreover, the old examination system had been abolished in 1905, even before the fall of the empire, so that there was no longer any career outlet for those with a purely classical training. But on the other hand, though modern Western-style education was considered essential, and though a certain number of modern schools had been established, it was not yet clear either what the place of their graduates as a whole in the life of the nation might be, or which of them were best adapted to fit young men of talent for successful careers.

Mao was therefore not alone in seeking his way with difficulty. Attracted by advertisements in the papers, he registered successively for a police and then a soap-making school. A friend who had become a law student directed Mao's attention to an advertisement of his own school that promised to teach students all about law in three years, after which they would instantly become 'mandarins'. Interested by this prospect, Mao wrote to his family asking for tuition money and painting a bright picture of his future 'as a jurist and mandarin'. Then he was persuaded by another advertisement that it would be preferable to become a commercial expert, and he enrolled in a higher commercial school. This time he actually attended classes for a month, and then left because most of the courses were taught in English, of which he understood very little.

After another six months of experiment at the First Provincial Middle School in Changsha, where he learned more traditional history, Mao decided that it would be more profitable to read and study independently. He therefore withdrew and spent the next half-year reading from morning to night in the Hunan Provincial Library. It was during this period that he read Yen Fu's versions of such classic works in the Western liberal tradition as *The Wealth of Nations* and *L'Esprit des lois*, as well as translations of Darwin, J. S. Mill, Rousseau, and Spencer. He also saw for the first time a map of the world, and made a study of the history and geography of Russia, the United States, England, France, and other countries.[4]

Although he found this period extremely fruitful, Mao was unable to continue, because his family refused to support such apparently aimless activity. Moreover, he had been thinking seriously of his 'career', and had decided that he was best fitted to be a teacher. Once more his eye was caught by an advertisement, this time for the Fourth Provincial Normal School in Changsha. Mao entered it in the spring of 1913; the following autumn the Fourth Normal School was combined with the First, and Mao and his fellow students were transferred to the latter, from which he graduated in the spring of 1918.

It would be hard to overestimate the importance of these five years in Mao's life. They were, first of all, the years during which he acquired a great part of his education, both classical and modern, Chinese and Western. But they were much more than that. They launched Mao on a brief career in teaching which served as a stepping-stone to a wider role. More important still, perhaps, the intellectually alert and politically conscious student body of the First Normal School, and of other boys' and girls' schools in Changsha, provided Mao with an ideal training-ground for his apprenticeship as a political worker. There he developed ideas and techniques which he was to apply later; there he acquired friends and comrades, some of whom were to follow his leadership until the final victory in 1949.

Neither of Mao's two most intimate friends at the normal school was to be at his side in the hour of triumph. One of them,

Ts'ai Ho-sen, had fallen in the revolutionary struggle many years before; the other, Hsiao Hsü-tung, had taken a different political direction and become Mao's bitter enemy.

Ts'ai Ho-sen, a native (like Mao's mother) of Hsianghsiang *hsien*, was the oldest of six children. His father had at one time been a minor official at the Chiangnan Arsenal in Shanghai. He was distantly related, through his mother, to Tseng Kuo-fan. He, Mao, and Hsiao Hsü-tung were considered the three most brilliant students in the school, and were highly conscious of the fact. They liked to call themselves *san-ko hao-chieh*, the 'three heroes' or 'three worthies', employing a term common in Mao's favourite novel, the *Romance of the Three Kingdoms*, which suggests the possession not only of strength and courage, but of intellectual and moral qualities as well.

Physical strength was, incidentally, one of Mao's constant preoccupations during his years in Changsha. He and his friends sought consciously to develop their bodily resistance through long walks, and through exposure to the elements.

While at the normal school, Mao made the acquaintance of another contemporary who was to figure extensively in his future experience: Li Li-san. At one point, Mao put an advertisement in the paper, signed with his pseudonym 'Twenty-eight-stroke Student' (28 being the number of strokes required to write the three characters composing his name), expressing his desire to meet young men concerned about the future of their country. He received, to use his own expression, 'three and a half replies'. The 'half reply' was that of Li Li-san, who listened to everything Mao had to say and then went away without expressing his own views. 'Our friendship never developed', commented Mao in 1936, with delicate irony. In fact, the two men were to clash violently a dozen years later regarding the strategy which could best assure the victory of the revolution in an agrarian country such as China.

The First Normal School was officially considered an institution of secondary education, and not of higher education. In a sense, therefore, Mao never acquired a university education. But the standards at the school were high, and in fact he learned as much as he might have done at a provincial university. Two teachers especially left their mark on Mao at the normal school.

The first was a teacher of Chinese, nicknamed 'Yüan the Big Beard', who ridiculed Mao's style copied from Liang Ch'i-ch'ao, and taught him to take as his model Han Yü, the famous T'ang dynasty essayist. 'Thanks to Yüan the Big Beard', declared Mao in 1936, 'I can today still turn out a passable classical essay if required.'[5]

The second teacher was Yang Ch'ang-chi, the professor of ethics. Yang taught only the senior classes, and so Mao did not become his student until 1915 or 1916. By that time, the intellectual world in China was already bubbling with the ferment of the 'May 4th Movement'. In the narrow sense, this term designates the student demonstrations which broke out on 4 May 1919 in Peking, and which were directed against the decision of the Paris Peace Conference to give the former German concessions in Shantung to Japan, instead of returning them to China. But the expression is commonly applied more broadly to the years from 1915 to 1921, and to the radical ideas which swept China during that period. On the eve of the 1911 revolution hardly anyone in China openly advocated the wholesale imitation of Western customs and ideas for their own sake; it was a question of adapting certain aspects of the West to China's needs. T'an Ssu-t'ung, who at the time of the Reform Movement of 1898 had advocated 'total Westernization', was one of the rare exceptions, and neither K'ang nor Liang had gone nearly so far. But in the years after 1911 there came a profound change, which is symbolized by the creation of the review *Ch'ing-nien* (*Youth*), soon re-named *Hsin Ch'ing-nien* (*New Youth*). The moving spirit of *Hsin Ch'ing-nien* was Ch'en Tu-hsiu, Dean of the Faculty of Letters at Peking University, the principal intellectual centre in the country, who was later to become the first secretary general of the Chinese Communist Party. Ch'en's message was one of complete and radical Westernization. The salvation of China, he maintained, lay in the hands of 'Mr Democracy' and 'Mr Science', who would sweep away the ignorance, superstition, and barbarism of the past and lay the foundations for a new, modern, secular state on Western lines.

Ch'en's dislike for almost everything in China's past did not mean that he was unconcerned about the future of his country. On the contrary, the often-quoted phrase, 'I would rather see the ruin of our national essence than the extinction of our race in the present and in the future because of its inaptitude for survival',* implies that his goal was precisely to save the nation by putting realism ahead of pride in the past. But there is no doubt that so sweeping a rejection of the Chinese heritage as was contained in Ch'en's writings of 1915–18 did offend against national pride to a degree hardly tolerable in the long run. Moreover, there remained the question of whether a whole people, and moreover a people of largely illiterate peasants, could be made to discard their entire heritage for a radically different tradition. At the same time, in judging Ch'en Tu-hsiu's historical role, it is essential to recall how heavy was the weight of tradition and authority in early twentieth-century China. Without the violent and devastatingly effective attack on the past led by Ch'en and his friends, it would hardly have been possible later to salvage a great deal from tradition without becoming captive to it again.

Yang Ch'ang-chi, Mao's ethics teacher, became an enthusiastic reader and partisan of *Hsin Ch'ing-nien* almost from the moment of its appearance. He urged all his students to read it, and it was through him that Mao soon found himself in touch with the mainstream of intellectual life in China. But at the same time, Mao received from Yang in his school courses an education in which Chinese and Western ideas were intimately combined. Yang was remarkably well qualified to give such an education, for to a solid classical training he had added some ten years' study abroad in Japan, England, and Germany. There he had become a disciple of Kant and T. H. Green (as well as an admirer of Samuel Smiles), without ceasing to profess the neo-Confucianism of Chu Hsi, the famous twelfth-century philosopher. In an article published in Changsha just before Mao entered his class, Yang Ch'ang-chi made a statement which undoubtedly sums up the spirit of his teaching:

Each country has its own national spirit, just as each person has his

* From his editorial for the first issue of *Ch'ing-nien*. For extracts, see Chow Tse-tsung, *The May Fourth Movement: Intellectual Revolution in Modern China* (Cambridge, Mass., Harvard University Press, 1960), pp. 45–7. (Chow's book is the basic and indispensable study of the May 4th Movement as a whole.) For another discussion of Ch'en Tu-hsiu's ideas, see Benjamin Schwartz, *Chinese Communism and the Rise of Mao* (Cambridge, Mass., Harvard University Press, 1958), pp. 12–27.

own personality. The culture of one country cannot be transplanted in its entirety to another country. A country is an organic whole, just as the human body is an organic whole. It is not like a machine, which can be taken apart and put together again. If you take it apart, it dies.[6]

Thus, while advocating the study of all aspects of Western thought and institutions and the intensive application of Western science and technology, he also urged students not to forget their national heritage. The utility of learning from the West, in his view, was not to find a new culture which could be substituted bodily for the Chinese one, but to prod China's own culture into movement once again.

Yang Ch'ang-chi's influence on Mao was certainly very profound. When speaking of him in 1936, Mao could not forget that he had later married Yang's daughter, whom he deeply loved, and who had been executed by the Nationalists in 1930. But there is every reason to believe that in describing Yang as 'a man of high moral character', who 'tried to imbue his students with the desire to become just, moral, virtuous men, useful in society',[7] Mao was moved not only by affection for his father-in-law but by sincere admiration for his teacher.

Yang guided Mao not only towards the radical Westernizers and iconoclasts of Hsin Ch'ing-nien, but towards another group, whose members were seeking the inspiration for a rebirth of their country within the Chinese heritage itself, in the writings of the great Hunanese scholar of the early seventeenth century, Wang Fu-chih. This group was the Ch'uan-shan Hsüeh-she, or Society for the Study of Wang Fu-chih,* which had been founded almost simultaneously with Hsin Ch'ing-nien by a group of distinguished literati in Changsha. Wang Fu-chih was one of the most radical among several remarkable figures who, at the time of the Manchu conquest, had refused to serve China's new foreign masters. These men had denounced the idealist perversion of the classics by Chu Hsi and the other Sung and Ming neo-Confucians, who to their mind had contributed to the loss of the nation's independence by turning scholars away from practical concerns toward self-cultivation and mystical practices. They also em-

---

* From Wang's style Wang Ch'uan-shan, Ch'uan-shan being the name of a mountain where he retired after the Manchu conquest.

phasized, against the grain of traditional attitudes if not of the classics themselves, the dignity of the military condition. In all of this, the members of the Ch'uan-shan Hsüeh-she found inspiration for the present. At Yang Ch'ang-chi's urging, Mao attended their meetings, and absorbed their point of view. ('Yüan the Big Beard' also encouraged him in this direction.)

Judging by his first literary effort, an article published in April 1917 in *Hsin Ch'ing-nien*, it would seem that Mao had thus far assimilated rather more from these neo-traditionalists than from the magazine in which his article appeared. Entitled 'A Study of Physical Culture', the piece recommended to his countrymen a system of exercises which Mao had devised, as a means of strengthening their bodies and characters alike. Mao was and has remained an energetic partisan of physical culture, and the subject of the article is accordingly not without significance. But far more interesting are the considerations with which he justified the importance of vigorous health and firmness of will in aiding the Chinese to withstand foreign pressure and save their nation. This aim is stated with the utmost clarity in the opening sentences:

Our nation is wanting in strength. The military spirit has not been encouraged. The physical condition of the population deteriorates daily. This is an extremely disturbing phenomenon. . . . If this state continues, our weakness will increase further. To attain our goals and make our influence felt are external matters, results. The development of our physical strength is an internal matter, a cause. If our bodies are not strong, we will be afraid as soon as we see enemy soldiers, and then how can we attain our goals and make ourselves respected?[*]

And in the paragraph dealing with the utility of physical education, Mao states explicitly: 'The principal aim of physical education is military heroism.'[8]

In this article we find, fully developed, two of the most characteristic traits in Mao's personality – a concern for the destiny of China and an emphasis on courage, strength, and the military ethos as the instruments of national salvation. In contrast, the theme of 'self-awareness' (to use Mao's own term of 1917) appears

[*] S. Schram, *The Political Thought of Mao Tse-tung* (London, Pall Mall Press, 1964), pp. 94–5. For a complete translation and a more detailed analysis of Mao's thought in 1917, see Mao Ze-dong, *Une étude de l'éducation physique* (Paris, Mouton, 1963).

as yet only in rudimentary form. 'When we speak of physical education', writes Mao, 'we should begin with individual initiative.' Already he has learned from the Chinese admirers of the nineteenth-century European liberal tradition (including Yang Ch'ang-chi) that the energy for making a nation rich and powerful is hidden within each member of society, and can be released only by stimulating individual initiative. But he has not yet reached the conclusion that in order to release these energies it is necessary to shatter the traditional patterns of thought and behaviour which limit the autonomy of the individual. On the contrary, he offers as a text to justify self-reliance a citation from the Confucian *Analects* ('What the superior man seeks is in himself'), and as an example of firmness of character, Tseng Kuo-fan. But although the intellectual world of Mao Tse-tung in 1917 appears as yet primarily Chinese and traditionalist, this does not mean that his reading of *Hsin Ch'ing-nien* and in general his contact with radical Westernizing ideas had been without effect. Even for a Chinese living in Peking or Shanghai who knew one or more Western languages, it was no easy thing, in the second decade of the twentieth century, to understand conceptions so profoundly alien to his own tradition. For a young Hunanese who knew no foreign languages and had never left his own province, the task was still more difficult. Nevertheless, Mao was making headway in it.

Mao's development toward a more progressive attitude was no doubt encouraged by another of his teachers who was later to become his faithful political disciple. This was Hsü T'e-li, a Hunanese of peasant origin who had studied in Japan, and who had there become a member of Sun Yat-sen's T'ung Meng Hui. Hsü was iconoclastic not only in his words but in his acts; he was one of two teachers who drew attention to themselves by coming to the school on foot. (Most of the other came in rickshaws; a few particularly affluent ones, including Yang Ch'angchi, were borne in sedan chairs.)

The modern side of Mao's personality during his student days in Changsha appears more clearly in his activities than in his one surviving article. This is only natural, for it is easier to form an organization to discuss new ideas than to reason coherently and at length in such an unfamiliar framework. As early as 1915, Mao had been elected secretary of the Students' Society at the Normal School, and in this and other contexts he showed himself increasingly active in promoting both the study of new ideas and the struggle against rigid and old-fashioned notions of discipline. One of his first and most characteristic initiatives was the creation of an 'Association for Student Self-Government', primarily with the object of organizing the students for collective resistance against such demands of the school authorities as seemed unreasonable.

Even as he was taking these first steps in the organization of resistance to the abuses of tradition, Mao Tse-tung remained profoundly imbued with the peasant environment of his childhood. He had an opportunity to demonstrate this when, in the spring of 1917, he commanded the defence of the First Normal School in the course of the wars among various military factions which continually ravaged the province. On this occasion, Mao, who headed the students' 'volunteer army', employed a device well known in the Hunanese countryside: he had the students cut young bamboos in such a way as to leave a sharp point which could be used to put out the eyes of any soldiers trying to climb over the school wall (none tried!).

On the occasion of the Moon Cake Festival in the autumn of 1917 Mao gave a new and even more striking demonstration of his traditional mentality. A group of students gathered on the hill behind the First Normal School to discuss ways of saving the country. Some proposed going into politics. To this Mao's reply was that one needed money and connexions to get elected. Others suggested exploiting their future positions as teachers to influence future generations – but Mao objected that this method would take too long. Asked to offer his own solution, he replied: 'Imitate the heroes of Liang Shan P'o.'*

Liang Shan P'o was the name of the mountain fortress on which the bandit heroes of Mao's favourite novel *Water Margin* had

---

* Information regarding this episode, and a wide variety of other facts regarding Mao's early life, were supplied in the course of interviews with Professor Pai Yü (or Yu Beh, as he transcribes it) of National Chengchi University in Taipei, who has kindly allowed me to make use here of his reminiscences. Professor Pai was a student in the First Normal School in Changsha, in a class two years behind that of Mao.

established themselves to fight for justice and order in an unjust and disorderly world. Exactly ten years later, Mao was to mount the Chingkangshan and begin an adventure not altogether dissimilar. But for the moment, at the end of 1917, he was about to take another step forward in the assimilation of modern and Western ideas. The most important concrete manifestation of this was his leading role in the formation of the Hsin-min Hsüeh-hui or New People's Study Society, which began its activities in the fall of 1917 and was formally constituted in the spring of 1918. This was one of the most radical student societies in all of China at the time; virtually its entire membership ultimately joined the Communist Party.

Other evidence confirms that it was during the winter of 1917–18 that the ideas and impressions which Mao had accumulated over half a dozen years' exposure to various forms of modern and Western thought began to coalesce into a world view infinitely more radical than that expressed in the article he had published a few months earlier in *Hsin Ch'ing-nien*. At this time he wrote in the margin of his textbook on ethics:

We must develop our physical and mental capacities to the fullest extent. . . . Wherever there is repression of the individual, wherever there are acts contrary to the nature of the individual, there can be no greater crime. That is why our country's three bonds must go, and constitute, with religion, capitalists, and autocracy, the four evil demons of the empire.*

Here we find not only a reference to 'capitalists', but a categoric rejection of the 'three bonds' (between prince and subject, father and son, husband and wife) which are the very heart of Confucian morality. At the same time, the new personality which emerges at this time shows a profound continuity with that of Mao's traditionalist phase. The admiration for the classics gave place to an enthusiasm for revolution and radical change, but the two traits of nationalism and of emphasis on the martial virtues remained at the heart of Mao's personality.

If Mao began to move towards a more modern outlook in 1917–18, this was no doubt primarily because he had reached the

---

* Schram, op. cit., p. 13. The textbook in question was a Chinese translation of a work by a second-rank neo-Kantian, Friedrich Paulsen.

appropriate moment in his own development. But it also cor-responded to a rapid acceleration of social and intellectual change in China as a whole, as the May 4th Movement began to gather momentum. And as these events unfolded, they were to carry Mao along with them towards his ultimate destiny. When he graduated from the First Normal School in the spring of 1918 he was merely a student from the provinces, albeit a politically conscious one. By the summer of 1919, when he returned to Hunan after half a year in Peking and threw himself into the political activity of all kinds that followed the May 4th student demonstrations, he had already set his foot on the path that would lead him shortly to a career as a professional revolutionary.

Notes

1. Snow, *Red Star Over China*, pp. 137–8.
2. ibid.
3. ibid., p. 139.
4. ibid., pp. 141–2.
5. ibid., pp. 143–4.
6. 'An Exhortation to Study', *Kung Yen*, Vol. 1, No. 1.
7. Snow, op. cit., p. 143.
8. Schram, *Political Thought of Mao*, p. 99.

Chapter 3
Mao Tse-tung at the Time of the May 4th Movement

Hitherto the radicals within the Chinese intelligentsia, in seeking a model for transforming the customs and mentality of their country, had fixed their eyes essentially on Western Europe. Some, such as Ch'en Tu-hsiu, were admirers of the French democratic tradition; others looked toward England. Still others, impressed by the strength of the Central Powers in the early years of the First World War, wrote articles in *Hsin Ch'ing-nien* praising 'German militarism'. Now, as Mao Tse-tung prepared to leave for Peking, a change was about to take place. A year after its occurrence, the Bolshevik revolution in Russia began to attract serious attention.

Mao Tse-tung himself has written that 'the salvoes of the October Revolution' brought Marxism to China.[1] This is an oversimplified view, but it contains a good deal of truth. Marxism was not completely unknown in China even before the October revolution; a fragment of the *Communist Manifesto* had been translated as early as 1906, and we have already referred to the polemics on 'socialism' that took place early in the century. But it is true that Marxism as a serious political force came to China only after the Russian revolution. It did not come all at once, as Mao has implied in affirming repeatedly that the May 4th period marked the watershed between the predominance of 'bourgeois' and of 'proletarian' leadership in China's political and intellectual life.[2] But its influence grew so rapidly that within a mere year or two the terms of the debate over China's future were completely altered.

It is characteristic of the manner in which this change came about that Li Ta-chao, who was to be, with Ch'en Tu-hsiu, one of the two principal founders of the Chinese Communist Party, hailed the historical and philosophical significance of the Russian revolution well before he had arrived at any serious understanding of its doctrinal basis. His first article on this theme appeared in *Hsin Ch'ing-nien* in July 1918, and a more famous one, 'The Victory of Bolshevism', in October, almost simultaneously with

Mao's arrival in Peking. The triumph of the October revolution, he declared, was not merely a Russian concern, but 'the victory of the spirit of all mankind' in the combat against militarism and autocracy. We can be particularly certain that Mao read these articles attentively, for thanks to an introduction from Yang Ch'ang-chi (who had just accepted a chair at Peking University) he found a job as librarian's assistant under Li, who was head of the university library, and later professor of history. Mao's salary of eight dollars a month allowed him only a frugal existence – he shared a room with seven other Hunanese students – but the beauty of the ancient capital was, in his own words, a 'vivid and living compensation'. He was also frequently invited to the home of Yang Ch'ang-chi, where he became better acquainted with his future wife, Yang K'ai-hui.

Mao's functions at the library were exceedingly menial, and in itself his job was merely a means of subsistence, offering him little contact with people or ideas. He tried to strike up conversations with some of the leading intellectual figures who visited the library, but in most cases found that they had 'no time to listen to an assistant librarian speaking southern dialect'.[3] But he ultimately found other ways of participating in the intellectual life of the university. On the one hand, he joined the Philosophy and Journalism Societies, so acquiring the right to attend courses. (This does not mean that, in the still highly status-conscious world of a Chinese university, he was admitted to full participation. Once, after a lecture, he tried to ask Hu Shih a question, but the latter, on learning that he was not a proper student but a mere librarian's assistant, refused to talk to him.) More important, both as a means of meeting people and as a step towards his ultimate political commitment, was his participation in the 'Marxist Study Group' which had been founded by Li Ta-chao the previous spring.*

Li Ta-chao himself was at this time very far from completely accepting Marxism or even adequately understanding it. The article, 'My View of Marxism', which he published in May 1919, in a special issue of *Hsin Ch'ing-nien* on the subject, presents a

* There is some doubt over whether this group was formally constituted before 1919, but it was probably functioning on an informal basis during the winter of 1918–19. See Chow Tse-tsung, *The May Fourth Movement*, p. 244.

narrowly determinist and completely undialectical account of the Marxist view of history, adding that such social philosophy, while useful, must be supplemented by other ideas placing greater emphasis on the creative role of the human spirit. It is not surprising that, with such a master, Mao should not have become committed while in Peking to Marxism, still less to Leninism, in any precise form. On the contrary, he admits in his autobiography that at this time he was, though 'more and more radical', also 'confused, looking for a road', and that he was much influenced by anarchism. He nevertheless declares that 'under Li Ta-chao' he 'developed rapidly toward Marxism' during the winter of 1918–19. To cap the confusion he adds that, during this first visit to Peking, Ch'en Tu-hsiu influenced him 'perhaps more than anyone else'.[4] And Ch'en leaned neither towards Marxism nor towards anarchism at this time, but continued to advocate parliamentary democracy and the scientific spirit as the answer to China's problems.

The contradiction among these various statements is perhaps more apparent than real. This was the period of most furious change in the Chinese intellectual and political scene, when even learned and mature scholars often altered their outlooks completely in the space of a few months. It is therefore hardly surprising that the ideas of a young man from the provinces, in contact for the first time with the life of the capital, should move rapidly in several directions at once. He was tempted by anarchism, because like all his generation he was bent on shattering the restraints imposed on the individual by traditional society. He was deeply influenced by Ch'en Tu-hsiu because for several years Ch'en had been his literary idol, and because Ch'en's uncompromising espousal of all that was young, uninhibited, and full of vitality answered the same urge for liberation. And he developed toward Marxism 'under Li Ta-chao', not only because Li was the architect of the study group in which Mao's own knowledge of the subject was enlarged, but because he closely resembled Li in his passionate commitment to the greatness of China.

The importance of this trait in Li's personality cannot be too strongly emphasized. It was, in fact, only in the autumn of 1918 that he joined the editorial board of *Hsin Ch'ing-nien*. Hitherto he had kept at a certain distance from the 'New Culture' movement, precisely because of the lack of interest in national values that appeared to him to characterize the outlook of Ch'en Tuhsiu.* His appearance among the leaders of the cultural revolution at this time should be taken as a symbol of the dual transformation which this movement was about to undergo: not only towards Marxism, as already indicated, but also towards nationalism. It was, of course, precisely the Leninist interpretation of Marxism, with its emphasis on the role of national liberation movements in the world revolution, which permitted the synthesis of these two tendencies. But the interpretation of Marxism which was ultimately to be developed by Li Ta-chao, and after him by Mao Tse-tung, was distinguished by the fact that, unlike Lenin, these men regarded the dignity of the Chinese nation and the greatness of its contribution to the world not merely as themes for anti-imperialist propaganda, but as values in themselves.

In the spring of 1919, however, as he prepared to leave Peking, Mao was not as yet propagating any variant of Marxism at all. To the 'curious mixture of ideas of liberalism, democratic reformism, and utopian socialism' which had characterized his mind at the time of his graduation from the Normal School in Changsha,[5] he had now added a certain number of Marxist and anarchist ideas, but the mixture had not yet produced a proper precipitate. Moreover, there were and have remained traditionalist elements in the mixture as well, despite his progressive assimilation of Marxism-Leninism. At the meetings of Li Ta-chao's Marxist Study Society Mao attracted attention by his efforts to combine socialism with the most diverse schools of

* Already in 1914–15 the two men had clashed in a celebrated literary battle, over whether the freedom and enlightenment of the people or the survival of the state were more important. Li had then maintained with great vehemence that to suffer even the most tyrannical state was better than to have no state at all and to become 'slaves without a country'. For extracts from the articles written by Li and Ch'en on this occasion, and for a comparison of their evolution from 1914 down to the death of Li Ta-chao in 1927, see Carrère d'Encausse and Schram, *Le Marxisme et l'Asie*, pp. 72–82, 280–314. For a detailed study of the intellectual development of Li Ta-chao, see Maurice Meisner's biography, to be published shortly by the Harvard University Press.

ancient Chinese thought. He is even reported to have admired Tseng Kuo-fan at this time to the point of taking him as his model. In identifying himself with Tseng, Mao was no doubt thinking not of the latter's specifically conservative traits, but of his example as a statesman who knew how to wield political and military power. And he was surely not indifferent to the fact that Tseng was a fellow Hunanese.

Political action during the summer of 1919 was to aid Mao Tse-tung in clarifying his ideas, and begin to give him at last a definite political outlook and political style of his own. He left Peking in late February or early March, and went first to Shanghai, to accompany some of his friends bound for France. They were going there thanks to a project for 'diligent work and frugal study' which had been launched in 1917 by a group of leading scholars and under which they would work half-time to pay for their education and subsistence. Mao's 'New People's Study Society' in Changsha had begun to take an interest in this scheme during the spring and summer of 1918, and the number of Hunanese students who went to Peking in order to prepare for their experience in France was particularly large. After his arrival in the capital Mao had continued to work actively in favour of the project, and helped draw up a plan for the training of candidates during their stay in Peking.[6]

Among those who took advantage of this scheme were Chou En-lai, and also Mao's intimate friend Ts'ai Ho-sen. In his autobiography Mao affirms that he never considered going to France himself, since he did not know enough about his own country. Other accounts by those who knew him at the time suggest that he did think seriously of going, but finally decided against it because he was not good at languages and was afraid he would be at a loss in a foreign country. In any case, he ultimately chose to remain in China, and merely accompanied some of his friends to the boat in Shanghai. From there he returned to Changsha. Characteristically, among the memories of the journey which stood out in his mind seventeen years later were his visits to landmarks of China's past: Confucius's grave or a walled city famous in the *Romance of the Three Kingdoms*.

On his return to Hunan, Mao Tse-tung found the province in considerable ferment. The governor, General Chang Ching-yao, was particularly brutal,* and the resentment caused by his repressive policies was further increased by the fact that he was a representative of the pro-Japanese Anfu Clique of General Tuan Ch'i-jui, the strongman then controlling the government in Peking. Ever since the Twenty-One Demands of 1915, resentment against Japan had been building up in China, and the explosion provoked by the news from Paris, where the Chinese delegation had meekly accepted the decision of the Peace Conference over Shantung, was all the greater since it was Tokyo that was benefiting from this particular act of imperialist dismemberment. The massive student demonstrations on 4 May 1919 were conducted to the accompaniment of anti-Japanese slogans, and the minister whose house was burned by the students was particularly associated with the acceptance of the Twenty-One Demands by the Chinese Government four years earlier.

The news from Peking found an immediate response in Hunan, despite the measures taken by Chang Ching-yao to suppress all anti-Japanese activity. In Changsha, as in Peking, the driving force of the movement during this first phase was furnished by the students. Mao Tse-tung played an active role in the creation, on 3 June, of a 'United Students' Association' of Hunan province, which issued a proclamation on the same day demanding the decapitation of the pro-Japanese politician whose house the Peking students had burned on 4 May.†

On 3 June, too, there took place mass arrests of students in Peking, which led two days later to protest strikes by both merchants and workers in Shanghai and soon in most other large centres in the country. For the first time in the history of modern China the progressive intellectuals had succeeded in inspiring and organizing a mass movement of other social categories. (The movement, however, remained confined to the cities; the peasantry

* His successor, General Chao Heng-t'i (Mao's old military commander in 1911–12, and his adversary in many political struggles of the May 4th period), told me in an interview in Taipei in June 1963 that Chang 'was not a man, but a wild beast'.

† This concluding sentence, cited by Li Jui (*Mao Tse-tung t'ung-chih ti ch'u-ch'i ko-ming huo-tung*, p. 95), does not appear in the manifesto as published at the time (*Hsüeh-sheng chiu-kuo ch'üan-shih*, Shanghai, October 1919), but it might well have been omitted for reasons of prudence in a book published under the reign of Tuan Ch'i-jui.

was entirely unaffected.) The succeeding weeks saw a proliferation of organizations of all kinds. In Hunan Mao Tse-tung continued to play a very active role, both in promoting new organizations and in launching demonstrations against the pro-Japanese policy of the government in Peking and of Chang Ching-yao. Two of the most significant organizations were the 'United Association for the Promotion of National Goods' (i.e. for the boycott of Japanese goods), and the 'Hunan United Association of all Circles' (i.e. social categories). The latter was inspired by precedents in Peking, Tientsin, and Shanghai, and was formed on 9 July at the initiative of the 'United Students' Association', with the participation of both workers and merchants. It is particularly significant as a manifestation of the solidarity among very wide and diverse strata of Chinese society felt at the time. It is also important as the background for Mao Tse-tung's most influential article of the May 4th period, entitled 'The Great Union of the Popular Masses', the first instalment of which appeared on 21 July.*

This article was published in Numbers 2, 3, and 4 of the *Hsiang River Review* (*Hsiang-chiang P'ing-lun*), a weekly magazine of the United Students' Association, founded on 14 July 1919, with Mao as editor. As part of his systematic attempt to demonstrate that Mao had already developed in his student days all the basic ideas which emerged from his later thought and activity, Li Jui endeavours to show that this 'great union of the popular masses' was really a kind of first cousin to the 'people's democratic dictatorship' put forward by Mao in 1949, in which workers and peasants were destined to play a leading role.[7] This is a completely unhistorical view. Mao does indeed mention the peasants (though they had as yet played no part in the May 4th Movement) and the workers, along with the students, the women, and various other categories, as participants in the 'great union'. He also evokes the victory of the Russian revolution, and hails the progress of the 'army of the red flag' throughout the world. But he does not distinguish between the red flag of the Bolsheviks and that of the May 4th Movement, which was as much the flag of Dewey as of Marx, nor is there a trace of Marxist categories in his analysis. Indeed, in another article published in *Hsiang-chiang*

* For extracts see Schram, *Political Thought of Mao*, texts IA and IVA.

P'ing-lun and cited by Li Jui himself, Mao declared that the science and democracy preached by Ch'en Tu-hsiu were exactly what China needed. Yet this article, even if it does not contain a prefiguration of the Marxist formulas which Mao was later to elaborate, does mark the appearance of an extremely important aspect of his thought which might be called his 'populist' tendency.

Mao's master Li Ta-chao had already begun to develop ideas on the importance and moral superiority of the peasantry not unlike those of the Russian *narodniks*. Mao was not yet 'populist' in this sense; he was not particularly interested in the peasantry at all. But he was, and has remained, 'populist' in a broader sense, in his feeling that the Chinese people as a whole are not merely an important historical entity, but a mighty progressive force. Indeed, this idea can be regarded as the bridge which led him from the relatively conservative and traditionalist nationalism of 1917 to a genuinely Marxist viewpoint. He had begun, as we have seen, with a concern for the survival of the Chinese nation as such. In this undifferentiated monolith he now carved out two blocks, the 'popular masses' on the one hand, and a tiny minority of militarists and profiteers who sided with the foreigners on the other. The succeeding decades were to witness many changes in Mao's definitions of the progressive or reactionary character of various social categories, but no retreat from his faith in the revolutionary mission of the Chinese 'people'.

Mao's work in editing the *Hsiang River Review*, and writing a large part of it himself, was not merely an academic exercise on the part of one intellectual for other intellectuals. The influence of his periodical was exceedingly wide: two thousand copies of the first issue were sold out in a single day, and thereafter five thousand copies of each issue were printed.[8] Although the vast majority of the Chinese people, especially in the countryside, were, of course, illiterate, the potential audience had been greatly increased by the introduction of a simple and easily comprehensible style of writing, which was one of the important conquests of the New Culture movement during its early period. The idea of doing away with the difficult and artificial literary language and replacing it by the vernacular had been put forward by Hu Shih in an article in *Hsin Ch'ing-nien* in January 1917, and by the summer of 1919 most political and literary writing was being done in a style relatively close to the patterns of spoken Chinese.* This 'literary revolution' carried out by the intellectuals within their own restricted circle was a necessary pre-condition for breaking out of that circle and speaking to the Chinese people as a whole.

Like the older and more famous scholars and intellectuals in Peking, Mao intended to make the fullest use of the instrument thus placed at his disposal for influencing public opinion. The impact of this new factor was not lost on the authorities, and it is not surprising that Chang Ching-yao closed the *Hsiang River Review* after the fifth issue, banning the United Students' Association at the same time. Nothing daunted, Mao took over the editorship of another student weekly published in Changsha, *Hsin Hunan* (*New Hunan*). Chang Ching-yao soon closed that too, so Mao was reduced to writing articles in the leading daily newspaper of Changsha, the *Ta Kung Pao*. Some of the most interesting of these were devoted to a theme which has always been close to his heart, that of equal rights for women. Following the suicide of a young girl in Changsha whose parents had forced her to marry against her will, he wrote no less than nine articles in thirteen days denouncing the restraints placed by the old society on the liberty of the individual, and hailing the 'great wave of the freedom to love'.†

The limitations on his literary activities led Mao to devote a greater portion of his time to practical political work. In November 1919 he reorganized the United Students' Association, and in December he organized a strike of the students in all the secondary and some of the primary schools of Changsha, directed against Chang Ching-yao.[9] This was no laughing matter, for Chang was not one to take opposition lightly. A few weeks earlier he had called a meeting of student representatives, which Mao had attended though himself no longer a student. There he had roundly abused the students for meddling in politics and not

* On the introduction of the vernacular or *pai-hua* style, see Chow Tse-tsung, op. cit., pp. 26–8 and 271–83.

† For extracts see Schram, op. cit., pp. 226–8. The views of Mao's future father-in-law, Yang Ch'ang-chi, on sexual mores were exceedingly radical. In an article published in 1915, he praised the English because, unlike the Chinese, they did not concern themselves with the sexual activities of widows. 'CZY Sheng' (pseudonym), in *Chia-yin Tsa-chih*, I, 6.

leaving relations with Japan to the government, and had ended by screaming at them: 'If you don't listen to me, I'll cut off your heads!' At this point one of the girl students had begun to cry, but Mao had told her to pay no more attention to Chang than to a dog barking.[10]

The strike of the students in December 1919 secured some concessions, but the situation in Hunan soon became too dangerous for those who, like Mao, had brought themselves to the attention of the authorities by their political activities. Moreover, the idea began to gain ground that Chang could only be driven from the province with outside help. Consequently, a certain number of representatives were sent to the principal centres, and Mao found himself once more in Peking.

By this time, however, he was no longer an unknown student, as he had been a year earlier. His articles in *Hsiang-chiang P'ing-lun* and *Hsin Hunan* had attracted favourable notice from three of the most important periodicals of the day: *Hsin Ch'ing-nien*, *Hsin Ch'ao* (*New Tide*), and Li Ta-chao's organ *Mei-chou P'ing-lun* (*The Weekly Critic*).* His stay in Peking was relatively brief, but it marked an important stage in his political development. Translations of all or substantial parts of several basic Marxist writings were just beginning to appear. November 1919 had seen the publication of the first section of the *Communist Manifesto* (the complete translation appeared in book form in April 1920), and of the major part of Kautsky's *Karl Marx's ökonomische Lehren*.† Mao read both these translations, as well as a history of socialism by Kirkup; the three items, he says, '. . . especially deeply carved my mind, and built up in me a faith in Marxism, from which, once I had accepted it as the correct interpretation of history, I did not afterwards waver.'[11] There may also have been a personal reason for Mao's going to Peking at precisely this time. Yang Ch'ang-chi died at the end of January, and Mao may have wanted to see and console Yang K'ai-hui.

From Peking Mao went once more to Shanghai, where he met

---

* For the enthusiastic praise of Lo Chia-lun, now a high dignitary on Taiwan, in *Hsin Ch'ao*, see Schram, op. cit., p. 104. See also *Hsin Ch'ing-nien*, Vol. 7, No. 1, p. 104, and *Mei-chou P'ing-lun*, No. 36, p. 4.

† For details regarding the date and form of publication of various Marxist writings at this time, see Chow Tse-tsung, op. cit., p. 299.

Ch'en Tu-hsiu and discussed with him the Marxist books he had been reading. Ch'en had been detained for six months as a result of the support which, as Dean of the Faculty of Letters, he had given to the May 4th student demonstrations; on his release in December 1919 he had sought refuge in Shanghai. During his imprisonment he had been turning more and more towards Marxism. 'Ch'en's own assertions of belief', Mao said in 1936, 'deeply impressed me at what was probably a critical period in my life.'[12]

In Shanghai Mao supported himself by working as a laundry-man.[13] But his fortunes were shortly to change. I P'ei-chi, an important member of the Kuomintang who had been his teacher at the First Normal School, was in Hengyang, under the protection of T'an Yen-k'ai, whose support he had managed to enlist in the campaign to drive out Chang Ching-yao. There Mao had the occasion to meet him and to become better acquainted, thanks to the leisure of an enforced retreat. When I P'ei-chi returned triumphantly to Changsha in June 1920, in the wake of the armies of T'an Yen-k'ai and Chao Heng-t'i, and took up a number of important positions, including that of director of the normal school, he appointed Mao in turn director of the primary school attached to the First Normal School. This position gave Mao considerable status in Changsha and facilitated his efforts to influence public opinion in the province. It also gave him for the first time a certain material basis for his existence and even a degree of comfort. This combination of status and financial stability influenced not only his professional and political activities, but also his personal life, for it undoubtedly made possible his marriage to Yang K'ai-hui in the following winter. Yang Ch'ang-chi had been a wealthy man, and the family would hardly have permitted his daughter to marry a young man who was not only radically non-conformist, and remarkably unkempt to boot, but had no visible means of support.

'By the summer of 1920', declared Mao in his autobiography, 'I had become, in theory and to some extent in action, a Marxist, and from this time on I considered myself a Marxist.'[14] Indeed, without abandoning his political work among the students, Mao was shortly to move forward to more specifically Marxist-inspired activities such as organizing labour unions. Once more this new phase in his life corresponded to a new stage in the development of the situation in China as a whole, for Mao's activities during the winter of 1920–21 are closely bound up with the preparations for the creation of the Chinese Communist Party, which was to hold its first congress in July 1921.

Notes

1. 'On the People's Democratic Dictatorship', *Selected Works* (Peking), Vol. IV, p. 413.
2. See in particular 'On New Democracy', *Selected Works* (Peking), Vol. II, pp. 339–84.
3. Snow, *Red Star Over China*, p. 148.
4. ibid., pp. 149, 151, 154.
5. ibid., p. 146.
6. Li Jui, *Mao Tse-tung t'ung-chih ti ch'u-ch'i ko-ming huo-tung*, pp. 83–6.
7. ibid., pp. 103–4.
8. ibid., p. 100.
9. ibid., pp. 115–17.
10. Reminiscences of Professor Pai Yü.
11. Snow, op. cit., p. 153.
12. ibid., p. 154.
13. Li Jui, op. cit., p. 120.
14. Snow, op. cit., p. 153.

Chapter 4
The Beginnings of the Chinese Communist Party

The Chinese Communist Party was from the beginning an indigenous phenomenon; it was composed of men who had become largely convinced of the value of Soviet theories and methods for solving China's problems before they were taken in hand by the emissaries of the Third International. But their gropings towards an effective revolutionary formula would not have crystallized into a concrete organization either so soon, or in the same way, had it not been for the efforts of Moscow. This Russian intervention offered certain advantages to the Chinese revolutionaries: theoretical and organizational guidance on the part of more experienced comrades, and material aid. But it also had two negative aspects. The Soviet leaders were not merely revolutionaries interested in the overthrow of capitalism throughout the world, they were also responsible for the destinies of the Russian state. And experience was to show that, when forced to choose, they frequently put the second category of interests before the first. Furthermore, most of the leading Bolsheviks were thoroughly European in experience and mentality, and even when their only desire was to give disinterested assistance to their Chinese comrades, their vision of Chinese realities was often so distorted that they gave very poor advice.

These two factors on the whole operated together and in the same direction. The excessive importance attached by Moscow to the urban working class led to the conclusion that, since the Asian proletariat was weak and immature, one would have to be satisfied for the moment with a 'national revolution' led by the bourgeoisie. Such a revolution, directed against the influence of Western imperialism without exciting too great a social upheaval in a neighbouring country, fitted in perfectly with Soviet foreign policy interests, which required above all a stable and friendly buffer in the East against any new attempt at intervention. This does not mean that Soviet advice to the Chinese Communists was always shortsighted and selfish. On the contrary, it was certainly very helpful at first in saving them from sectarian errors.

But in the long run, the priority given to Soviet foreign policy objectives, with Stalin's naïve conviction that he understood China perfectly and had everything completely under control, was to lead the Chinese revolution to repeated catastrophes, which the Chinese leaders, and Mao in particular, have never forgotten.

Lenin had been among the first of the European revolutionaries to grasp the full significance of the movement for colonial emancipation. His ideas on the proper tactics for harnessing the national revolution in Asia to the overall goal of world revolution had been taking shape gradually ever since the Japanese victory over Russia in 1905 had called forth new stirrings of hope and rebellion throughout the East. But it was only in July 1920, at the Second Comintern Congress, that his views on the 'national and colonial question' were formulated in a set of theses and formally adopted by the International.

In 1920 the International was not the disciplined instrument of Soviet policy which it later became, and Lenin was obliged to defend his conclusions in the course of very lively exchanges. His principal adversary was the Indian Communist M. N. Roy; their debate was in many respects a remarkable prefiguration of the current Sino-Soviet controversy. On the one hand, Roy considered, as the Chinese do today, that the Soviet leaders were prepared to go much too far in conciliating the bourgeoisie, and did not insist sufficiently on the leadership of the proletariat in the Asian revolution. On the other hand, he affirmed categorically that the key to the world revolution lay in Asia, and that the European proletariat would never accomplish anything until a prior upheaval in the colonies had shattered the foundations of the capitalist order and prepared the ground for its overthrow. Lenin had long fought the tendency of European Social Democrats, including even such radicals as Rosa Luxemburg, to minimize the revolutionary capacities of the non-European peoples and consider the latter as mere passive objects awaiting deliverance at the hands of the European proletariat. He was not prepared, however, to follow Roy to the opposite extreme, and told the Indian with some asperity that he was going too far. Finally, the congress adopted a position which was basically that of Lenin, with some limited and largely verbal concessions to Roy. Over tactics, it was laid down that in the colonial countries the International (and the local Communists, if any) should collaborate not with 'bourgeois-democratic' movements (as Lenin had written in his original draft) but only with 'national-revolutionary' movements. Lenin made quite clear in his report to the congress that this change was purely formal, since by the nature of things any nationalist movements in the colonies would be of a bourgeois-democratic character. The mutual dependence of the European and Asian revolutions was affirmed, but no clear statement was made on their relative importance.*

Although this position was laid down in July 1920, it had virtually no impact in China before January 1922. The Chinese Communists had only just begun to assimilate the rudiments of Marxism, and as yet they had little notion of the differences among the disciples and interpreters of Marx. For the moment they were inspired not by Lenin's ideas, but by orthodox Marxist conceptions, which attributed the entire responsibility for revolution to the urban proletariat, and made no provision for an alliance with the bourgeoisie to struggle for the common goal of national independence.

Although the principal explanation for this lag lies no doubt in the difficulty of assimilating so foreign an ideology, another is the slowness with which the Soviet leaders undertook to propagate their faith in China. As early as 1918 Lenin and Chicherin had begun efforts to establish relations both with the conservative government in Peking and with Sun Yat-sen in the south,† but it was only in early 1920 that the first emissary of the International, Grigorij Voitinskij, arrived in China. With the aid of Soviet diplomats then in Peking, he succeeded in making contact with Li Ta-chao, who gave him an introduction to Chen Tu-hsiu in Shanghai. The result was first the creation in March at Peking of a Society for the Study of Marxist Theory, and then the

---

* On the discussions of the Second Comintern Congress, see Carrère d'Encausse and Schram, *Le Marxisme et l'Asie*, pp. 40–48.

† On this aspect of things see, in particular, Allen S. Whiting, *Soviet Policies in China* (New York, Columbia University Press, 1954), pp. 25–41, which contains an exhaustive study of the various versions of the famous Karakhan declaration of July 1919, promising to return the Chinese Eastern Railway to China without compensation.

establishment of nuclei of an actual Communist Party, at Shanghai in May, and at Peking in September.*

In September or October, shortly after receiving word from Peking that a Communist group had been established there, Mao Tse-tung took the initiative in forming one at Changsha. At the same time he began efforts to establish a branch of the Socialist Youth Corps, which finally bore fruit in the last days of December 1920. (The Youth Corps had appeared first at Shanghai, in August.) In Hunan both of these organizations were recruited in large part from among the members of the Hsin-min Hsüeh-hui.[1]

These two organizations, like the Society for the Study of Marxism which he had founded in September, were created by Mao in imitation of what had been done elsewhere by his masters Li Ta-chao and Ch'en Tu-hsiu. Certain other undertakings of the same period show more originality. One of these was the establishment at the end of July of a Cultural Book Society, with the purpose of making available in Hunan the radical books and periodicals which were appearing in increasing numbers in Peking and Shanghai. For this good cause, Mao persuaded the new governor T'an Yen-k'ai, a former member of the imperial Hanlin academy proud of his calligraphy, to write the characters on the sign in front of the store.[2] Another and more singular activity was a campaign for the autonomy of Hunan province. In a China torn by civil strife, the idea of provincial autonomy was propagated at the time by people of varied political tendencies, and in fact T'an Yen-k'ai himself had espoused it on his return to Changsha in the summer of 1920. According to the official historiography, Mao took up the autonomist cause with the aim of exploiting the situation in a double sense. On the one hand, T'an Yen-k'ai's intention was clearly not to establish genuine popular self-government, but to consolidate his own sphere of influence in the face of the rival military clique

* There is some dispute over whether the Shanghai group was merely one group among others (seven in all by the beginning of 1921) or *the* Chinese Communist Party in embryo, and consequently over whether the founding of the party should be dated May 1920 or July 1921, when the First Congress was held. (Chow Tse-tsung, op. cit., pp. 248–9.) I prefer the second date, but the concrete steps in the establishment of the party seem to me to be more important than the superior dignity attached to one of them as constituting *the* foundation of the party.

in Peking. By organizing – as Mao proceeded to do – a campaign in favour of popular participation in the writing of the new provincial constitution, it would be possible to expose the bogus character of T'an's democracy. At the same time, to the extent that the future constitution contained guarantees of civil liberties, a more favourable climate would be created for revolutionary work in general. There is no doubt some truth in this, but it is doubtful that it is the whole truth. In August 1921, Chao Heng-t'i, who had replaced his former chief as the ruler of the province in the previous November, put through a provincial constitution which came into force on 1 January 1922. By the summer of the next year, Mao had forgotten that he had ever supported provincial autonomy even as a tactical manoeuvre, declaring in an article attacking Chao's oppressive régime in Hunan that he had 'always opposed' provincial self-government.* This repudiation of his previous attitude was no doubt motivated in large part by the criticisms of his Hunanese particularism to which he had been exposed in the interval.

During the winter of 1920–21 Mao Tse-tung began a brief career as a labour organizer. (This was only a part-time activity and he retained his post as head of the primary school in Changsha.) In the process, he developed an 'orthodox Marxist' viewpoint of the type which we have already noted as characteristic of the Chinese Communist Party as a whole at this time. In a letter to his friend Ts'ai Ho-sen in France he described the world as made up of one third 'capitalists' and two thirds 'proletarians', without any reference at all to national differences.[3]

By the spring of 1921 there were six 'small groups' of Communists in China itself (at Shanghai, Peking, Changsha, Wuhan, Canton, and Tsinan), and one formed by the Chinese students in Paris. After some preliminary discussions,† the First Congress of

* *Hsiang-tao chou-pao*, No. 36, 15 August 1923, pp. 270–71. On Mao's autonomist activities in general, see Li Jui, op. cit., pp. 121–31.

† The account by Boris Shumjatskij in the organ of the Far Eastern Secretariat of the Comintern (*Narody Dal'nogo Vostoka*, No. 1, 1921) mentions a conference which had taken place 'recently' in Central China, attended by representatives of the various Communist groups, and which had concluded that 'a united political party of the revolutionary proletariat, that is to say, a communist party', must be created. Shumjatskij added that these words had now been transformed into reality, since the First Congress

the party met at Shanghai in July 1921. Among those attending were two delegates from each of the six groups existing in China at this time, plus one representative of the Chinese in Japan, thirteen in all. Among those best known later were, in addition to Mao himself, Chang Kuo-t'ao from Peking, Tung Pi-wu from Wuhan, and Ch'en Kung-po (Wang Ching-wei's successor as head of the pro-Japanese puppet government in 1944) from Canton. Neither Li Ta-chao nor Ch'en Tu-hsiu was able to attend. The former remained in Peking; the latter had been called to Canton in December to head the education board of the government just set up by Sun Yat-sen under the protection of the war-lord Ch'en Chiung-ming. The congress was also attended by two Comintern delegates. Voitinskij had been joined recently by a Dutchman, Henricus Sneevliet, who went under the name of 'Maring' and was to play a major role in determining the future orientation of the Chinese revolution.

The work of the congress was conducted under dramatic circumstances, which have often been described.* The first sessions were held at a girls' school then closed for the holidays in the French concession in Shanghai, beyond the reach of the Chinese authorities. Eventually the comings and goings of the delegates attracted the attention of the police; but the intrusion of a suspicious character (evidently a police spy) gave them prior warning, and all but two of the delegates escaped just in time. (The others were soon released after questioning.) For greater security it was therefore decided to leave the city, and hold the final session on a boat on South Lake, near Chiahsing in Chekiang province, in the guise of a holiday excursion.

---

of the Chinese Communist Party was coming to an end as he wrote his article. The conference 'in Central China' may be the meeting in May 1921 in Shanghai referred to by Tung Pi-wu (in *Red Dust: Autobiographies of Chinese Communists*, edited by Nym Wales, p. 39), attended, according to him, by both Li Ta-chao and Ch'en Tu-hsiu.

* For the most complete and meticulous account of the First Congress, including a discussion of the evidence on such points as the names of those who attended and the exact date of the meetings (the official date for the foundation of the party, 1 July, is open to considerable doubt; the congress probably took place later in the month), see Ch'en Kung-po, *The Communist Movement in China* – an essay written in 1924, edited and with an introduction by C. Martin Wilbur (New York, Columbia University, 1960; reproduced for private distribution by the East Asian Institute).

The documents adopted by the First Congress comprised a particularly blatant expression of the 'orthodox Marxist' mentality. The goal set forth in the programme was to 'overthrow the capitalistic classes' with the 'revolutionary army of the proletariat', and to 'adopt the dictatorship of the proletariat' in order to abolish classes. A further decision on the objects of the party dealt almost exclusively with problems of organizing labour unions. Nowhere was there any reference to the collaboration with bourgeois nationalists advocated by Lenin, or even to national goals as distinguished from class ones. On the contrary, it was affirmed: 'Towards the existing political parties, an attitude of independence, aggression, and exclusion should be adopted. Our party should stand up on behalf of the proletariat, and should allow no relationship with the other parties or groups.'[4]

It was also decided that regular monthly reports should be submitted to the Third International, and that a representative should be sent if required to the Far Eastern Secretariat in Irkutsk.[*] In view of this close link with the Comintern, and the fact that representatives of Moscow attended the congress, it is surprising that the party's line was not closer to that of Lenin. But it appears that Maring and Voitinskij did not attend the final session on the lake at which the resolutions were adopted, and in any event the representatives of the International may well have felt that the main objective was to get a party organization established, even if it were made up of people of varied and heterodox ideological tendencies (there would always be time later to take them in hand).

It seems clear that there was some discussion on the problem of relations with other parties, if not in general terms then at least over collaboration with Sun Yat-sen. Disagreement on the issue was reportedly such that an 'anti-imperialist, anti-militarist' manifesto prepared for the congress was never published, because many of those present considered that Sun was no better than the northern war-lords.[†]

[*] Several sources affirm that there was not even a clear decision to affiliate with the Third International, but on this point I agree with C. Martin Wilbur, who presents the evidence in his introduction to Ch'en Kung-po, op. cit., pp. 25–6.

[†] On this see Wilbur's introduction to Ch'en Kung-po, op. cit., pp. 22–3, and also Ch'en's own essay. In reminiscences written in 1943, when he was

There is no conclusive evidence as to which side Mao took in these debates. His authorized biographer, Li Jui, claims that Mao opposed both the idea that the proletarian dictatorship was an immediate task of the party, and the 'erroneous extreme-left view-point hostile to the acceptance of intellectuals in the party'.* This would appear to put him on the side of a non-sectarian policy, and by implication on the side of collaboration with the nationalists. Although this fits in very well with the long-term tendency of Mao's thought and action, directed towards the 'great union of the popular masses' of China, his letters to Ts'ai Ho-sen of the previous winter make it seem much more likely that at this time he was going through a sectarian phase and probably approved the exclusive accent on work among the proletariat which characterizes the documents of the First Congress.

Whatever Mao's attitude in the summer of 1921, it was soon to be modified, like that of the party as a whole, in the direction of acceptance and even enthusiastic espousal of collaboration with the 'bourgeois nationalists'. But before any definite policy could be developed along these lines it was necessary for Moscow to decide *which* nationalists it wanted to support. In December 1921 Maring visited not only Wu P'ei-fu, the pro-English war-lord dominant in Central China, but Chao Heng-t'i, the governor of Hunan, on his way to Sun Yat-sen's headquarters at Kweilin, near Canton.[5] This hesitation was no doubt inspired not only by foreign policy considerations and the desire to deal with the real masters of the country, but by the confused and unsystematic character of Sun Yat-sen's political action and his lack of any solid organizational base. The conversations of December 1921 at Kweilin convinced both the participants that they had some-thing to gain from each other, and the ground was prepared for the further discussions which other Soviet emissaries were to carry on with Sun in the course of the following year; but

---

collaborating with the Japanese, Ch'en Kung-po further stated that the representative of the International tried to get the decision against co-operation with other parties reversed. ibid., p. 53.

* Li Jui, *Mao Tse-tung t'ung-chih ti ch'u-ch'i ko-ming huo-tung*, p. 150. The programme adopted at the First Congress contained the statement that the party 'absolutely cuts off all relations with the yellow intellectual class . . . '. Ch'en Kung-po, op. cit., p. 106.

Moscow still hesitated between Sun and his military protector Ch'en Chiung-ming.

Meanwhile Mao Tse-tung had returned to Changsha as secretary of the Chinese Communist Party for Hunan Province. The party had only some sixty or seventy members in all of China, so there was much to be done to create a serious and effective organization. Nevertheless, in addition to his work in the party and in the labour unions, Mao found time for several new initiatives in the cultural domain. Perhaps the most characteristic was the creation, in August 1921, of the 'Self-Study University', aimed at providing students with an opportunity to read and reflect independently, as Mao himself had once done, in an environment which provided the stimulus of occasional lectures as well as discussions with teachers and fellow students. The university was housed in the premises of the Ch'uan-shan Hsüeh-she.* This is no accidental detail; on the contrary, it may be regarded as a symbol of the whole future course of the Chinese revolution, as it was to take shape years later under Mao's direction. For the 'Self-Study University', though it placed great emphasis on modern ideas and especially on Marxism, also gave full weight to traditional Chinese learning, including in particular such sceptical, materialist, and nationalist thinkers as Wang Fu-chih himself. Many future cadres of the Chinese Communist Party were students at this school, and Mao Tse-tung, in the course of his participation in the discussions conducted there, continued to lay the foundations of that 'Sinification of Marxism' which was to be his greatest theoretical and practical achievement.

Another of Mao's initiatives, which illustrated, like the 'Cultural Book Society', his genius for exploiting respectable people and institutions for radical ends, was the establishment in Hunan late in 1922 of the 'mass education movement', launched in 1921 on a broad scale by leaders of the Chinese YMCA with American backing. Under Mao's guidance this enterprise had, like the original movement, the object of teaching illiterate adults a basic thousand-character vocabulary; but instead of using the textbooks employed elsewhere, Mao had a special set prepared. These began by affirming that all the riches of society were produced by the peasants and workers, and went on to explain

* See Chapter 2, pp. 40–41.

Communism, the Russian revolution, and the alliance of workers and peasants. Mao's instructions to those responsible for the movement in various localities were to make contact in each village with the local gentry, and obtain in this way respectable backing for the programme.[6]

Interesting and characteristic as these initiatives in the field of education were, Mao's central preoccupation was, of course, the work in the party itself, and above all in the labour unions. During 1922 a wave of labour unrest was rising throughout the country. In January Mao organized a campaign against Chao Heng-t'i, following the execution of two anarchist labour leaders in Changsha. Together with two other Hunanese who were later to play key roles in the history of the Chinese Communist Party, Li Li-san and Liu Shao-ch'i, he also helped lead a series of strikes in the whole province, of which the most important were the miners' strike at Anyüan and the strike on the Canton–Hankow railroad in September. Mao also worked among the stonemasons, the printers, and men of various other trades.*

While Mao was thus occupied, the political context in which he worked was about to undergo a radical change. In January 1922 a Chinese delegation, including representatives of both the Communist Party and the Kuomintang, attended the First Congress of the Toilers of the Far East in Moscow and Petrograd. There they were roundly lectured by Zinoviev, who conjured his Chinese comrades not to look down on those 'publicans and sinners', the bourgeois nationalists, but to make common cause with them.[7] Thus instructed, the Chinese Communist Party proceeded, at its Second Congress in July 1922, to abandon its sectarian attitude of the previous year and to adopt a resolution espousing an alliance with the nationalists. Mao has claimed that he went to Shanghai to attend this congress, but lost the address, could not find any comrades, and was obliged to return to Changsha.[8]

At this point dramatic events occurred which obliged the Soviets to decide which 'nationalist' they wanted to support in China. Ch'en Chiung-ming turned against Sun Yat-sen, who narrowly escaped arrest and arrived in Shanghai once more a

---

* Li Jui, op. cit., pp. 160–211. On the history of the Chinese labour movement in general, see Jean Chesneaux, *Le Mouvement ouvrier chinois 1919–1927* (Paris, Mouton, 1963).

refugee, without any base of operations whatsoever. There he was visited first by Dalin, the emissary of the Youth International, and then by Maring; but despite his precarious position, he refused to accept their proposal for a united front of the usual type between the Chinese Communist Party and the Kuomintang, fearing apparently that his new allies would prove difficult to control. He was prepared, however, to admit Communists to membership in the Kuomintang as individuals, and thus cooperation could be established between the two parties.

This proposal appeared much less shocking to Maring than it might have done to most other emissaries of the International, for he had participated in a roughly similar type of 'bloc from within' in the Netherlands East Indies, where he had played a major role in the development of the social-democratic movement. There the socialists had cooperated intimately with a nationalist revolutionary organization of religious colouring called Sarekat Islam, and the cooperation had been interrupted just as it seemed about to yield results. Here was a chance to make a new attempt at the utilization of the same tactics. Maring seized it with alacrity, called a special plenary meeting of the Central Committee of the Chinese Communist Party, dismissed the objections from a majority of his Chinese comrades, and obtained formal endorsement of the decision.*

Thus the Chinese Communist Party was launched on the adventure of collaboration with the Kuomintang to carry out the 'national revolution' in China, an adventure which was to bring it seemingly so near to triumph but end in defeat and the total destruction of the party's urban base.

* On the very obscure question of whether or not Maring acted in this instance on his own initiative, or on direct instructions from Moscow, see Conrad Brandt, *Stalin's Failure in China* (Cambridge, Mass., Harvard University Press, 1958), pp. 30–32; E. H. Carr, *A History of Soviet Russia. Socialism in One Country*, Vol. III, Part II (Macmillan, 1964), pp. 689–92. My own view, expressed in the text, is the same as that of Brandt and Carr, namely that Maring acted on his own responsibility and secured Moscow's approval only *ex post facto*. But the evidence is not sufficient to demonstrate this conclusively.

Notes

1. Li Jui, *Mao Tse-tung t'ung chih ti ch'u-ch'i ko-ming huo-tung*, pp. 146–53.
2. ibid., pp. 138–43; Chow Tse-tsung, *The May Fourth Movement*, p. 249.
3. Schram, *Political Thought of Mao*, pp. 25 and 214–16.
4. Ch'en Kung-po, *The Communist Movement in China*, pp. 106–9.
5. Letter of 19 December 1921, from Chao Heng-t'i to T'an Yen-k'ai, Institute of Modern History, Academia Sinica, Taiwan.
6. On the 'Self-Study University', see Li Jui, op. cit., pp. 153–9; on the mass education movement in Hunan, ibid., pp. 166–8, 181–2.
7. See Allen Whiting, *Soviet Policies in China*, Ch. V, and Carrère d'Encausse and Schram, *Le Marxisme et l'Asie*, pp. 79, 296–7.
8. Snow, *Red Star Over China*, p. 156.

Chapter 5
Collaboration with the Kuomintang

Whether or not the Soviet leaders gave their blessing in advance to the singular *form* in which Maring agreed to establish co-operation between the Chinese Communist Party and the Kuomintang, there is no doubt that the *substance* of this policy was in harmony with the climate which prevailed in Moscow. The lack of confidence in the revolutionary capacities of the Asian peoples, which characterized nearly all of the leading Bolsheviks, found particularly blatant expression at the Fourth Comintern Congress in November 1922. After denouncing the Chinese Communists for their lack of influence among the workers and taunting them with their 'Confucian' mentality, Karl Radek enjoined them to recognize that in China neither the establishment of socialism, nor even the unification of the country under a democratic republic were as yet on the order of the day.[1] In other words, they were to help in building up Sun Yat-sen's position in the south, while Moscow pursued Soviet diplomatic interests in dealings with Wu P'ei-fu in Peking, and with Chang Tso-lin in Manchuria.

Although all the Chinese Communists were obliged to obey the orders of Moscow, certain of their leaders accepted the new line more readily than others. Ch'en Tu-hsiu, who had sought in Marxism above all a new and more effective method for modernizing the country, was one of the most reticent, and only resigned himself to the necessity of the alliance with the bourgeoisie when Wu P'ei-fu's massacre of the striking railway workers on 7 February 1923 convinced him that the proletariat alone could not fight against the armed force of the 'militarists'.*

* This act of repression, which took thirty-five lives, came as the climax of an episode lasting nearly a year, during which the Chinese Communists had given their support to Wu P'ei-fu against his rivals among the war-lords in exchange for a free hand to organize the workers in the area controlled by him. This policy, later denounced as opportunist, appears to have been the joint invention of Maring and Li Ta-chao. See Brandt, *Stalin's Failure in China*, pp. 24–5, who puts the responsibility entirely on Maring and the Comintern, and Chesneaux, *Le Mouvement ouvrier chinois*, p. 277, who attributes the initiative to Li Ta-chao.

Li Ta-chao, who had never abandoned his intense concern with the independence and grandeur of China, embraced with enthusiasm the policy of collaborating with Sun Yat-sen, as did Mao Tse-tung. Indeed, the new policy could hardly have failed to receive Mao's support, for the anti-imperialist objective of the alliance with Sun struck a profoundly responsive chord in his personality.

It would be an oversimplification and an injustice to accuse Mao of mere xenophobia; but it is certain that he could not forgive the foreigners who had forced their way into his country and humiliated its inhabitants. An incident which took place a little later, in 1924, illustrates this aspect of his outlook. A former schoolmate who had been studying abroad for several years ran into him one day in Shanghai. Mao, who was dressed in worn Chinese clothes, looked disdainfully at his friend's Western-style suit and said to him: 'You'd better change your clothes.' 'Why?' asked the other. 'I'll show you,' replied Mao, and took him to see the famous sign at the entrance to a municipal park in Shanghai, 'Chinese and dogs not allowed'.*

The bitter resentment inspired by such indignities finds abundant expression in Mao's writings during 1923. He denounced the subservience of the Peking Government in Rabelaisian terms,† and berated his compatriots because they did not detest the Anglo-Saxons as much as they did the Japanese. 'Do the Chinese people only know how to hate Japan, and don't they know how to hate England?' he exclaimed in disgust. 'Don't they know that the aggression of the English imperialists against China is even more atrocious than that of the Japanese imperialists?'[2] As for America, she was 'the most murderous of hangmen'.[3]

More surprising at first glance than this hostility to the arro-

---

* Reminiscences of Professor Pai Yü. Professor Pai also recounts an incident from Mao's Changsha days which illustrates his elemental nationalism. At a soccer game between the First Normal School and Yale in China (a preparatory school frequented by the sons of Chinese closely linked to the West), Mao's tall form suddenly rose in the midst of the crowd of spectators, and he bellowed out: 'Beat the slaves of the foreigners!' – *yang-nu*, a contemptuous expression applied to those considered too subservient to Western interests.

† 'If one of our foreign masters farts', he wrote in August 1923, 'it's a lovely perfume.' Schram, *Political Thought of Mao*, p. 143.

gant 'foreign masters', against whom the anti-imperialist alliance was to be directed, was Mao's affirmation, in July 1923, that the merchants were the leaders of the national revolution:

The present political problem in China is none other than the problem of the national revolution. To use the strength of the people to overthrow the militarists and foreign imperialism, with which the former are in collusion to accomplish their treasonable acts, is the historic mission of the Chinese people. This revolution is the task of the people as a whole. The merchants, workers, peasants, students and teachers should all come forward to take on the responsibility for a portion of the revolutionary work; but because of historical necessity and current tendencies, the work for which the merchants should be responsible in the national revolution is both more urgent and more important than the work that the rest of the people should take upon themselves.[4]

If the struggle against imperialism corresponded to the nationalist streak in Mao's character, the idea of a joint struggle of the whole people against the 'militarists' appealed to the populist tendency he had already manifested in 1919. Only the vocabulary had changed; instead of the 'great union of the popular masses', he called for a 'united front'.[5] If, at the same time, he recognized that the leadership of such a front must belong to the bourgeoisie, his attitude was no doubt shaped by a mixture of conviction and opportunism. At the Third Congress of the Chinese Communist Party in June 1923 a similar line had been laid down. 'The Kuomintang must be the central force in the national revolution and assume the leadership of the revolution', declared the manifesto adopted on this occasion.[6] At the Third Congress it was also decided that the Communists should relinquish control over the labour movement to the Kuomintang. Chang Kuo-t'ao claims that, on this occasion, Mao Tse-tung originally supported trade union independence, but switched his vote once the opposite viewpoint had triumphed, in order to advance his career in the party.*

Chang Kuo-t'ao is a highly partisan witness, who is quite

* Brandt, op. cit., pp. 36–7. The decision regarding the unions, which was apparently taken under Comintern pressure, was reversed two months later.

capable of distorting the facts to suit his own purposes,* and he is systematically hostile to Mao. This is, however, no reason to discount his testimony altogether. Mao no doubt *was* concerned, at the Third Chinese Communist Party Congress, to improve his position in the party, just as his article recognizing the leadership of the merchants in the national revolution may have been designed to ingratiate himself with the Kuomintang. (In both of these objectives he succeeded: shortly after the Third Congress of the Chinese Communist Party, where he had been elected to the Central Committee, he was appointed head of the Organization Bureau, replacing Chang Kuo-t'ao; and in January 1924, at the First Congress of the Kuomintang, he was elected an alternate member of the Central Executive Committee.) But at the same time, Mao was probably moved not merely by the desire to play a role, but by a deep-rooted conviction that it was both possible and necessary to unite the overwhelming majority of the Chinese people in a defence of their common interests. And he was certainly conscious of the weakness of the Chinese Communist Party, which still only had some four hundred members.

The reorganization of the Kuomintang, carried out under the guidance of Stalin's envoy Borodin, who arrived at Canton in October 1923, was a paradoxical undertaking. In the only previous example of such cooperation 'from within', Maring's experience in Indonesia, the Sarekat Islam had been a loosely-organized mass movement which could easily be taken in hand by a determined minority. Similarly, the Kuomintang, in the years before 1924, could hardly be called a party at all, as Borodin, whose task it was to make a party out of it, frequently remarked.†

Now, on the contrary, it was proposed to transform the Kuo-

* See Chapter 6, p. 118, the discussion of his version of the Nanchang Uprising.

† According to the testimony of Vera Vladimirovna Vishnjakova-Akimova, who served as Borodin's secretary and interpreter for more than a year towards the end of his mission in China. See her memoirs, *Dva Goda v Vosstavshem Kitae (1925–1927)* (Moscow, Izdatel'stvo 'Nauka', 1965), p. 179. Mme Vishnjakova-Akimova's book is an exceptionally important contribution to the history of the Soviet missions to China in the 1920s; it contains, in particular, entirely new information regarding the identity of certain military advisers which we shall have occasion to cite below.

mintang itself into a highly efficient organization based on Leninist principles of democratic centralism. Moreover, Sun Yat-sen was also to be provided with a modern army trained by Soviet instructors.* And Sun, though he had concluded by the end of 1923 that there was no alternative to reliance on the Soviet Union, had no intention whatever of allowing himself to become the puppet of Moscow. In terms of Soviet foreign policy objectives this strengthening of the Kuomintang organization was entirely desirable, for Moscow thereby gained a more effective ally. But in terms of the future of the Chinese revolution, it was fraught with danger for both partners. If they succeeded in gaining control of the Kuomintang organization at its middle and lower levels, the Communists could manipulate the whole of the Kuomintang for their own ends. If they failed to do so, they might be crushed by the machine which they themselves had helped to create.

For the moment Mao Tse-tung threw himself into his task of organizing cooperation with the Kuomintang so enthusiastically that he was soon looked upon with suspicion in his own party. Since his election to the Central Committee in June 1923 he had taken up residence in Shanghai, where the headquarters of the Chinese Communist Party was located. In January and February 1924 he visited Canton briefly to attend the First Congress of the Kuomintang. Here, for the first time, shortly after his thirtieth birthday, he emerged into view as a significant political figure at a forum attended by Nationalist leaders from all over China. The place which he had already won for himself in the hierarchy of the Communist Party is attested by the fact that he was one of the three Communist representatives delegated to serve on the 19-member committee for examining the new Kuomintang Party Constitution.† He also intervened frequently in the debate to underscore the Communists' firm support for the principles of Sun Yat-sen.

* The Whampoa Military Academy opened unofficially on 1 May 1924, and was solemnly inaugurated on 16 June by Sun Yat-sen; it was headed by Chiang Kai-shek; Chou En-lai ran the political department (see note on page 85).

† *Ko-ming wen-hsien*, Vol. 8, Taipei, 1955, p. 1152. This reference to Mao was very kindly called to my attention by Professor C. Martin Wilbur of Columbia University. The other two Communist representatives were Li Ta-chao and T'an P'ing-shan.

As already noted, the First Congress elected Mao an alternate member of the Kuomintang Central Executive Committee. In this capacity, he attended the meetings of the central party organization until his departure for Shanghai on 7 or 8 February. He left behind him a remarkably revealing statement of his political thinking at that time, in the form of four resolutions, which were discussed in his absence at the fourth meeting of the party centre on 9 February 1924.

These resolutions display in full measure the exceptional grasp of organizational problems which has been one of Mao's greatest assets throughout his political career. On the one hand, he protests against the plethora of high-level cadres in Canton. The central and provincial KMT party organizations, he declares, are 'hollow organizations'; the 'real organizations', which constitute 'the decisive organ for directing the activities of party members', are the bureaus in the cities and at an intermediate level in the countryside (*hsien* or regional bureaus). But at the same time, Mao emphasizes that the Kuomintang should not spread its human and financial resources too thin. In the course of the coming year, he proposes that 70 per cent of the party's resources should be devoted to eight or nine major centres where the movement is already well developed, and the remaining 30 per cent to eleven or twelve other places where there are real prospects for making a good start. The rest of the country should be neglected for the moment 'in order to avoid dispersing our energy'.

Mao's desire to decentralize the work of the Kuomintang did not reflect an interest in the peasantry. The localities in which he proposed to develop the party's efforts were primarily large cities. Nor is Changsha included in the list. The most likely explanation for this omission is that Mao wished to avoid giving the impression of favouring his native province. The text of the pertinent resolution includes only nine of the twenty localities in which Mao proposed an active effort during 1924, the others being represented by the Chinese equivalent of 'etc'. In Mao's mind, Changsha was probably included among the cities thus left un-named. That he was not indifferent to the progress of the revolution in Hunan is shown by the fact that in April 1924 he sent his friend Hsia Hsi, a former member of the New People's

Study Society which Mao had organized during his student days, to set up a Kuomintang organization in Changsha.*

Returning to Shanghai shortly after presenting these resolutions, Mao combined high office in the Chinese Communist Party with membership in the Shanghai Bureau of the Kuomintang. There his colleagues were Wang Ching-wei and Hu Han-min, soon to emerge as leaders respectively of the left and right wings of the party. By the end of the year he had, according to his auto-biography, become ill, and had to return to Hunan for a rest.[7] He may in fact have been tired from overwork, but there is little doubt that his illness was at least partly diplomatic. He was under heavy attack from those in the Chinese Communist Party opposed to an excessive emphasis on cooperation with the Kuomintang and to sacrifice of their own party's independence; his Hunanese comrade Li Li-san derided him as 'Hu Han-min's secretary'.[8] The double strain of hard work and hostile criticism proved too much, and Mao took refuge in his native village of Shao-shan.

Once he got there he found that conditions in the countryside were not nearly so restful as they had been in the past. The peasantry, which had played virtually no role in the upheaval of the May 4th period, was at last awakening to political activity. Hitherto this development had received more attention from the Kuomintang than from the Chinese Communist Party, though individual Communists in the Kuomintang apparatus had played an important part in it. As early as 1922 P'eng P'ai, a landlord's son who had joined the Communist Party, began organizing the peasants at Haifeng and Lufeng in his native province of Kwang-

---

* For the text of Mao's four resolutions, see the report of the fourth meeting of the central party organization in *Chung-kuo Kuo-min-tang Chou-k'an*, No. 9, 24 February 1924, p. 11; for Hsia Hsi's mission, see Hsia's letter, transmitted by the Shanghai Executive Bureau to the twenty-third session of the Central Executive Committee on 21 April, ibid., No. 31, 18 May 1924, p. 5. These materials were very kindly called to my attention and a microfilm supplied by Mr Roy Hofheinz, whose work on Chinese Communist rural politics is referred to in the Acknowledgements. Mr Hofheinz also put forward the above explanation of Mao's failure to mention Changsha in his plan of work for 1924, and contributed greatly, in the course of a discussion of these documents, to my interpretation of Mao's position in 1924.

tung.* When the Kuomintang established a Peasant Department in its Central Executive Committee in February 1924, it was P'eng who became the first secretary of the Department, and who took charge of the Peasant Movement Training Institute created in July 1924. P'eng P'ai soon left Canton to resume work among the Kuangtung peasantry, but other Communists continued to exercise a decisive influence in the Kuomintang Peasant Department.† The leaders of the party, however, showed little interest in the peasants. In August 1923 Ch'en Tu-hsiu had affirmed that, in a country of smallholders such as China, 'over half the peasants are petty bourgeois landed proprietors who adhere firmly to private property consciousness. How can they accept Communism?' And in December 1924 he still held that the proletariat was not only 'the principal force in the social revolution in the capitalist and imperialist countries', but 'must also play the role of commander in the national revolution of the countries oppressed by capitalist imperialism'. As for the peasantry, it was merely one of several allies of the proletariat, all of them 'inclined to compromise'.[9]

This disdain for the peasantry was not characteristic of the Comintern at the same period. On the contrary, the Peasant International or Krestintern had been founded in October 1923, and held its First Congress in 1924. And in a directive of May 1923 to the Chinese Communist Party, the International affirmed that the peasantry was 'the central problem in our whole policy'.[10] The divergence between Moscow and the Chinese Communists over the revolutionary qualities of the peasantry was a manifestation of their continuing disagreement over collaboration with other classes and parties in general. The Chinese had been obliged by Moscow to accept the alliance with Sun Yat-sen, but many of them continued to regard him with suspicion. Nor was their anxiety altogether without foundation. Sun was a strange and volatile character. After asking Borodin to write the text of a

---

* Haifeng and Lufeng (often designated collectively as Hailufeng) are located on the sea coast, north-east of Canton.

† On P'eng P'ai's activities, both in the Peasant Department and among the peasantry in Hailufeng and elsewhere, see the two articles by Eto Shinkichi in the *China Quarterly*, Nos. 8 and 9, October–December 1961 and January–March 1962, pp. 160–83 and 149–81.

manifesto for the First Congress of the Kuomintang, calling for a 'national revolution' directed against imperialism, he had become alarmed at the reaction of the party's conservative wing and, in the midst of the debates, proposed to his Soviet adviser the abandonment of this document and its replacement by the programme for a national government which he himself had composed. The latter text was characterized by what Borodin called the 'empty phraseology' of the pre-1924 Kuomintang. It took Borodin several hours to persuade Sun to give up this sudden whim and go ahead with the adoption of the document which had been agreed upon as the ideological foundation of Chinese-Soviet cooperation.*

In late 1924 Sun accepted an invitation from Tuan Ch'i-jui, head of the pro-Japanese Anfu clique, who had just regained power in Peking, to come north and discuss terms for the peaceful unification of the country. On the way he stopped in Japan and made speeches extolling Sino-Japanese friendship. In the face of these manoeuvres the organ of the Chinese Communist Party published early in January 1925 an article declaring that Sun Yat-sen had 'lost the confidence of the people', and that his attitude constituted 'an obstacle to the national revolution'.[11]

All of these reproaches were of course buried when Sun died at Peking in March 1925, and his name became the symbol of a firm policy of collaboration with the Communists for the good of China. But the tension persisted between Stalin, for whom the Kuomintang was merely an abstract entity promising a profitable alliance, and the Chinese Communists, who had to live with this ally and watched its growing political and military might with anxiety.

In this triangular relationship of Moscow, the Kuomintang, and the Chinese Communist Party, the position of Mao Tse-tung during the crucial years 1925-7 was on the whole closer to that of the Kuomintang than to that of either Stalin or Ch'en Tu-hsiu. This fact is obviously a source of extreme embarrassment to Mao

* See Borodin's own contemporary note regarding his conversation of 23 January 1924 with Sun Yat-sen, published in A. I. Cherepanov, *Zapiski Voennogo Sovetnika v Kitae. Iz Istorii Pervoj Grazhdanskoj Revoljutsionnoj Vojny (1924–1927)*, Vol. I (Moscow, Izdatel'stvo 'Nauka', 1964), pp. 67–72.

himself and to the historians now writing in Peking, who (as we shall shortly see) have distorted the record of his activities during this period. Actually there is nothing in the least discreditable about it, unless one considers it discreditable for an Asian revolutionary to be more concerned about the destiny of his country than about social revolution for its own sake. Mao agreed with Stalin whenever Chinese and Russian national interests appeared to coincide; when they did not, he took an independent line. He supported Chiang Kai-shek's plan for a Northern Expedition to unify the country well before Stalin accepted it, because national unity was a matter of vital importance to him. Subsequently he recommended breaking with the Kuomintang and relying on a radical agrarian revolution led by the Communists before Stalin became resigned to such a rupture, because his ultimate concern was not the security of Russia's Siberian frontier, but the salvation of the Chinese revolution.

Mao's discovery of the revolutionary force among the peasantry dates, as already indicated, from his retirement to Shaoshan in the winter and spring of 1925. At that point the peasant movement in Kwangtung had already attained considerable proportions. During Chiang Kai-shek's first 'Eastern Expedition' against Ch'en Chiung-ming in February and March 1925, the peasants of Hailufeng not only welcomed his troops with open arms, but provided them with precious assistance in the form of scouts, porters, etc. Reports of this 'Peasants' Association' had reached Hunan, and the peasants there began to organize spontaneously themselves.

This embryonic peasant movement in Hunan received a powerful impetus from the events which followed the incident of 30 May 1925, itself one of the great watersheds in the Chinese revolution. On that afternoon a column of workers and students, demonstrating against the killing of a Chinese worker by a Japanese foreman, was fired on by the police of the International Settlement in Shanghai, under the orders of a British police officer. Ten of the demonstrators were killed, and fifty wounded. The following afternoon a meeting at the Chamber of Commerce decided on a general strike of all students and shopkeepers, as well as of all workers in foreign-owned factories. Further shootings by the police during the succeeding days merely strengthened the determination of the population, and by the middle of June the movement had spread to Canton. There once again, on 23 June, the British and French police fired into the crowd, killing fifty-two people. The result was a total boycott of all traffic between Kwangtung and Hongkong which was to last sixteen months.*

Only six years earlier the peasantry had remained unstirred by the great anti-Japanese movement which swept the coastal cities. Now they soon manifested their hostility to the indignities inflicted on the Chinese by the foreigners, and this increased political awareness reflected itself in the progress of the peasant associations. 'Formerly I had not fully realized the degree of class struggle among the peasantry,' declared Mao to Edgar Snow, 'but after the May 30 Incident ... and during the great wave of political activity which followed it, the Hunanese peasantry became very militant. I left my home, where I had been resting, and began a rural organizational campaign.'[12]

One of the Soviet officers who served as an adviser in China at this time has recently affirmed in his memoirs that Mao came to Borodin's house early in 1924 'to consult him before returning to Hunan, where he was going to organize peasant associations'.[13] This is an obvious attempt to emphasize the Soviet role in the Chinese revolution, and the facts cannot be true as stated. Mao did not return to Hunan until later in 1924, and he did not have the idea of organizing peasant associations when he went there. But it is possible that Mao may have absorbed some ideas from Borodin, who declared during 1925, in lectures to Kuomintang members, that the success of the Chinese revolution would depend entirely on organizing the peasants for a solution of the land question.[14]

Whatever the sources of his inspiration, once he began his work, in the summer of 1925, Mao showed characteristic skill in adapting the slogans of the day to the mentality of his peasant auditors. 'Down with the militarists!' was immediately clear and

---

* Regarding this period see Chesneaux, op. cit., pp. 371–444. The demonstration of 30 May which triggered off the whole series of events had been organized in large part by the Chinese Communist Party; one of the most active in this respect had been Ts'ai Ho-sen, Mao Tse-tung's intimate friend from Changsha.

convincing; the Chinese peasantry had suffered too long from the depredations of armies of every hue to have any difficulty in grasping this slogan. 'Down with the imperialists!' was a bit more puzzling; the Chinese term for 'imperialism' (literally 'emperor-countryism') did not mean much to the peasant who heard it for the first time. Mao therefore explained that it meant 'Down with the rich foreigners!'[15]

Mao's activity in organizing the peasant associations eventually came to the attention of his old acquaintance Chao Heng-t'i, and he was forced to flee once more to Canton. He arrived there apparently sometime in October or November, after nearly a year's absence.*

In Canton Mao found a tense and complicated situation. The Fourth Congress of the Chinese Communist Party in January 1925, which he had missed because of his illness, had adopted a relatively sectarian line on relations with the Kuomintang, obviously reflecting the suspicions inspired by Sun Yat-sen immediately before his death. The 30 May incident had led to a momentary improvement in relations between all social groups and political tendencies in China, as they closed ranks against the atrocities of the foreigners, but its ultimate effects were profoundly divisive. The foreign interests whose factories were closed as a result of strikes supported by their Chinese competitors (the

---

* The date of Mao's arrival in Canton marks the beginning of a period of over a year during which the official historiography has grossly distorted the chronology of Mao's life, with the object of concealing the fact that he collaborated with Chiang Kai-shek's Kuomintang longer than any other leading Communist equally in view. This rewriting of history was begun by Mao himself in 1936, when he related the story of his life to Edgar Snow, but Mao was considerably more honest than many of those now writing in Peking. As regards the precise point at issue here, in his autobiography he indicates that his activity in organizing the Hunan peasants (which, as we have seen above, began in earnest only after 30 May) lasted 'a few months' before Chao Heng-t'i put an end to it (Snow, *Red Star Over China*, p. 157). This is compatible with the date given above for his return to Canton. Secondly, Mao's poem 'Changsha', obviously written in the autumn, is dated 1925 in the most recent edition of his poems, which would indicate that he remained in Hunan at least until September. Finally, a source which is not always accurate, but frequently contains useful information, says that Mao arrived in Canton in November (*Hsien-tai shih-liao*, Shanghai, 1934, Vol. IV, Part 3, p. 343, cited in Jerome Ch'ên, *Mao and the Chinese Revolution*, p. 100).

Chinese merchants and industrialists even contributed to a fund for the maintenance of the strikers and their families) soon saw to it that the Chinese-owned factories were obliged to shut down too, so that the Chinese capitalists found themselves paying compensation at a time when they had no income. And, more generally, the well-to-do Chinese were not at ease in a situation where they found themselves in acute conflict with their foreign colleagues and with the diplomatic and military representatives of all the powers, while the influence of the trade unions and of the Communist Party was manifestly growing by leaps and bounds. One result was the appearance in the Kuomintang, during the summer and autumn of 1925, of a 'new right wing' hostile to the divisive activities of the Communists, which proposed to expel the latter from the Kuomintang and establish a looser and less dangerous form of cooperation between the two parties. In August 1925 the strongest leader in the left wing of the Kuomintang, Liao Chung-k'ai, was assassinated, and Hu Han-min, one of the leaders of the 'new right', was so far suspected that it was necessary to send him off to Moscow.* The situation became so serious that, according to his own later affirmations, Ch'en Tu-hsiu proposed at the plenum of the Chinese Communist Party in October 1925 that the Communists withdraw from the Kuomintang, but he was overruled by the representative of the International. It is certain in any case that this plenum adopted a very hostile attitude towards the Kuomintang.[16]

Mao Tse-tung, on his return to Canton, does not seem to have been greatly disturbed by all this tension. On the contrary, it is at this point that he began to play a really major role in the apparatus of the Kuomintang. In this he was faithfully reflecting the position of Stalin, who by no means shared the hostility and suspicion of Ch'en Tu-hsiu and many of the other Communist leaders towards Chiang Kai-shek and the tendencies emerging in

* On Tai Chi-t'ao and the 'new right wing' see Brandt, op. cit., pp. 56–61; C. Martin Wilbur and Julie How, *Documents on Communism, Nationalism and Soviet Advisers in China* (New York, Columbia University Press, 1953), pp. 206–12. It was called the 'new right' in opposition to the 'old right' of the diehards, who had opposed the entry of the Communists into the Kuomintang from the very beginning, in 1923. While Hu Han-min was unquestionably suspicious of the Communists, the accusations of complicity in the murder of Liao Chung-k'ai may in fact have been unjust.

the Kuomintang. He was also, no doubt, seeking a suitable framework for applying his new ideas on the revolutionary capabilities of the peasantry, and the Kuomintang continued to display infinitely more interest in the peasantry than did the Central Committee of the Chinese Communist Party.

It was the Peasant Movement Training Institute, founded in 1924, which offered Mao his principal opportunity for influencing the course of the agrarian revolution. When the Institute began its fifth session on 1 October 1925, the number of students from Hunan suddenly jumped to nearly forty per cent of the total. Many of these were Communists recruited by the Hunan provincial organization of the party,[17] and Mao had probably played a role in encouraging this development. This hypothesis is strongly supported by the fact that his own brother, Mao Tse-min, who was later to be his close political collaborator, was one of the Hunanese students.[18] Mao was probably not in charge of this session, but he may well have given some lectures, and he certainly took an interest in it. He did serve as principal of the sixth training session, which began the following May, but by that time decisive events had intervened in the triangular relationship of Moscow, Chiang Kai-shek, and the Chinese Communist Party.

The first act in this complicated drama was provided by the Second Congress of the Kuomintang, which took place in January 1926. In addition to his interest in the peasant movement, Mao had taken over, on his return to Canton, the principal responsibility for the Propaganda Department of the Kuomintang Central Executive Committee. His title was originally secretary, and later deputy chief, but as the nominal head, Wang Ching-wei, was absorbed by his duties as head of the government, Mao in fact ran the department.* He therefore presented the report on propaganda at the Second Congress.

In the resolution on propaganda adopted by the Congress, which summarized the main ideas of Mao's report, appeared the following statement:

The success of a party depends necessarily on the fact that it has a

* The statement in his autobiography that he became 'chief of the Agitprop department' (Snow, op. cit., p. 157), though formally erroneous, is thus accurate in substance. Mao's position was parallel with that of Chou En-lai at Whampoa, who ran the political department of the Academy, though nominally he was only deputy chief.

centre of gravity. The centre of gravity of the Kuomintang is hidden among the countless masses of the exploited peasantry. The Propaganda Department must ceaselessly draw the attention of party members to this point, and direct them to rely more on this centre of gravity.[19]

Although Mao's interest in the peasant question at this point is undeniable, his rather numerous interventions in the debates of the congress reflected above all a strong concern for the problems of organization and discipline which were to be at the root of developments during the next few months. The gravest problem which confronted the congress in this respect was how to deal with the so-called 'Western Hills' faction, whose members had met in late November and early December in front of Sun Yat-sen's tomb in the Western Hills district of Peking and adopted resolutions expelling the Communists from the Kuomintang. Mao recommended leniency. The rightists, he urged, should not be punished for the 'Western Hills' meeting itself, but only for any subsequent breach of discipline. He also moved an amend-ment (adopted by the congress) doubling the delay extended to three of the worst offenders in which to mend their ways, 'because we still hope that they will return to the revolutionary path'.[20]

Whether Mao was merely expressing his personal view, or whether he spoke for at least a sizeable faction of the Chinese Communist Party, his intervention on this point is important and characteristic. In thus stressing comradely solidarity with men whom, although conservative, he regarded as sincerely committed to the struggle for the unity and independence of China, Mao clearly demonstrated that he attached (at least for the moment) a higher priority to the national revolution than to the social revolution.

Equally significant was the discussion of the resolution on party affairs, during which one of Chiang Kai-shek's collaborators at the Whampoa Military Academy proposed a three-point pro-gramme for allaying anxiety over the intentions of the Communists in the Kuomintang. All Communists who entered the Kuomin-tang must declare their membership in the Chinese Communist Party; their activity within the Kuomintang must be entirely open; and no Kuomintang member should join the Communist party without the authority of the local Kuomintang organiza-tion. First Chang Kuo-t'ao and then Mao Tse-tung replied on behalf of the Communists, and for once they adopted exactly the same position. These were splendid ideas in theory, but under the conditions existing in most of China it was impossible to declare openly that one was a Communist without being arrested and executed. To oblige part of the revolutionary forces to expose themselves to such blows, said Mao, 'would not be advantageous to the future of the national revolution'.[21]

The congress ultimately adopted a compromise proposed by Ch'en Kung-po, providing for a meeting between the executive bodies of the two parties to seek methods for reassuring the right about the activities of the Communists. Mao had lost his seat on the Central Committee of the Chinese Communist Party when he failed to attend the Fourth Congress, but he was now once more elected an alternate member of the Kuomintang Central Executive Committee. He received 173 votes, compared with a maximum of 248 for Wang Ching-wei, T'an Yen-k'ai, Hu Han-min, and Chiang Kai-shek, and 192 for Li Ta-chao.[22] Shortly after the congress he was appointed editor of a new Kuomintang organ, the *Political Weekly* (*Cheng-chih chou-pao*), which began publication at the beginning of March.

Although the problem of Communist influence in the leading organs of the Kuomintang was to play a part in the test of strength shaping up between Chiang Kai-shek and Moscow, the crucial issue was that of the Northern Expedition. A military expedition to overthrow the war-lords in Peking and establish a unified national government had been Sun Yat-sen's dearest ambition in the last years of his life. He had accepted Soviet assistance in large part because he had become convinced that it was the only way to create a strong military force of his own capable of undertaking this task. Now, not only his own ambition, but a concern for the future of the country impelled Chiang Kai-shek to carry out Sun's desires.

The Soviet viewpoint was altogether different. In March 1926, at the very moment when the decisive test was about to arise, the Politburo of the Soviet Communist Party met in Moscow to discuss a report on Far Eastern policy which had been prepared by a special commission presided over by Trotsky. Under the guiding hand of this apostle of 'permanent revolution', a document had been prepared dealing almost exclusively with the possibilities of diplomatic manoeuvring. Alarm had been raised in Moscow by recent incidents on the Chinese Eastern Railway. The Soviets now proposed to buy their security from Japan at the expense of Chinese national interests, by recognizing the *de facto* autonomy of the pro-Japanese Chang Tso-lin in Manchuria. They also entertained the idea of resuming dealings with Wu P'ei-fu in Central China, despite his massacre of the railway workers three years earlier. As for the revolutionary government in Canton, Rakovskij, the Soviet Ambassador in Paris, was to be asked to sound out the possibilities of obtaining recognition of its autonomy as well, to balance that of Chang Tso-lin. To the passage of the report dealing with Canton, the following clause was added at Stalin's own suggestion:

The Canton government should in the present period decisively reject the thought of military expeditions of an offensive character, and in general of any such proceedings as may encourage the imperialists to embark on military action.[23]

These deliberations took place on 25 March, five days after Chiang Kai-shek's coup on 20 March, of which the news had not yet reached Moscow. Obviously Mao could not approve Chiang's action in arresting a large number of Communists among the political commissars in the army, and confining the Soviet advisers to their residences. But to the extent that Chiang's motive was to put pressure on Moscow in order to obtain support for his projected Northern Expedition, he and Mao were in substantial agreement. On 30 March 1926, ten days after Chiang's coup, Mao participated in a meeting of the Kuomintang's Committee on the Peasant Movement, of which he was a member, and proposed a motion summarized as follows in the minutes:

Inasmuch as there is a close link between politics and the movement of the popular masses, and in view of the peasant movement now going on in many provinces, we should devote careful attention to the peasant movement in the provinces ... which will be crossed by the Northern Expedition in the future.[24]

His position on this occasion is doubly remarkable: as a manifestation of his imagination in foreseeing the effect which the Northern Expedition was in fact to have in stimulating the peasant movement, and as a confirmation of his independence from Moscow, which at the same date was firmly opposed to any such offensive by Canton. It is therefore not surprising that, once Chiang had made his peace with the Soviets by reinstating all the military advisers save two or three whom he particularly disliked,* and by carrying out a symbolic offensive against the right wing of the Kuomintang to balance his coup against the left, Mao should have gone on working in collaboration with him. What is more surprising is that he should have continued to occupy a post of such importance as that of principal of the Peasant Movement Training Institute during the entire period from May to October 1926. For part of the price exacted by Chiang in return for restoring good relations with Moscow, in addition to Stalin's support for the Northern Expedition, was a 'reorganization' of the Kuomintang designed to limit Communist influence within the party. The number of Communists who could belong to any leading organ was limited to one third, and Communists were barred from holding top posts such as the directorates of departments. Under this provision T'an P'ingshan was obliged to give up his leadership of the Organization Department, and Lin Tsu-han to quit the leadership of the Peasant Department. Mao himself lost his post as deputy head of the Propaganda Department. And yet he remained in charge of the Training Institute. Whether this can be explained by considerations of opportunism, or whether the key factor was a com-

* Among the latter was Nikolai Vladimirovich Kuibyshev (the brother of the well-known Soviet leader Valerian Vladimirovich Kuibyshev, then Commissar for Economic Affairs), who went under the pseudonym of Kisanka (generally misspelled Kisanko in Western sources). Kuibyshev had replaced Vasilij Konstantinovich Bljukher, who went under the pseudonym of Galin (generally misspelled Galen), as head of the Soviet military mission to Canton during the summer of 1925. At Chiang's request, he was now recalled to Russia and replaced in his turn by Bljukher, in whom Chiang had more confidence. Kuibyshev may have earned Chiang's dislike by his insistence on centralized administration of the whole Nationalist Army, a policy clearly designed to maximize Russian influence. It is also reported that Kuibyshev was a partisan of a plan proposed by Wang Chingwei for forming an anti-Chiang alliance in order to force Chiang not to yield to the demands of the anti-Communist faction in the Kuomintang. And in his diary Chiang complained that 'Kisanka' had 'ridiculed' him. Thus there were probably both personal and political reasons for the rupture. See Wilbur and How, op. cit., pp. 212–17, 267–8, etc.; Vishnjakova-Akimova, op. cit., pp. 190, 207, 236–7, etc. Bljukher was to be a victim of Stalin's purges in 1938.

mon bond of nationalism between Mao and Chiang, the fact is there.*

It is true that Mao's activities at the Peasant Movement Training Institute would not benefit Chiang and the Kuomintang. The young men whom Mao trained as rural agitators would, in the short run, help assure the Chinese Communist Party of control over the revolutionary process at the grass roots; after the rupture of 1927 many of them would accompany Mao when he led his first little guerrilla band into the mountains. But did Mao foresee this result? One who had known him well since Changsha days recalls how Mao told him in the spring of 1926 that he 'had a way' to remedy the weakness of the Communists in their dealing with Chiang Kai-shek, and has suggested that this 'way' involved preparing the cadres for a future guerrilla war at the Peasant Movement Training Institute.[25] It is probable that Mao, who could hardly have been entirely confident of Chiang's ultimate intentions after the March coup, did indeed take as one of his objects the creation of an independent power base for the Chinese Communist Party in the countryside to balance Chiang's military might. But it is also likely that in the summer of 1926 he still looked forward to a long period of collaboration, if not with Chiang Kai-shek, then with other elements of the Kuomintang.

The curriculum of the sixth session of the institute, which lasted from 3 May to 5 October 1926, was on the whole similar to that of the other sessions. Like his predecessors, Mao continued the heavy emphasis on military training (128 hours out of a total of 380), which corresponded both to the needs of the situation and to a fundamental bent in his own personality. In addition to his other duties he gave the lectures on the peasant question in China (twenty-three hours) and on methods of teaching in the countryside (nine hours) himself. He also participated personally in a highly characteristic programme which he in-

* It is attested by a document published in Peking in 1953 (*Ti-i-tz'u Kuo-nei ko-ming chan-cheng shih-ch'i ti nung-min yün-tung*, pp. 20–32), though this does not prevent Li Jui from trying to conceal the date of the sixth session in an article published in the very same volume. For Mao's appointment as principal at a meeting of the Peasant Movement Committee on 16 March, see *Chung-kuo Nung-min*, No. 4, 1926.

troduced for the first time, involving independent study of books and articles by the students, followed by the submission of written answers to questions which were corrected by Mao and one other teacher.[26]

The sixth session of the Peasant Movement Training Institute opened almost simultaneously with the Second National Labour Congress and the Second Provincial Peasant Congress, which began on 1 May. On 2 May Chiang Kai-shek presented a report to a joint session of the two congresses, entitled 'The Great Union of the Workers, Peasants, and Soldiers'. Recalling the aid given to his armies by P'eng P'ai's peasants during the eastern expedition of 1925, Chiang affirmed: 'The armed workers and peasants play a more important role in the revolution than the army.'[27] His words probably made a more favourable impression on the peasants present than did those of Liu Shao-ch'i, who declared three days later that the workers must 'take the peasants by the hand' and lead them forward.*

Liu's position, which was in harmony with that of the Central Committee of the Chinese Communist Party, helps explain why, in some respects, Mao was more at ease within the Kuomintang than within his own party. Among the Kuomintang members with whom he cooperated were, however, a certain number of Communists who shared his belief in the peasantry. Li Fu-ch'un, the brother-in-law of Ts'ai Ho-sen, who headed the Political Training Institute of the Kuomintang, invited Mao to lecture to his students on the peasant question,† and in September Mao took the whole student body of the Peasant Movement Training

---

* The expression he used – *t'i-hsi* – also means 'to carry in the arms like a child'. I am assuming that the article published in *Cheng-chih Chou-pao*, No. 14, 5 June 1926, corresponds in fact to Liu's report at the congresses. I have taken the date of the report from *Hsin Ch'ing-nien*, No. 5, p. 778.

† Li Jui, *Mao Tse-tung t'ung chih ti ch'u-ch'i ko-ming huo-tung*, p. 249. In line with his systematic distortion of the chronology of Mao's activities, Li Jui claims that this collaboration took place during the sixth session of the Peasant Movement Training Institute, in the autumn of 1925. As already explained, these two indications are incompatible; the events in question must have occurred either during the *fifth* session, in the autumn of 1925, or during the sixth session, but in that case in the summer of 1926. The latter hypothesis seems more probable. If this is correct, Li, too, continued to occupy a high post in the Kuomintang after the 'reorganization' of May 1926.

Institute on a two-week visit to Haifeng, where they were given lectures by P'eng P'ai and shown an agrarian revolution in action.[28]

Despite his continued intimacy with the Kuomintang, Mao was not entirely out of favour with the Chinese Communist Party. Indeed, in the course of the summer the reaction of Ch'en Tu-hsiu and his colleagues to Chiang's 'reorganization' led to a new role for Mao in the party hierarchy. At the second plenum in July 1926 a resolution was adopted to provide for a loosening of the ties between the two parties, without going so far as to abolish the 'bloc from within' or the principle of dual membership.[29] A natural corollary of this policy was to intensify the work of the Chinese Communist Party itself among the peasantry. (Hitherto the peasants had been regarded as petty bourgeois who could best be left to the Kuomintang.) A Peasant Department was therefore created under the Central Committee of the Chinese Communist Party, and Mao was appointed to head it.[30] But the resolution providing for greater independence from the Kuomintang was shortly annulled in Moscow, where reliance on Chiang Kai-shek continued to be the cornerstone of Stalin's policy, and it is not clear whether the separate Peasant Department of the Chinese Communist Party ever really existed except on paper. In any event all Mao's essential work with the peasantry up to July 1927 was carried out under the auspices of the Kuomintang.

During the summer of 1926, while Mao was training his peasant agitators in Canton, Chiang Kai-shek's Northern Expedition made triumphant progress. By August Hunan was entirely in the hands of the revolutionary army, and Chiang declared in Changsha:

Only after the overthrow of imperialism can China obtain freedom. . . . In the present world revolution, there is the Third International, which can be called the general staff of the revolution. . . . If we want our revolution to succeed, we must unite with Russia to overthrow imperialism. . . . If Russia aids the Chinese revolution, does that mean she wants to oblige China to apply Communism? No, she wants us to carry out the national revolution. If the Communists join the Kuomintang, does this mean that they want to apply Communism? No, they do not want to do that either, they want to apply the Three People's Principles. I am persuaded that the Communists who have joined our party do not, at the present time, want to apply Communism, but want rather to carry out the national revolution. . . . The Chinese revolution is part of the world revolution. We want to unite the partisans of the world revolution to overthrow imperialism. . . .[31]

It is easy to see how a Communist such as Mao, who had demonstrated his commitment to the national revolution by rallying to the support of the Northern Expedition, would be more favourably impressed by this language than most of his comrades. There is, incidentally, a striking parallel between Chiang Kai-shek's words just cited, and Mao's language in the article written in February 1926, which has become the first item in the currently accepted canon of his *Selected Works*. Mao, too, was discussing the problem of how the Kuomintang should lead the national revolution. And in opposing the 'two huge banners' of the Third International and of the counter-revolution, he was not taking an uncompromising class line or affirming that the proletariat and/or the Chinese Communist Party should lead the revolution,* but merely repeating slogans that Chiang himself would still not hesitate to use six months later. This has not, of course, prevented the historians in Peking from presenting his 1926 article as ' he earliest and most clear-cut Marxist-Leninist document in China' which 'correctly solved all the fundamental problems' of the Chinese revolution.[32] In fact, Mao was still only an apprentice in the study of Marxism-Leninism, and the original version of the article contains a number of theoretical blunders.[33] He acquired an adequate grasp of Marxist theory only some ten years later, when the pattern of his rise to power had crystallized in practice.

In the practical experience which was to shape Mao's later theoretical conceptions, few episodes are more important than those during the winter of 1926–7. At the close of the session of the Peasant Movement Training Institute in October Mao apparently went briefly to Shanghai, to occupy his post as director of the Peasant Department of the Communist Party.† But he was soon on the move again. He may have inspected the peasant

* The only references to 'proletarian leadership' were inserted in 1951. For an example of this rewriting, see Schram, op. cit., p. 177.

† Snow, op. cit., p. 158. Mao's chronology is so hazy that it is hard to know when, or whether, he went to Shanghai.

movement in the provinces immediately north and south of Shanghai in November;* in December he was in Changsha, addressing the first provincial peasant congress.[34] And then he embarked on the investigation of the peasant movement in the province as a whole which was to form the subject of his famous report.

The problem of the ideological orthodoxy of this document, regarding which so much ink has been spilt,† can be summed up in a few words. The importance attached by Mao to the peasantry was not itself heretical, inasmuch as Lenin himself had spoken at the Second Comintern Congress of peasant soviets and of a peasant-based revolution in the East.‡ He fell into error, by orthodox standards, in failing to mention the leading role of the proletariat,§ but inasmuch as the report was addressed to the Kuomintang, it would have been tactless to do so. In any case, the importance of the report lies not in its Marxist ideological content, which is primitive, but in the revolutionary passion which animates it, and the historic moment it records.

At the time of the May 4th Movement Mao Tse-tung had written and spoken with great eloquence of the need for a cultural revolution as the only means of liberating the energy to achieve China's national goals. Then for several years he had become so preoccupied with his work, in the Kuomintang apparatus, for the national revolution, that his interest in the need for changing the nature of Chinese society was largely neglected. Suddenly, in the Hunanese countryside, he found himself face to face with

* The article, 'The Bitter Sufferings of the Peasants in Kiangsu and Chekiang and Their Movements of Resistance', published in *Hsiang-tao* on 25 November (extracts in Schram, op. cit., pp. 178–9), may have been based on personal observation, or on the reports he received in Shanghai. The first hypothesis seems more likely, for the account has the vividness of first-hand impressions.

† See, in particular, the exchange between Karl August Wittfogel and Benjamin Schwartz, in *China Quarterly*, Nos. 1 and 2, 1960.

‡ He had even suggested, in a note recently published, that the composition of the Communist Party itself should be adapted to the conditions in peasant countries. On this, and on Lenin's ideas regarding a peasant-based revolution in general, see Carrère d'Encausse and Schram, *Le Marxisme et l'Asie*, pp. 45–6.

§ The references to proletarian or Communist direction were added in 1951, as in the case of Mao's 1926 article. See Schram, op. cit., p. 184.

an outpouring of revolutionary energy such as he had scarcely ever imagined:

> In a very short time, several hundred million peasants in China's central, southern, and northern provinces will rise like a tornado or tempest – a force so extraordinarily swift and violent that no power, however great, will be able to suppress it. They will break through all the trammels that now bind them and push forward along the road to liberation.[35]

The liberation that Mao saw at work was of two kinds: cultural and social. The first, which involved primarily the overthrow of superstitions and conventions – from ancestor worship to the authority of the husband over the wife – might have been acceptable to a large part of the Kuomintang. The second type of liberation most decidedly was not. To be sure, Mao spoke frequently in his report of attacks directed exclusively against 'local bullies and bad gentry', and not against the landlords as such. But in other places his language was embarrassingly clear. 'A revolution', he wrote, 'is . . . an act of violence whereby one class overthrows the authority of another. A rural revolution is one in which the peasantry overthrows the authority of the feudal landlord class.'[36]

If the Kuomintang had been a 'peasant party', as not only Stalin but Trotsky believed,* its leaders should have welcomed the triumph of their class over its oppressors. Unfortunately, most of the officers of the 'revolutionary army' had issued either from the landlord class itself or from well-to-do urban families related to the landlords by ties of blood. The remarkable sympathy of the Kuomintang for the peasant movement in the years 1924–6 had been due in large part to the fact that the landlords whose interests the peasants menaced were those supporting Ch'en Chiung-ming in Kuangtung. But now the revolutionary tempest unleashed by the Northern Expedition was sweeping across the very areas from which the advancing armies recruited most of their officers. A clash was inevitable – unless the Communists, who had trained most of the peasant agitators, even if

---

* In a note written in September 1926, Trotsky described the Kuomintang as the 'peasants' own party', comparable to the Russian Social-Revolutionaries (Brandt, op. cit., p. 217).

they had done so in the name of the Kuomintang, agreed to call off the rural revolution.

That is exactly what Stalin ordered them to do. As early as October 1926, when the problem first began to become acute, he had sent a telegram urging the Chinese Communists to restrain the peasant movement.* Two months later, at the Seventh Plenum of the Executive Committee of the International, he tried to find a formula for eating his cake and having it too.

The core of Stalin's position on this occasion was reliance on the 'revolutionary army' of Chiang Kai-shek. From the exaggerated scepticism towards the Northern Expedition which had characterized his attitude before it began, he had swung over to an equally excessive optimism over the nature and intentions of the Kuomintang forces. To those who claimed that, in the areas where the revolutionary army had established itself, a 'certain disillusionment' had set in, he replied that 'the same thing' had happened in the Soviet Union during the civil war. Thus an explicit parallel was established between Chiang's army and the Red Army in 1919. Once this premise was accepted, it seemed easy to solve the problem of reconciling the need to gain the sympathies of the peasants with the no less urgent need to avoid alarming the Kuomintang by a radical agrarian policy: it would suffice to 'satisfy the urgent demands' of the peasant masses through the revolutionary army, as well as through the apparatus of the 'people's revolutionary government'. (The Communists were invited to join this government and push it towards the left.) In any event, there was no other practical possibility, for 'some tens of thousands' of Chinese revolutionaries (the Chinese Communist Party had now grown to these dimensions) could not possibly organize the 'ocean of peasants' except through the Kuomintang's governmental and military apparatus.[37]

Was this crass misrepresentation of the nature of the Kuomintang forces and of the Communist capacity to manipulate them due primarily to ignorance; or to utter cynicism over the fate of the Chinese revolution, provided that the interests of Russia herself were safeguarded? The answer, no doubt, is a mixture of the two. The roots of Stalin's policy appear with striking clarity

---

* He admitted as much himself later. See his speech of August 1927, *Works*, Vol. 10, p. 18.

in a speech of January 1925, in which he analysed the contribution which four 'allies of the Soviet power' could make to 'our régime, our *state*'. The first and most important ally was the proletariat of the developed capitalist countries, but unfortunately the 'state of the revolutionary movement' in Europe did not offer much hope from that quarter in the immediate future. The second ally, 'the oppressed peoples in the underdeveloped countries', constituted 'an immense reserve for our revolution', but were 'very slow in getting into their stride'. Consequently it was necessary to rely on the third 'ally', the contradictions among the capitalist countries (i.e. on diplomatic manoeuvring), and above all on the support of the fourth ally, the Russian peasantry.[38]

This egoism, plus the overweening confidence of a man who believed that he understood everything about a large and complex country very different from his own because he was the guardian and the prophet of Marxist orthodoxy, goes far to explain why Soviet policies in China during this period ended in total defeat. This does not mean that, with more realistic policies, it would have been easy, or even possible at all, to deprive Chiang Kai-shek of the fruits of his triumph and establish a régime under Communist control. Added evidence of just how difficult an undertaking this was is provided by the experience of those within the 'Left Kuomintang' itself, who shortly endeavoured to establish civilian control over Chiang and his armies.

The coup of 20 March 1926 had given a new impetus to the polarization of the Kuomintang into left and right factions. During the last months of 1926, while the party and government authorities were moving northwards in the wake of the army and making plans for the establishment of the revolutionary government in Central China, the tension between Chiang and the left wing continued rapidly to increase. It was unquestionably fostered by the Chinese Communists, who did not share Stalin's confidence in Chiang Kai-shek, but it also grew out of personal rivalries between Chiang and the leaders of the left, and out of the natural anxiety of the civilian authorities in the face of an all-powerful military commander who made his own strategic decisions without consulting them. By January 1927 relations had reached the point of open split between the 'Left Kuomintang' in Wuhan, and Chiang with his partisans established in Nan-chang. Moreover, Borodin and the other Soviet advisers had changed their tactics and were openly mobilizing support against Chiang because of the latter's clearly proclaimed intention to move eastwards against Shanghai, instead of driving immediately to-wards Peking. In the middle of March 1927 the Third Plenum of the Kuomintang Central Executive Committee, meeting in Wuhan, reorganized the party, military, and governmental hierarchies in such a way as to subordinate Chiang to the party authorities. Wang Ching-wei, in exile in France since the spring of 1926, was named to head both the party and the government; in his absence, T'an Yen-k'ai occupied the leading role.

Apart from these organizational measures, the third plenum took concrete steps at last to put into practice Stalin's counsel of the previous November on the entry of a minority of Com-munists into the government. Five new ministries were created, and Communists were appointed to head two of them; T'an P'ing-shan was named Minister of Agriculture and Su Chao-cheng Minister of Labour. Mao Tse-tung, who had just returned from his inspection trip in Hunan, participated in the plenum in his capacity as alternate member of the Kuomintang Central Ex-ecutive Committee.[39]

Although he was not one of those appointed to a ministerial post, Mao played an active part in the discussion of the peasant issue at the third plenum. On 15 March he defended a set of 'Regulations for the Repression of Local Bullies and Bad Gentry' that provided for the death penalty or life imprisonment for those who indulged in counter-revolutionary activities. In a revolutionary situation, he declared, 'peaceful methods cannot suffice to overthrow the local bullies and bad gentry'. They should be dealt with by revolutionary methods, 'preferably by the direct action of the peasants themselves'.*

* Chiang Yung-ching, *Pao-lo-t'ing yü Wu-han cheng-ch'üan* (*Borodin and the Wuhan Government*) (Taipei, Chung-kuo Hsüeh-shu chu-tso chiang-tsu wei-yüan-hui, December 1963), p. 259. This work is naturally most hostile towards the Communists, but Mr Chiang has had access to the archives of the Kuomintang, containing the manuscript reports on the proceedings of the third plenum, and of other meetings in which Mao took part in April 1927. His book thus sheds totally new light on Mao's position at this time.

It is clear that Mao was still caught by the revolutionary tempest he had witnessed in Hunan, and his stand on the land question continued to be influenced by this experience during the ensuing weeks. On 2 April 1927 the Kuomintang Central Executive Committee in Wuhan established a 'Central Land Committee' of five members, including Mao and T'an P'ing-shan.[40] This committee first held several meetings of its own members, at one of which Mao summarized the essence of what he had learned in Hunan:

> What we call land confiscation consists in not paying rent; there is no need for any other method. At the present time, there is already a high tide of the peasant movement in Hunan and Hupei, and on their own initiative the peasants have refused to pay rent, and have seized political power. In solving the land question in China, we must first have the reality, and it will be all right if legal recognition of this reality comes only later.[41]

On 19 April the first in a series of 'enlarged meetings' of the Land Committee was held, attended by a number of political workers active in the peasant movement, as well as by leaders of the Wuhan Government. Here Mao justified the confiscation of the landlords' land in terms of the national revolution. The revolutionary forces could move forward, he said, only if an 'army of production' supplied them with food, and land reform was the best way to enlist the support of the peasants in this respect.[42]

In order to arrive at a definite policy on land reform, the Land Committee appointed a group of five people, of whom Mao was the leading figure, to prepare a 'Draft Resolution on the Land Question'. This text provided for the confiscation of all land from 'local bullies and bad gentry, corrupt officials, militarists and all counter-revolutionary elements in the villages.'[43] These criteria in themselves, though somewhat ambiguous, were relatively moderate. But a 'Land Survey' also carried out by Mao and his group of five contained the following much more radical statement: 'All rich peasants, small, middle, and big landlords, possessing over thirty *mou* [four and a half acres], representing altogether 13% of the population, are uniformly counter-revolutionary.'[44] It will be recalled that Mao's own father had owned 22 *mou* of land; he was thus fixing the criterion for a counter-revolutionary attitude at a very low level indeed.

It is in this spirit that Mao explained his draft to the committee, his presentation illuminating his approach to revolutionary action, with its mixture of national and social aims:

What we are establishing now [he stated] is political confiscation, that is, confiscation of the land of local bullies and bad gentry, militarists, etc. This is the first step. If we go a step farther and speak of the confiscation of the land of all those who do not cultivate it themselves, but rent it out to others, then this is an economic measure.* Such economic confiscation is no longer a problem in Hunan; there, the peasants themselves have already divided up the land. . . . The militarists in Hunan are exploiters of the peasants. The National [i.e. Wuhan] Government, after establishing itself in Hunan, has also been unable to eliminate this exploitation completely.

Here Mao explained at some length that it was the need to draw from the peasantry the resources necessary for the prosecution of the war that had perpetuated this exploitation, and that a much larger sum could be obtained at less hardship to the population if the land were confiscated, since then the peasants, having no rent to pay, would be able to pay considerably heavier taxes with ease. He also drew explicitly the conclusion implicit in the 'Land Survey', but not stated openly in the 'Draft Resolution on the Land Question' – that the land of the rich peasants, like that of the landowners, should be confiscated. Finally, he underscored heavily the great difference in local conditions, and the consequent need for a flexible policy. In Hunan, where the peasants had already begun dividing up the land, 'political confiscation' would be insufficient; conversely, in other provinces, where the peasant movement was less advanced, 'economic confiscation' would be premature.[45]

At the meeting of the Enlarged Land Committee where Mao presented his resolution on 22 April, Wang Ching-wei expressed his anxiety at this text, which called 'for political confiscation in name, but for economic confiscation in fact'. In his view, political

* The terms 'economic' and 'political' confiscation were used at the time in two different senses, to designate either the *method* or the *aim* of confiscation; i.e., 'economic confiscation' meant either confiscation by 'economic' means, such as by refusal to pay rent, or confiscation with the economic aim of changing the agrarian structure, as opposed to the political aim of consolidating the revolutionary forces. Here Mao uses it in the second sense.

confiscation should be extended only to the large landlords. This criticism from Wang can hardly have surprised Mao. A much unkinder cut was the attack by Hsia Hsi, who was not only a Communist but had been a close friend and political associate of Mao's ever since the days of the 'New People's Study Society' and other groups formed by Mao in Changsha during his student days. Hsia claimed that the resolution was 'full of contradictions', would lead to 'an immediate struggle between the poor and rich peasants', and was generally impracticable.[46]

Another Communist, P'eng Tse-hsiang, attacked the resolution as insufficient, and demanded 'unconditional confiscation' of all land immediately. Replying to those critics who complained that he went either too far or not far enough, Mao began by stating that if only there were four or five provinces such as Hunan, it would be possible to proceed immediately to a complete solution of the land question. But under existing circumstances, this would be premature. He then proceeded to re-state his basic idea, according to which facts were more important than legal formulas: 'It suffices that people do not pay rent to the landlords, and you have a system in which the land question is resolved; legal recognition can come after the *fait accompli*.' Once the time was ripe for 'economic confiscation' in the whole country, another resolution could be prepared, providing for a complete solution to the land question.[47]

On 20 April 1927, at the Fourth Enlarged Meeting of the Land Committee, a set of 'Regulations for Protecting the Land of Revolutionary Military Men' was adopted which exempted the land even of the biggest landowners if they had relatives in the army. For the rest, it was generally agreed that Mao's system of 'economic confiscation' – that is to say, of direct action by the peasants in refusing to pay rent – was not a suitable method, and that 'political confiscation' – within the narrow limits permitted by the exemption for military men – was the best solution. It was also decided to invite Ch'en Tu-hsiu, Ch'ü Ch'iu-pai, and Chang Kuo-t'ao to attend the sessions.[48]

On 26 April Borodin also participated in the meeting, and recommended a differentiated policy. In areas controlled by the Wuhan Government, there should be only limited and orderly confiscation; in other regions, on the contrary, sweeping and radical slogans should be employed to excite the peasants. Ch'en Tu-hsiu expressed general agreement with Borodin's views. The resolution on the solution of the land question finally adopted on 6 May provided only for the confiscation of the land of large landlords, and specifically exempted both small landlords and 'revolutionary' military men. It also provided, as demanded both by Mao and by Borodin, for the organization of local self-government by the peasants.[49]

Such were the April discussions. In his autobiography Mao affirmed that on this occasion he 'discussed the proposals' of his 'thesis' (presumably the draft resolution written under his direction), which 'carried recommendations for a widespread redistribution of land'.[50] But at the same time, he gives the impression that the decisions adopted corresponded closely to his own position, whereas we have seen that this was not the case.*

The problem is further complicated by the fact that, in the course of the month of April, Mao was also appointed a member of the five-man Standing Committee of the Provisional Executive Committee of the All-China Peasant Association. Three of the five members of this Standing Committee were the same as those of the Land Committee: Mao himself; T'an P'ing-shan, the Minister of Agriculture in the Wuhan Government; and Teng Yen-ta, an important leader of the Kuomintang left. But the Standing Committee of the Peasant Association was under the chairmanship of T'an Yen-k'ai, a close political associate of Wang Ching-wei who, like Wang himself, had hitherto been regarded as a member of the Kuomintang left, but was becoming increasingly reticent in the face of the revolutionary action by the peasants. Mao himself claims that this organization was formed under the auspices of the Chinese Communist Party, but it is clear that a body headed even nominally by a leading figure of the Kuomintang could hardly have been created by the Communists.† In any event, its function was to consist primarily in

* For other accounts of the April discussions, see Robert C. North and Xenia J. Eudin, *M. N. Roy's Mission to China. The Communist-Kuomintang Split of 1927* (Berkeley, University of California Press, 1963), p. 74, and also Vishnjakova-Akimova, op. cit., pp. 349–50.

† The above list of leaders is based on the signatures at the end of a manuscript document from the Isaacs Collection at the Hoover Institution. (Directive dated 13 June 1927.) Eto Shinkichi cites a Japanese source which

restraining the peasants in order to prevent the disruption of the alliance between the Communists and the Kuomintang left. This was the more urgent because the Wuhan Government relied for its military support on the Hunanese war-lord T'ang Sheng-chih, who had rallied to the Kuomintang cause at the time of the Northern Expedition. T'ang affected a very liberal attitude, and had begun as early as October 1926 to cultivate good relations with the Communists.[51] But this did not change the fact that his officer corps was in large part the emanation of the Hunanese landlords, and consequently most directly affected by the events which Mao had been observing with so much enthusiasm a few weeks before. Ho Chien, the commander of the Thirty-Fifth Army under T'ang Sheng-chih, declared openly and frankly in the course of the April discussions on the land question that a great many officers and men, especially in his own army, owned land, and that this fact should be taken into account in deciding on an agrarian policy.[52]

Despite all these warning signals, Moscow was still insisting on collaboration not only with the Kuomintang left in Wuhan, but with Chiang Kai-shek. Chiang very soon solved this problem himself by his celebrated massacre of the Shanghai workers in April 1927, after their insurrection had delivered the city over to him. In the shadow of this disaster, the Chinese Communist Party held its Fifth Congress in Wuhan at the end of April, and endeavoured to devise a policy which would make it possible to rely on the shield of T'ang Sheng-chih's military might, and yet carry forward the agrarian revolution. Not surprisingly, the attempt failed, and they got neither security nor revolution. But they did get experience, and from that experience some of them were to draw the conclusion that they could manage their own affairs at least as well as Stalin.

---

also identifies T'an Yen-k'ai as chairman (*China Quarterly*, No. 9, 1962). For Mao's account, see Snow, op. cit., p. 159. Chiang Yung-ching does not mention this body, but it is probable nevertheless that it was formed in the course of the April discussions, perhaps as a means for ensuring that the practical application of the resolutions adopted did not distort them in a radical sense.

Notes

1. Carrère d'Encausse and Schram, *Le Marxisme et l'Asie*, p. 265.
2. *Hsiang-tao Chou-pao*, No. 38, 29 August 1923, pp. 286–7.
3. Schram, *Political Thought of Mao*, p. 266.
4. ibid., p. 140.
5. ibid., p. 142.
6. *Hsiang-tao Chou-pao*, No. 30, 20 June 1923, p. 228.
7. Snow, *Red Star Over China*, p. 156.
8. Brandt, *Stalin's Failure in China*, p. 37.
9. Ch'en Tu-hsiu, *Hsiang-tao Chou-pao*, No. 34, 1 August 1923, and *Hsin Ch'ing-nien*, No. 4, 1924, p. 22.
10. *Strategija i taktika Kominterna v natsional'no-kolonialnoj revoljutsii na primere Kitaja* (Moscow, 1934), pp. 114–16.
11. Article by P'eng Shu-chih, close friend of Ch'en Tu-hsiu, in *Hsiang-tao Chou-pao*, No. 98, 7 January 1925, pp. 817–18.
12. Snow, op. cit., p. 157.
13. Cherepanov, *Zapiski Voennogo Sovetnika v Kitae*, p. 84.
14. Borodin, *Kuo-chi cheng-chih chi Chung-kuo ko-ming ti ken-pen wen-t'i*, third edition (Canton, January 1927).
15. Li Jui, article in *Ti-i-tz'u kuo-nei ko-ming chan-cheng shih-ch'i ti nung-min yün-tung* (Peking, Jen-min Ch'u-pan-she, 1953), p. 264.
16. Wilbur and How, *Documents on Communism*, pp. 92, 234–6; Brandt, op. cit., p. 60.
17. Li Jui, in *Ti-i-tz'u kuo-nei ko-ming chan-cheng shih-ch'i ti nung-min yün-tung*, p. 265.
18. Mao Tse-min's name is included in the complete list of 113 students of the fifth session published in *Chung-kuo Nung-min*, No. 2, 1926.
19. *Chung-kuo Kuomintang ti-erh-tz'u ch'üan-kuo tai-piao ta-hui hui-i chi-lu*, April 1926, published by the Central Executive Committee of the Kuomintang, p. 136.
20. ibid., p. 179.
21. ibid., pp. 165–7.
22. ibid., pp. 145–6.
23. Carr, *Socialism in One Country*, pp. 769–72.
24. *Chung-kuo Nung-min*, No. 5, 1926.
25. Souvenirs of Professor Pai Yü, cited in greater detail in S. Schram, 'The "Military Deviation" of Mao Tse-tung', *Problems of Communism*, No. 1, 1964, pp. 50–51.

26. Report on the session in *Chung-kuo Nung-min*, No. 9, reprinted in *Ti-i-tz'u kuo-nei ko-ming chan-cheng shih-ch'i ti nung-min yün-tung*, pp. 20–32.

27. *Chung-kuo Nung-min*, No. 6/7, 1926.

28. *Ti-i-tz'u kuo-nei ko-ming chan-cheng shih-ch'i ti nung-min yün-tung*, p. 23.

29. Brandt, op. cit., pp. 82–3; Wilbur and How, op. cit., pp. 278–81.

30. Snow, op. cit., p. 158.

31. *Chiang Chieh-shih Ch'üan-shu* (*The Complete Works of Chiang Kai-shek*) (n.p., 1927). The preface, dated 15 January 1927, is signed with a pen name, 'Tung-ya wu-wo-tzu' ('East Asian Altruist').

32. Ho Kan-chih, *A History of the Modern Chinese Revolution* (Peking, 1958), p. 108.

33. Schram, op. cit., pp. 28–30, 143–7.

34. Li Jui, op. cit., pp. 259–61.

35. Schram, op. cit., pp. 179–80.

36. ibid., p. 182.

37. Stalin, *Works*, Vol. 8, pp. 384–8, passim.

38. ibid., Vol. 7, pp. 25–9.

39. K. A. Wittfogel, *China Quarterly*, No. 2, 1960, p. 17, citing *People's Tribune*, 18 March 1927.

40. Chiang Yung-ching, *Borodin and the Wuhan Government*, p. 278.

41. ibid., p. 282.

42. ibid., p. 284.

43. ibid., p. 286.

44. ibid., p. 289.

45. ibid., pp. 289–90.

46. ibid., pp. 291–2.

47. ibid., p. 293.

48. ibid., pp. 297–8.

49. ibid., pp. 298–304.

50. Snow, op. cit., p. 158.

51. Wilbur and How, op. cit., pp. 417, 394, etc.

52. Chiang Yung-ching, op. cit., p. 294.

Chapter 6
The Years in the Wilderness

Chiang Kai-shek's break with the Communists, and his merciless repression of the workers' movement in Shanghai, effectively ended the policy of collaboration between the Chinese Communist Party and the Kuomintang; but it would be over five months before Stalin agreed to drop the principle, and even longer before certain elements in the Chinese Communist Party broke completely with the past. In Stalin's case the principal motive was to safeguard the myth of his omniscience. In the spring of 1927 the lines were being formed for the decisive battle with the opposition of Trotsky and Zinoviev within the party. If Stalin could be shown to bear the responsibility for the ruin of the Chinese revolution, it would be a powerful blow to his prestige. The Opposition, which had hitherto never treated China as a major issue, therefore launched a full-scale attack. Stalin's defence was to declare that he had always foreseen Chiang's defection, and that it did not really matter, for the same policy would now be shown to work in dealings with the authorities in Wuhan.

The reaction of the Chinese Communists was different, and more complex. They had fewer illusions than Stalin about the 'Left Kuomintang' in Wuhan, though some of them remained unduly optimistic on this score. But at the same time they would be exposed to the full fury of repression in the event of a break with the Kuomintang. Those of them like Mao who still had the good fortune to go on living peacefully under the régime of Wang Ching-wei and T'ang Sheng-chih could see only too well, from the example of their comrades in Shanghai and elsewhere, what would probably be their fate if it came to a test of strength. Almost simultaneously with Chiang's blow in Shanghai, Chang Tso-lin in Peking had staged a raid on the Soviet Embassy, and a number of Communists, including Li Ta-chao, who were arrested on this occasion, were shortly executed by strangulation. Repression began also in Canton. In such a situation there were two natural reflexes: to cling to the alliance with the Wuhan Kuomintang at any price, however many compromises might be required; or to break radically with bourgeois nationalism in any form, and to make a desperate effort to carry out a revolutionary uprising which would overthrow the power of the Kuomintang and enable the Communists to avenge their fallen comrades. The party was to try these two policies successively. After both of them had failed Mao Tse-tung began his long and tortuous effort to devise and apply a method combining revolutionary violence and tactical flexibility, which finally opened the way to victory.

The dislocation of the alliance between the Chinese Communist Party and the Kuomintang took place in three stages: a pause while Stalin decided what to do next, accompanied by strenuous efforts to conciliate the Kuomintang leaders in Wuhan; an attempt to continue the cooperation with Wang Ching-wei and T'ang Sheng-chih, and at the same time to establish Communist control at the grass roots, which finally led in July to the rupture with Wuhan; and a last, forlorn effort to use the 'flag of the Kuomintang' in agreement with a tiny rump of Left Kuomintang leaders who were still prepared to cooperate with the Communists.

The Fifth Congress of the Chinese Communist Party took place in Wuhan two weeks after Chiang Kai-shek's Shanghai coup. The debates were characterized by a remarkable degree of confusion, engendered not only by the inherently contradictory nature of Stalin's policy, but by the fact that the two Soviet representatives on the spot interpreted the Moscow line in widely different ways. Borodin, whose official title was that of adviser to the Kuomintang Government, was still the most powerful emissary of Moscow in China, but he had recently been joined by an official Comintern delegate, the Indian Communist M. N. Roy. As he had been doing ever since the Second Comintern Congress in 1920, Roy ostensibly defended the prevailing orthodoxy, but in fact distorted it in the direction of greater sectarianism and intransigence towards the Communists' bourgeois allies. Borodin, on the other hand, was inclined to go even further than Stalin himself in conciliating the Left Kuomintang leaders, and Ch'en Tu-hsiu supported him in this.

One of the most curious and interesting manifestations of the divergence between the two men was to be found in Borodin's

'Northwestern Theory', and Roy's rebuttal of it. This theory called for a temporary retreat of the Communist Party into the hinterland, where the forces of reaction were not so strong as they were in the great coastal cities. At first sight this sounds like what Mao actually did later, but in fact it implied something totally different: not the creation of an independent armed force under Communist control, but the replacement of one military protector by another. More and more anxiety was felt about the attitude of T'ang Sheng-chih; it would therefore be better to move out of his reach and to rely on the benevolence of the 'Christian general', Feng Yü-hsiang. A corollary of this idea, advanced by Borodin and Ch'en Tu-hsiu, was that the Communists should support the plans of the Wuhan Kuomintang to take Peking, thus opening a pathway to Feng's sphere of influence in the north.

Roy defended the opposite view. Instead of 'broadening' the revolution in this way (and in fact merely weakening it), the Communists should 'deepen' it by returning to the original base in Kwangtung, where the revolutionary ferment was furthest advanced, and also by promoting agrarian revolution.*

Agrarian reform was clearly the touchstone of the direction in which the Wuhan régime was heading, for only agrarian reform could change the relationship among the classes in the country-side, and ultimately provide a real basis for 'proletarian hege-mony'. On this question, as in all of its deliberations, the congress endeavoured to follow the directives of the Seventh Comintern Plenum of November–December 1926, despite the total change in the situation which had ensued. On the agrarian question the instructions were not very precise; Stalin had advised the Chinese Communists to go ahead with land reform, but only within the limits of what was acceptable to the Kuomintang. Concretely, it had been stated in the theses of the seventh plenum that only the lands belonging to the reactionary militarists and landlords who were waging civil war against the Kuomintang Government should be confiscated. In the course of the debates at the Fifth

---

* North and Eudin, *M. N. Roy's Mission to China*, passim; also Brandt, *Stalin's Failure in China*, pp. 121–2. Moscow, questioned by telegram, blandly replied that the Chinese comrades should do both things at once.

Congress of the Chinese Communist Party, three main currents of opinion emerged on this issue:

1. The congress should simply take over the formulation of the seventh plenum, providing for the confiscation only of the land owned by 'counter-revolutionaries'.

2. They should maintain the principle of 'political' confiscation, but invert the formulation so as to put the burden of proof on the landlords; i.e., instead of confiscating only the land of 'reactionaries', they should confiscate all the land of large landlords except when the latter were 'revolutionary', as demonstrated by their relationship to officers in the revolutionary army.

3. All land should be confiscated immediately.[1]

What was Mao's role in this debate? In his autobiography Mao claims the proposals for 'a new line in the peasant movement' contained in his report on the peasant movement in Hunan were not even discussed by the Fifth Congress because of Ch'en Tu-hsiu's opportunism.[2] In fact, the original version of Mao's report contained no specific recommendations regarding land reform,* but, as we have seen in the previous chapter, Mao did elaborate, on the basis of his Hunanese experience, the distinctly radical proposals which he presented, shortly before the Communist Party Congress, to the Kuomintang Land Committee. In view of his assertion then that all those who owned more than 30 *mou* were counter-revolutionaries, he was closer to the third than to the second of the positions listed above. (It was the second position, corresponding roughly to the policy of 'political confiscation' adopted by the Kuomintang, which was finally approved by the congress.)

Mao thus belonged, in April 1927, on the left of the Chinese Communist Party – though not on the extreme left, for he had been criticized by other Communists, during the April meetings of the Kuomintang Land Committee, for not going far enough. But if we are to understand the originality of his position at the time, and its implications for the future, we must consider not only whose land he wanted to confiscate, but how he proposed to go about it. At the April meetings of the Land Committee he had recommended direct action by the peasants themselves,

* There was, however, one reference to land reform. See Schram, *Political Thought of Mao*, pp. 184–5.

either in the form of 'economic confiscation' – i.e., simply refusing to pay rent – or by actually re-distributing the land through the local organizations of the peasant associations. Presumably he defended a similar line before the Communist Party Congress, only to discover that Ch'en Tu-hsiu and the majority of the party leaders were no more prepared than Wang Ching-wei and T'an Yen-k'ai to trust their fate to the 'disorderly' activities of the peasants. He was apparently realistic enough not to go on indefinitely defending a policy which had been condemned both by the Central Committee and by the representatives of the International. (Even Roy had finally thrown his support to the compromise solution of exempting the land of small landlords and revolutionary officers from confiscation.) Mao therefore abandoned temporarily his support for direct action by the peasants. But carrying out social and economic change by the action of the masses at the grass roots has remained his constant concern.

After a few days, Mao became discouraged at the atmosphere of the congress and stopped attending on pretext of illness. The Peking historians claim that it was because Ch'en Tu-hsiu prevented him from expressing his views.[3] It may also have been out of pique at the fact that he had been replaced by Ch'ü Ch'iu-pai as head of the Peasant Department of the Chinese Communist Party.[4]

The Fifth Congress was marked by great apparent cordiality between the Communists and the leaders of the Kuomintang. In fact the Comintern representative, Roy, spelled out so bluntly and tactlessly the aims of the Communists that one is led to wonder whether the nationalist leaders did not decide then and there that a split was inevitable. On 4 May Roy made a report to the congress on the theses of the Seventh Comintern Plenum regarding the course of the Chinese revolution. Wang Ching-wei, the president of the Wuhan Government, came especially to this session in order to hear him. Before an audience including Wang and other Kuomintang luminaries, Roy made statements such as this:

The Communist Party is entering the government, because this is a revolutionary government. At the present stage of the revolution, hegemony in the struggle belongs to the proletariat, and it is participating in the national-revolutionary government in order to use the state machinery as an instrument for achieving hegemony.*

In clear language, this meant: 'We are entering your government in order to take it over.' Following this speech Wang Ching-wei arose and said that he agreed entirely with Roy, that the petty bourgeoisie (which the Kuomintang leadership was supposed to represent) must indeed advance towards socialism together with the proletariat.[5] But one can reasonably have doubts over whether he was perfectly sincere.

Although Roy was particularly tactless in talking in this way to Wang and the other 'petty bourgeois' leaders of the Kuomintang, the substance of his position was very similar to that of Stalin, as embodied in a resolution of the Eighth Comintern Plenum adopted on 30 May. In order to facilitate the transformation of the existing 'three-class bloc'† into a workers' and peasants' dictatorship, and the establishment of 'proletarian' (i.e. Communist) hegemony within the Wuhan Government, the Comintern called on the Chinese Communists to promote a reorganization of the Kuomintang, so as to turn it into a loose federation of mass organizations which could be more easily manoeuvred. The agrarian revolution should be carried forward, and the army transformed, through simultaneous action 'from above', by the government, and 'from below', by the masses.[6]

Such advice, which amounted to requesting the Kuomintang politicians and generals to dig their own mass grave so that their Communist allies could push them into it, was astonishing enough. But it appeared almost prudent when compared with the precise instructions which Stalin sent the Chinese Communists at the same time:

Without an agrarian revolution, victory is impossible. Without it, the

* North and Eudin, op. cit., p. 227. Roy's speeches to the congress were studded with similar remarks; for example, on 30 April, he explained that Communist hegemony should be achieved through the creation of Communist factions in all the Kuomintang state and party organs which would in fact control everything. ibid., pp. 205–6.

† According to Stalin, the 'four-class bloc' (workers, peasants, petty bourgeoisie, and national bourgeoisie), which had existed until the betrayal of Chiang Kai-shek and the national bourgeoisie, had henceforth been transformed into a 'three-class bloc'. See his interview with the students of Sun Yat-sen University, *Works*, Vol. 9, pp. 249–55.

Central Committee of the Kuomintang will be converted into a wretched plaything of unreliable generals. Excesses must be combated – not, however, with the help of troops, but through the peasant unions. . . .

Certain of the old leaders of the Central Committee of the Kuomintang are afraid of what is taking place. They are vacillating and compromising. A large number of new peasant and working class leaders must be drawn into the Central Committee of the Kuomintang from below. Their bold voices will make the old leaders more resolute, or throw them into discard. . . .

It is necessary to liquidate the dependence upon unreliable generals immediately. Mobilize about 20,000 Communists and about 50,000 revolutionary workers and peasants from Hunan and Hupei, form several new army corps, utilize the students of the school for military commanders, and organize your own reliable army before it is too late. . . . It is a difficult matter, but there is no other course. . . .[7]

The full irony of this document can be appreciated only in the light of the concrete situation in China at the time. Though T'ang Sheng-chih himself continued to pose as a 'revolutionary military man', some of his subordinates had already begun to manifest by their actions the anxiety about the consequences of agrarian ferment which had been voiced by Ho Chien at the April meetings. In the middle of May, General Hsia Tou-yin, commander of the Fourteenth Independent Division stationed at Ich'ang in Western Hupei, received orders from Chiang Kai-shek in Nanking to march on Wuhan. Though he had originally been under the command of T'ang Sheng-chih, Hsia proceeded to rise in revolt against the Wuhan Government on 18 May. The situation was saved only by the intervention of a hastily assembled force of military cadets and others, commanded by the Communist General Yeh T'ing, which halted the attackers when they were only some twenty-five miles from Wuhan. Meanwhile, tension had been increasing between the worker and peasant organizations in Hunan, largely animated by Kuomintang cadres who were also members of the Communist Party, and the local military forces. On 21 May, Colonel Hsü K'e-hsiang, commanding the Thirty-Third Regiment of Ho Chien's Thirty-Fifth Army, stationed in Changsha, acting on direct orders from General Ho, suddenly turned on the Communists and their supporters. Several hundred peasant militiamen were machine-gunned before Changsha, leading Communists were arrested, and the provincial labour union and peasant association were dissolved. This action was directly linked to that of Hsia Tou-yin, for the latter's advance had cut the railway between Wuhan and Changsha and precipitated events by starting all sorts of rumours. Hsü and Ho likewise transferred their allegiance from T'ang Sheng-chih and the 'Left Kuomintang' to Chiang Kai-shek, who rewarded Colonel Hsü by appointing him to a divisional command.

In the face of these open attacks on the Communists and on the policies they represented, the best Borodin was able to obtain after lengthy discussion was the appointment of a commission of inquiry, including himself, T'ang Sheng-chih, Ch'en Kung-po, and two of T'ang's staff officers. But halfway to Changsha they were turned back by Hsü K'e-hsiang, who threatened to execute them, and only Ch'en Kung-po was allowed to proceed. T'ang Sheng-chih did indeed condemn – in gentle terms – the anti-Communist activities of his former subordinates in Hunan, but the protestation remained entirely platonic.*

How could Stalin's orders be applied? When they received his telegram, the members of the Central Committee of the Chinese Communist Party were unanimous in the opinion that talking to the Kuomintang about land reform was like 'playing a lute to entertain a cow'.[8] They therefore decided to act only on the other half of the instructions: restraining the 'excesses' of the peasants. This was Mao's job, as the most active Communist member of the Executive Committee of the All-China Peasant Federation. He carried it out faithfully, though undoubtedly with much mental

* The most complete and thoroughly documented account of Hsia Tou-yin's and Hsü K'e-hsiang's offensives against the Communists is that of Chiang Yung-ching, *Borodin and the Wuhan Government*, pp. 311–54, making use both of the archives available to him and of the re-miniscences of Hsü K'e-hsiang and Ho Chien published recently in Taipei. A number of details in the above two paragraphs come from this source. See also North and Eudin, op. cit., pp. 96–100.

By a singular coincidence, it was precisely at this moment that the Kuomintang leaders at last put into practice the decision taken in principle two months earlier to hand over two ministries to the Communists, and solemnly installed T'an P'ing-shan as Minister of Agriculture – no doubt in order to give him more prestige for repressing the peasants. See the ironic account of Vishnjakova-Akimova, *Dva Goda v Vosstavshem Kitae*, pp. 354–5.

anguish.* A few days before the receipt of Stalin's famous telegram cited above, the Peasant Federation had called for patience
following the massacre of 21 May. Throughout the month of June
Mao and his colleagues continued to exercise a restraining hand,
though they also denounced the calumnies against the peasant
associations spread by the landlords, and affirmed that the
Wuhan 'National Government' could not be consolidated unless
the 'reactionary feudal forces' in the countryside were annihilated.[9]

In continuing to participate in this Kuomintang-sponsored
organization, and also continuing to direct the Peasant Movement Training Institute, of which a branch had been established
in Wuhan,[10] Mao Tse-tung was merely carrying out the policy of
the Chinese Communist Party. On 4 June the Central Committee had issued a proclamation warning the Hunanese peasants
against 'encroachments' on the members of soldiers' families.
'All such juvenile actions', affirmed the Central Committee,
'present the reactionary militarists and bureaucrats with an excellent opportunity to excite the soldiers against the peasant
movement.† To forbid such juvenile actions is an important task
of the peasant associations.' The peasants must therefore 'help
the people's government re-establish its political power in Changsha'.[11] Unfortunately, the 'people's government', contrary to the
illusions of Stalin, had not the slightest intention of disciplining
the 'counter-revolutionary' militarists. On the contrary, Wang
Ching-wei was already approaching Feng Yü-hsiang to negotiate
a reconciliation with Chiang Kai-shek.[12] The immediate motive
of Wang's decision was M. N. Roy's celebrated gesture in com-
municating to him Stalin's telegram of 30 May. Unlike Borodin
and the Chinese Communist leaders, Roy had been in favour of
trying to apply conscientiously Stalin's instructions for a simultaneous revolution from above and from below. Outvoted before
the Central Committee, he had then tried to force the hands of

---

* Mao's friend Ts'ai Ho-sen later claimed that, in late May and early
June 1927, Mao was carrying out open propaganda in favour of the union
of all true leftist forces against the Kuomintang leadership in Wuhan
(North and Eudin, op. cit., p. 114). But it seems very doubtful that Mao
worked openly against the Kuomintang at this time; he swung decidedly to
the left only late in August. (See pp. 120–22.)

† In other words, the peasants had brought Hsü K'e-hsiang's onslaught
on themselves.

his comrades by obtaining Wang's approval of Stalin's plans for establishing complete Communist control of the Kuomintang. This was certainly not a very rational undertaking, but it would be absurd to overestimate the consequences. Roy's action may have hastened by a week or two the rupture between the Communists and Wuhan, which finally occurred on 15 July, but the rupture was inevitable in any case. Even such last desperate gestures as the resignation of T'an P'ing-shan as Minister of Agriculture at the end of June, accompanied by apologies for his failure to 'put the peasants on the right track', could not influence the course of events.

Summing up the events of 1927 nine years later, Mao Tse-tung put the greatest blame on Ch'en Tu-hsiu, but he was also severe in his judgement of the two Comintern representatives. He thought that 'Roy had been a fool, Borodin a blunderer, and Ch'en an unconscious traitor.'[13] He did not mention Stalin, but the Russian leader's role in the catastrophe was assuredly not absent from his mind. A quarter of a century later, in the course of their polemics with Khrushchev, the Chinese affirmed: 'While defending Stalin, we do not defend his mistakes. Long ago the Chinese Communists had first-hand experience of some of his mistakes.'[14] This 'experience' most certainly encouraged Mao in his subsequent efforts to seek a new and specifically Chinese road to power.

Following the rupture with Wuhan, the disintegration of the alliance between the Communists and the Kuomintang moved from its second phase – that of simultaneous action from above and from below – into its third and last phase: that of the attempted exploitation of the 'Kuomintang flag', although none of the leaders of the Kuomintang enjoying any real power was still prepared to collaborate with the Communists. This period lasted only a little over two months, from the middle of July until the end of September, but it is of decisive importance in Mao's career. For it is during these months that Mao was sent by the Central Committee to organize an armed uprising in Hunan at the time of the autumn harvest. This action ended in defeat, but in a sense its echoes have still not died. For out of the survivors Mao put together a little guerrilla army which he led up into the Chingkangshan, the first stop on the road to Peking.

The Autumn Harvest Uprising in Hunan was part of a larger strategic design which had taken shape during the last half of July. Ch'en Tu-hsiu, depressed by the failure of the policy which had been partly his but mostly Stalin's, had withdrawn to Shanghai, but many members of the Central Committee had remained in the Wuhan area. Police repression was apparently not yet very efficient, and they were able to move about and meet with considerable freedom. As might be expected in such a dramatic situation, they held almost continuous sessions for some three weeks. By the time the new Comintern representative, Lominadze, arrived on the spot about 20 July (Borodin and Roy had both left in the wake of the disaster), the Central Committee had already decided on a movement in the autumn to reduce rents in the countryside. Lominadze approved, but added that the movement should have the expropriation of the land as its central point, though this expropriation should be confined to large and medium landlords. (Thus the substance of the agrarian policy adopted in November 1926 was left untouched.)

Meanwhile a group of Communist leaders, including Li Li-san, T'an P'ing-shan, and Ch'ü Ch'iu-pai, were conferring in the vicinity of Kiukiang, and decided to recommend to the Central Committee a plan for a revolt by units of the nationalist army stationed at Nanchang.* Informed of this proposal, both the Central Committee and Lominadze indicated their approval. After some discussion it was decided that the best course, once the uprising had been carried out, would be to return to Kwangtung and join up with P'eng P'ai's peasant insurgents. Chou En-lai was appointed secretary of the Front Committee which was to be set up to direct the uprising, and dispatched to Nanchang on 25 July.

Chou En-lai passed through Kiukiang on 26 July, and found several Communist leaders, including Li Li-san, Teng Chung-hsia, and T'an P'ing-shan in conference on the aims of the revolt. In accordance with Stalin's line at the time it had been decided to use the Kuomintang flag, and to issue all proclamations in the

---

* Except where otherwise indicated, the account of the Nanchang Uprising is based on the documents from the confidential inner-party organ of the Chinese Communist Party published by C. Martin Wilbur in *China Quarterly*, No. 18, 1964.

name of the Kuomintang Central Executive Committee. There was, however, a disagreement among those present over agrarian policy. T'an P'ing-shan, who apparently still regretted his failure, as Minister of Agriculture, to 'put the peasants on the right track', proposed that there should be no attempt, after the uprising, to confiscate any land at all, so as not to provoke division among the troops. Chou succeeded in rallying them all to the position of the Fifth Congress providing for the confiscation of land from big landlords only. (A big landlord was defined as one who owned over 500 *mou*, an enormous estate for that part of China.)

The military leaders of the uprising were to be Yeh T'ing and Ho Lung.* The former was already a Communist; the latter was as yet a representative of the extreme left wing of the Kuomintang, though he joined the Communist Party not long after the uprising. Ho Lung's prestige was enormous among the officers under his command, and his influence was counted on to sway a large number of them to the side of the uprising. His importance was also recognized by his more conservative colleagues, who made repeated attempts to win him to their side.

Ho Lung commanded the Twentieth Army, and Yeh T'ing the Twenty-Fourth Division of the Eleventh Army. Their superior, General Chang Fa-k'uei, the commander of the Second Front Army, did not seem to be firmly committed to the anti-Communist position of Chiang Kai-shek and Wang Ching-wei. This was, at least, the illusion cherished by certain among the Chinese Communist leaders, and above all by the Soviet military adviser, Galin. On 26 July, at a meeting of the Central Committee, Galin reported on a discussion he had just had with Chang Fa-k'uei. The latter, said Galin, had agreed to return to Kwangtung, the goal of the conspirators. This being the case, Galin was in favour of carrying out the uprising in concert with Chang, and breaking with him only on arrival in Kwangtung. In this way, he argued, the insurgents would have a force of 30,000 troops, instead of the 5,000 to 8,000 that Ho Lung and Yeh T'ing could rally for an uprising without Chang Fa-k'uei's support. This conclusion was approved unanimously by those present.

* Chu Te, now presented, in accounts published in Peking, as the principal artisan of the uprising, is mentioned only in passing in the contemporary sources.

Lominadze, who continued to favour the uprising, then produced a telegram from Moscow stating that if the project had no chance of success, it would be all right to abandon it and send the comrades to work among the peasants instead. Chang Kuo-t'ao was chosen to carry this information to the conspirators in Nanchang. In the course of his journey he sent a telegram ordering them to do nothing before his arrival, but on reaching Nanchang on 30 July he found the leaders there firmly resolved to go ahead with the uprising whatever he might say. In order to win them to his own very prudent position, he communicated Moscow's message in a distorted form; instead of saying that it would be all right to abandon the uprising if there were no hope of success, he told his comrades that the orders were to go ahead only if there was a guarantee of success.

The effect of this argument was nil. Chou En-lai threatened to resign as secretary of the Front Committee if the project was not pursued. The next morning Chang Kuo-t'ao finally agreed to bow to the will of the majority, but he still pleaded for two or three days' further delay in order to see whether an agreement could not be reached with Chang Fa-k'uei. Inasmuch as it had just been learned that the latter was conferring on near-by Lu Shan with Wang Ching-wei over measures for repressing the Communists, the others rejected this suggestion, and the uprising took place in the early hours of 1 August.

It has been claimed that the Nanchang Uprising, the anniversary of which is now celebrated in China as Red Army day, was carried out against Stalin's orders. The source of this legend is Chang Kuo-t'ao himself, who has given this version of events in interviews with various Western historians,* in order to conceal the fact that he did not faithfully execute the orders of the International at the time. But though Chang's story as it stands is false, it reflects an important psychological truth: the catastrophic failure of Stalin's policy had undermined confidence in Moscow to such an extent that a group of Chinese Communist leaders were prepared to disobey what they thought were Stalin's orders.

The subsequent history of the Nanchang Uprising was a series of blunders. The 10,000 or 11,000 troops who had rallied to the in-

* See, in particular, Brandt, op. cit., pp. 143–5.

surrection moved south on 4 and 5 August, but throughout the retreat to Kwangtung they made no attempt whatever to agitate among the peasantry or carry out any agrarian reforms, even to the extent of confiscating land from the largest landowners. When they finally took Swatow at the end of September, they showed themselves to be such zealous defenders of law and order (taking over bodily the existing police force and shooting several people for looting) that the local populace exclaimed: 'These are the troops of another Chiang Kai-shek!' The whole undertaking was a military adventure from beginning to end.

'Military adventurism' was the charge also levelled against Mao Tse-tung for his leadership of the Autumn Harvest Uprising in Hunan, but it was a different kind of adventure, though Mao's taste for military action certainly played a role in it. In the eyes of the Central Committee the uprisings in Hunan, Hupei, Kiangsi, and Kwangtung represented the basic plan, to which the revolt of the troops in Nanchang had been a later and secondary addition. And although Mao's name is not mentioned in the contemporary documents, he apparently participated in some, if not all, of the meetings of the Central Committee in late July and early August at which this policy was elaborated.

The most decisive of these was held on 7 August, and has gone down in history as the 'August 7th Emergency Conference'. It was called an emergency conference because the full quorum necessary for a plenary session of the Central Committee was not present. This did not, however, prevent the group from electing a new Central Committee, on which Mao was given a seat, and from which the absent Ch'en Tu-hsiu was excluded. 'Opportunism' was roundly denounced, without naming its source in Moscow. The new guiding spirit of the party was Ch'ü Ch'iu-pai, an intellectual with a remarkable knowledge of Russian literature. He was soon to develop extremely radical tendencies and lead the party into what is now called, in the official historiography, the 'first Left opportunist line'. But for the moment the policy laid down at the August 7th Conference was a highly moderate one, entirely in conformity with the line of the International. The banner of the Kuomintang (the 'true revolutionary Kuomintang') was once more to be used; only the land of the big landlords was to be confiscated; soviets were not yet to be formed, since to do so would amount to declaring an open struggle with the Kuomintang.

Shortly after the conference Mao Tse-tung was sent to Hunan as special commissioner to take charge of the uprising there, and to correct certain erroneous tendencies which had manifested themselves among the Hunanese comrades. Chief of these was 'military opportunism', which consisted in relying essentially on the intervention of organized armed forces of various kinds, from workers' militias to bandits, rather than on the spontaneous uprising of the peasant masses. The Central Committee was soon to conclude that Mao was even more guilty of military opportunism than the comrades he had been sent to put on the right track. But apart from this defect, which could be characterized as 'rightist' in Communist terminology, Mao manifested certain other eccentricities which would have to be called 'leftist'. The chief of these was a plea for the immediate establishment of soviets. On 20 August, a week or ten days after his arrival in Hunan, he wrote to the Central Committee:

A certain comrade has come to Hunan announcing that a new instruction from the International proposes the immediate establishment of soviets of workers, peasants, and soldiers in China. On hearing this, I jumped for joy. Objectively, China has long since reached 1917, but formerly everyone held the opinion that we were in 1905. This has been an extremely great error. . . .*

Mao's impression that the International had changed its policy probably resulted from distorted echoes of a modification which Stalin had effectively made in his position at the end of July. Without admitting that the time was as yet ripe for soviets – for this would have justified the criticism of the Opposition, which had been calling for soviets since March – he decreed that it was in order for the propagandists of the Chinese Communist Party to familiarize the masses with the *idea* of soviets. Thus the ground

* This letter, and most of the information in the remainder of this chapter regarding inner-party correspondence and directives in the period August 1927–March 1928, comes from the confidential inner-party bulletin *Chung-yang t'ung-hsin (Central Newsletter)*, later renamed *Chung-yang cheng-chih t'ung-hsün*. For a longer citation from Mao's letter of 20 August, see my article: 'On the Nature of Mao Tse-tung's "Deviation" in 1927', *China Quarterly*, April–June 1964, pp. 55–6.

would be prepared for the actual formation of soviets as soon as the revolutionary tide in China began rising once more. But Stalin did not expect such a tide for at least six months or a year, and in the meantime it was strictly forbidden to organize soviets, as distinguished from merely talking about them.

Mao's enthusiastic espousal of what he believed to be a shift in the International's line on this point was accompanied by a leftward movement on two other basic questions. Since his arrival in Hunan, Mao declared in his letter, he had at last come to understand that the Kuomintang was not an effective instrument for appealing to the masses. 'We really cannot use the Kuomintang flag', he wrote. 'If we do, we will only be defeated again. Formerly, we did not actively seize the leadership of the Kuomintang, and let Wang Ching-wei, Chiang Kai-shek, T'ang Sheng-chih and the others lead it. Now we should let them keep this flag, which is already nothing but a black flag, and we must immediately and resolutely raise the red flag.' Secondly, Mao came out firmly in favour of 'a complete solution of the land question', involving the confiscation and redistribution of all land without exception.

The arguments in Mao's letter of 20 August made no impression whatever on the Central Committee, which replied by ordering him categorically to follow the official line: he was not to establish soviets, was to use the Kuomintang flag, and was to promote confiscation of land only from large landlords. But at the same time that they condemned Mao's 'left' deviations, Ch'ü Ch'iu-pai and his colleagues were themselves embarking on a leftist adventure of a different sort. A series of resolutions adopted about 20 August sketched out quite clearly the 'putschist' line which was to be further elaborated and formally adopted at the plenum of November 1927.

This line found concrete expression in the plan for the uprising in Hunan and Hupei, elaborated by the Central Committee and communicated to Mao late in August. The accent in these instructions was on violence and terror as a means of revolutionary agitation. Prior to the uprising, the Communists were ordered to assassinate a few of the more notorious landlords in rural areas, so as to exalt the revolutionary spirit of the peasantry. In the course of the uprising itself, they were to kill off completely all bad gentry, local bullies, and reactionaries in the countryside, as well as all government officials in the cities. Only thus could the 'common people's political power' be established. The revolt should begin with the cities, and in order to take these, such bandits, remnants of various armies, and workers' or peasants' militias as were available could be employed, but only as auxiliary forces. The uprising both in the cities and in the countryside was to be primarily the work of the people themselves in arms.

This vision of the revolution as a ceaseless effusion of spontaneous mass violence clashed head on with the strategic conceptions which Mao was beginning to develop, in which organized military force played a leading role. To be sure, he had not yet formulated the notion of stable revolutionary base areas surviving for a long period in hostile surroundings. For the moment, his conflict with the Central Committee concerned only the best way to achieve the initial successes which would serve as the starting-point for a nation-wide revolutionary upheaval. Very rapidly, Mao was led to regard the military units at his disposal as the most effective instrument for striking a blow on behalf of the revolution.

All together, four regiments were available to Mao at the time of the Autumn Harvest Uprising. The First Regiment, the most important component in his forces from the standpoint of military efficiency, was the former guards regiment of the Left Kuomintang Government in Wuhan. This unit, whose officers were virtually all Communists, had left the city following the rupture of 15 July, with the intention of participating in the Nanchang Uprising, but en route they had learned that the forces of Yeh T'ing and Ho Lung had already moved south from Nanchang, and they had consequently veered into Hunan.

The Second Regiment was the most reassuring from the standpoint of Marxist orthodoxy, for it included a thousand-odd miners from Anyüan, as well as peasants from P'inghsiang and Liling. The Third Regiment was made up of worker and peasant volunteers from P'ingchiang and Liuyang, and of peasant militiamen from Hupei, and the Fourth Regiment consisted of reorganized troops formerly under the command of General Hsia Tou-yin, who had revolted against the Wuhan Government in May.

Mao's plan was that the First and Fourth Regiments, which constituted the right flank, should advance on Changsha from the north-east. The Second Regiment, on the left, was to advance from Anyüan, south-west of Changsha, take P'inghsiang and Liling, and then concentrate at Liuyang. The Third Regiment, constituting the central column, was to attack Tungmenshih, which lies eastward and slightly north of Changsha, and then take Liuyang, on 12 September, in a joint action with the Second Regiment. After this, the four regiments would converge on Changsha, beginning their assault on the provincial capital on 15 September. 'Several hundred armed comrades' and several thousand workers were to act in concert with the attackers, and thus the city would be taken by a combined action from within and from without.

Informed of this plan, the Central Committee declared, in a letter of 23 August (the same one which refuted Mao's views regarding soviets, land reform, and the Kuomintang flag), that it was 'correct in principle' to take Changsha as the main starting-point of the insurrection, but that Mao's way of going about this was full of errors. To begin with, he was accused of attaching undue importance to military force, lacking faith in the revolutionary strength of the masses, and thereby turning the uprising into a 'mere military adventure'. The Central Committee then proceeded to spell out for his benefit a plan for mass uprising in three areas lying to the east, south, and south-west of Changsha.

To this, Mao replied with some asperity that the Central Committee was practising 'a contradictory policy consisting in neglecting military affairs and at the same time desiring an armed insurrection of the popular masses'. He added that he regarded the organized military units under his command as mere 'auxiliary forces' to 'make up for the insufficient force of the workers and peasants'.

Undoubtedly Mao was more acutely conscious than the Central Committee of the 'insufficient force' of the masses. Hunan had suffered more than any other province from the repression of the spring and summer. Statistics reported in the inner-party bulletin revealed that nearly half of the 130,000 persons killed, wounded, or missing in the country as a whole (these figures apparently do not include Shanghai) were from Hunan. Prior to the 21 May massacre at Changsha there had been over 20,000 Communist Party members in the province as a whole, some 3,000 of them in Changsha. Now there were less than 5,000 altogether, and only 1,000 in the capital. This was also no doubt the reason for Mao's concentration on the Changsha area, despite the fact that the Central Committee berated him for not making preparations in Southern Hunan as well. His forces were limited; he did not think it prudent to spread them too thin.

The uprising began on 9 September. The main railroad lines into Changsha were immediately cut by sabotage, and no trains reached the city in the course of the ensuing week. On the 10th, the Second Regiment rose at Anyüan and attacked P'inghsiang according to plan. Liling was taken on the 12th, and a 're-volutionary committee' established, which proclaimed the confiscation of the land, and the restoration of the labour unions and the peasant associations. Attacked by a superior force, the Second Regiment was obliged to evacuate Liling, but attacked Liuyang on the 15th. The result was a disastrous defeat in which the regiment lost two thirds of its men. A further reverse occurred when, in the course of the joint attack by the First and Fourth Regiments on P'ingchiang, the Fourth Regiment, composed of recruits from the Kuomintang, returned to its original allegiance and struck at the First Regiment from the rear. The latter force – the former guards regiment – escaped with the loss of over a hundred men. For its part, the Third Regiment took Tung-menshih as planned, and then proceeded to attack Liuyang, but since the Second Regiment was not present at the rendezvous, it could not take the city unaided.

In the face of this situation, Mao decided, on 15 September, to call off the attack against Changsha as hopeless.* He assembled the remnants of the First and Third Regiments – a thousand men

* In the contemporary documents, the decision is attributed to the Provincial Committee of the Chinese Communist Party, and the secretary of that committee, P'eng Kung-ta, is made to bear a considerable share of responsibility. But Mao, as special commissioner and member of the Central Committee, outranked P'eng in the party; moreover, it is stated that he was the 'heart' of the Provincial Committee (of which he was also a member) during the uprising. I have therefore simplified matters by attributing the decision to him.

in all – and led them away from the provincial capital toward the countryside.*

These developments were reported to the Central Committee by a certain 'comrade Ma'†, who came in person to denounce the cowardly betrayal of the party authorities in Hunan, and to assert that if the insurgent forces turned and attacked Changsha victory was assured. On the basis of this information the Central Committee sent peremptory orders to the Hunan committee on 19 September, demanding that the assault on Changsha be carried out immediately. It is doubtful whether Mao ever received this letter; in any case he did not act upon it, and, after reorganizing his forces and instituting a system of political representatives in each unit, he led them up into the mountain range known as the Chingkangshan, on the border between Hunan and Kiangsi, where the first base was established in October.

By a curious coincidence, on the very same day – 19 September – that the members of the Central Committee bitterly denounced Mao for his 'betrayal' before Changsha, they adopted a resolution incorporating his point of view on all three of the basic policy issues at stake in the exchanges of late August: the immediate formation of soviets, the abandonment of the Kuomintang flag, and the confiscation of all land without exception. Naturally, the fact that Mao had been right too soon on these issues did not prevent him from receiving severe punishment at the next plenum of the Central Committee in November; he was relieved of his posts both on the Central Committee and on the Hunan Provincial Committee because of his failure to press the agrarian revolution, and his 'military opportunism'.

Mao did not even learn of his condemnation until many months later. Nor was he informed of the political line adopted by the November plenum, which put forward the idea of 'permanent' or 'uninterrupted' revolution as a general theoretical principle –

---

* The above account of the Autumn Harvest Uprising is based primarily on two sources: the materials published at the time in the inner-party organ *Chung-yang t'ung-hsin*, which have the advantage of being contemporary accounts, but are strongly slanted against Mao and the other leaders of the uprising in Hunan; and the account in *Hu-nan chin-pai-nien ta-shih chi-shu* (Changsha, Hu-nan Jen-min Ch'u-pan-she, 1959), pp. 538–40.

† Apparently the pseudonym of a foreigner, probably a Russian.

thus providing the opportunity for Stalin to attack the Chinese leadership as Trotskyite once its policy had failed. The practical expression of this principle was to be an uninterrupted series of armed uprisings until final victory was achieved. For the moment these uprisings would have to take place in the countryside, but their purpose would not be to establish stable base areas, as Mao was in the process of doing, but to keep the masses in a state of revolutionary excitement. Although insurrection in the cities could come only later, when the ruling class had been weakened by the day-to-day economic struggle, the intervention of the urban proletariat would ultimately be decisive. Meanwhile the workers were to be given secret military training to prepare them for their role, and in the immediate future these para-military organizations should be used to kill off the 'yellow' labour leaders, rob banks, attack police stations, etc.

A directive to the Communists of Hunan and Hupei in early December 1927 spelled out the tactics which Mao was expected to apply with greater precision. They consisted in 'sealing off' a *hsien* here and there (it was not explained how this feat could be achieved), establishing a revolutionary government, killing off as many bad gentry, reactionaries, and big landowners as possible, and carrying out a thoroughgoing social revolution.* In the cities, as well as the policy of terror and assassination already mentioned, the masses should be mobilized to 'attack the foreigners' (*ta wai-kuo-jen*).

Like the resolutions of the November plenum, this directive reached Mao, if at all, only in the spring of 1928. Meanwhile he was engaged in building up his position on the Chingkangshan according to quite different tactics. Far from carrying out, as the Central Committee had ordered in November, the 'workerization' (*kung-jen-hua*) of the cadres of the party and of the army, he had united with two bandit chiefs, Yüan Wen-ts'ai and Wang Tso, who were already established in these remote and inaccessible regions. In this way, he added six hundred men to the

---

* The model in this respect was considered to be P'eng P'ai's Hailufeng soviet régime, established in November 1927, which was to be crushed in February 1928 because of its extreme and terroristic policies, and because it did not have an adequate military force at its disposal. cf. Eto Shinkichi's articles in *China Quarterly*, cited previously in the footnote on p. 79.

thousand who remained of his own forces. He also gained a hundred and twenty rifles for his meagre arsenal. According to Mao himself, Yüan and Wang, who were each made regimental commanders (Mao being army commander), were 'faithful Communists' as long as Mao remained on the Chingkangshan. Later, when left alone, they returned to their bandit habits, and were killed by the peasants.[15]

The episode of Wang and Yüan has in fact very broad implications. It reflects the accent on the human will, rather than on objective factors, which still characterizes Mao's version of Marxism. A little later, commenting on the presence of an extremely high percentage of *éléments déclassés* in his army, Mao affirmed that the only remedy was to intensify political training, 'so as to effect a qualitative change in these elements'.[16] The notion that rural vagabonds can be transformed by suitable training into the vanguard of the proletariat is a striking reflection of that extreme voluntarism which has culminated in the idea that 'the subjective creates the objective'.*

But the presence in Mao's army of singularly unproletarian elements, suitably purified by indoctrination, did not merely reflect the difficulty of securing soldiers and officers of proper class origin, though this is naturally how Mao himself presented the matter to his superiors in the party. It also corresponded to a fundamental bent of his own temperament and imagination. From childhood he had admired the bandit heroes of the popular Chinese novels, and though his vision of the world was no longer circumscribed by the horizon of the peasant rebel, neither did his Marxist convictions lead him to repudiate the enthusiasms of his youth. His articles written in 1926 on the classes of Chinese society had offered striking confirmation of this. In them, Mao had given a colourful description of the five categories which constituted the *éléments déclassés* – soldiers, bandits, robbers, beggars, and prostitutes – and praised their capacity as revolutionary fighters, with none of the reservations he has seen fit to insert in the current orthodox edition of his works. He had also referred sympathetically to the secret societies which served as 'their mutual aid societies in the political and economic struggle'.[17] There is little doubt that this attitude of his persisted during the time he was on

* See Chapter 10.

the Chingkangshan, and that he accepted the support of bandits and similar outcasts from the existing society not merely with resignation, but with a certain sympathy. Moreover, like so many outlaws throughout Chinese history, Yüan and Wang were leaders of the secret societies. The society in question was related to the Ko Lao Hui, to which Mao attributed the principal merit for the 1911 revolution in Hunan, and this fact probably increased his confidence in the two men.*

Mao's sympathy for these old-fashioned rebels was by no means shared by the Central Committee, in its Shanghai refuge. On the contrary, the reports reaching them on Mao's activities inspired grave misgivings. They therefore wrote to Chu Te, requesting him to join forces with Mao, and to correct certain errors of which the latter had been guilty. The chief of these was precisely that Mao was behaving like the bandit heroes of *Water Margin* in their mountain fortress of Liang Shan P'o, sallying forth to carry out heroic exploits on behalf of the masses, instead of rousing the masses to undertake a spontaneous armed insurrection.[18]

Although Chu Te was apparently regarded by the Central Committee as more reliable than Mao at this time, he himself had fallen into an error of a different kind, denounced in the same letter. This error, indeed, was nothing less than returning to a policy of collaboration with the Kuomintang. Following the defeat of the forces from Nanchang at Swatow in late September, Chu Te had gathered together some two thousand of the survivors and marched them hither and thither about Kwangtung and the neighbouring provinces. Finally, in order to preserve his little army, he agreed to be 'reorganized' as a unit of the Kuomin-

---

* The fact that Wang and Yüan were leaders of the Triad Society, historically linked with the Ko Lao Hui, is given in a report on the 'Chu–Mao army' dated 1 September 1929 and published in January 1930 in a bulletin on military affairs issued by the Military Department of the Central Committee of the Chinese Communist Party: 'Kuan-yü Chu-Mao Hung-chün ti li-shih chi ch'i chuang-k'uan ti pao-kao', *Chung-yang chün-shih t'ung-hsün*, No. 1, 15 January 1930. This report makes it possible to verify a certain number of points in Mao's own story of his adventures, as told in his report to the Central Committee and in his autobiography; to correct some errors; and to add further details. See also the volume of reminiscences, *Hui-i Chingkangshan-ch'ü ti tou-cheng* (Peking, Kung-jen Ch'u-pan-she, 1955), pp. 1–23.

tang forces under the command of the Yünnanese general Fan Shih-sheng, who gave him in exchange considerable assistance in the form of supplies. The Central Committee was much disturbed by the news of this alliance, and ordered Chu abruptly to break it off, taking with him as large a portion as possible of Fan's troops. On receipt of these instructions he did finally break with Fan, in January 1928, and went to Southern Hunan.*

Chu Te's behaviour was not so eccentric as might appear at first sight. From one end of China to the other, in half a dozen provinces, various groups of Communists were still endeavouring to work through the Kuomintang. The phenomenon was so widespread that on 31 December the Central Committee was obliged to issue a circular letter ordering all party organizations to cease such practices immediately and expel all comrades who refused to resign from the Kuomintang.[19]

These assaults on collaboration with the Kuomintang were issued only after the failure of the uprising known as the 'Canton Commune' in the middle of December. This futile adventure, ordered by Stalin (in complete disregard of the consequences to Chinese revolutionaries) because he wanted a victory to exploit in his continuing struggle with the Russian Opposition, led in the first instance to an even more radical insistence on a policy of 'uninterrupted uprisings'. Within two months, however, Moscow had drawn other conclusions from the event. The Ninth Plenum of the Comintern, in a resolution of February 1928, roundly condemned the 'putschist' line of Ch'ü Ch'iu-pai, just as the opportunist line of Ch'en Tu-hsiu had been denounced in the previous summer. In both cases Stalin had been blind to these deviations until they led to obvious failure, and remained blind to his own responsibility.

* The letter of 21 December mentioned above was followed by another and even blunter missive of 27 December; see *Chung-yang cheng-chih t'ung-hsün*, No. 16, pp. 88–9. The fact of the alliance is confirmed and some details regarding it are given in the report mentioned previously in *Chung-yang chün-shih t'ung-hsün*, No. 1, 15 January 1930. In his autobiography as told to Agnes Smedley, Chu Te mentions his contacts with Fan Shih-sheng, but presents them as though Fan had taken the initiative, and as though they consisted merely in the supplying by Fan, at Chu Te's request, of a few hundred uniforms and some ammunition, without any obligations on Chu Te's part. *The Great Road* (Calder, 1958), pp. 213–14. This version is not compatible with the contemporary sources.

Ch'ü himself, who had been summoned to Moscow for the occasion, was not allowed to return to China, but remained in Moscow until 1930 under the pseudonym of 'Strakhov'.

The Comintern resolution of February 1928 reached the Central Committee in Shanghai only at the beginning of June, and was not received by Mao until 2 November.[20] Meanwhile the forces on the Chingkangshan had lived through nearly a year of dramatic episodes, provoked in great part by the intervention of various emissaries who were inspired by one form or another of the 'putschist' line already condemned by Moscow.

In fact, Mao's own policies were not so totally at variance with the Ch'ü Ch'iu-pai line as he would now have us believe. The terrorist methods advocated by the November plenum were employed to a certain extent in the Chingkangshan base area. In the course of the first winter a large number of guerrilla uprisings 'to massacre the landlords' (and not merely to 'overthrow' them, as the current version of the *Selected Works* has it) were undertaken. Some forms of political organization established during this early period, including the local soviets, were, according to Mao, his own 'clumsy inventions', but others were based on the practices and terminology of the 'Canton Commune', about which Mao and his comrades had read in the newspapers.*

On agrarian policy Mao regarded all peasant proprietors (and not merely the richer ones, as in the current version of his works) as potential enemies; only the poor landless peasants were reliable. Moreover, he justified the execution (and not merely the 'punishment') of several such peasants as a necessary measure to break their resistance to land reform.†

But though Mao was not quite so moderate in 1928 as he now

---

* *Hsüan-chi*, 1947 edition, Supplement, p. 55; cf. *Selected Works* (Peking), Vol. I, p. 75. In the current edition all mention of the influence of the Canton Commune, and even the term 'people's committees' copied from it, has been eliminated. This flagrant example of the sacrifice of Chinese lives to Stalin's whim is something the Chinese Communists would rather forget.

† *Hsüan-chi*, 1947 edition, Supplement, pp. 71–2; *Selected Works* (Peking), Vol. I, pp. 87–8. In the current edition, Mao has introduced into this early text the categories of 'middle' and 'rich' peasants, which were to be employed in his later writings and directives on agrarian policy. At the time, he used the term 'peasant proprietors' (*tzu-keng-nung*), which he had also employed in 1926. See Schram, *Political Thought of Mao*, pp. 172–4.

pretends, he was much too moderate for the taste of the party authorities in Hunan, to whom he was theoretically responsible in the absence of all contact with the Central Committee. This was particularly so with the urban petty bourgeoisie. Mao wrote of his experiences in this respect with a certain wry humour:

In March the representative of the Hunan Special Committee arrived in Ningkang [the capital of the Chingkangshan area] and criticized us for having done too little burning and killing, and for having failed to carry out the so-called policy of 'turning the petty bourgeois into proletarians and then forcing them into the revolution'; whereupon the leadership of the Front Committee was reorganized and the policy was changed. In April, after the whole of our army arrived in the border area, there was still not much burning and killing, but the expropriation of the middle merchants in the towns and the collection of compulsory contributions from the small landlords and rich peasants in the countryside were rigorously enforced. The slogan of 'All factories to the workers', put forward by the Southern Hunan Special Committee, was also given wide publicity. This ultra-left policy of attacking the petty bourgeoisie drove most of them to the side of the landlords, with the result that they put on white ribbons and opposed us.[21]

The 'reorganization' referred to by Mao did not merely involve a change in policy; it also cost him his dominant position in the party in the border area. Hitherto the 'Front Committee' appointed by the Central Committee to direct the Autumn Harvest Uprising in Hunan, with Mao as secretary, had continued to function as supreme party authority in the Chingkangshan area. Now it was transformed into a 'Divisional Party Committee', with a representative of the Southern Hunan Special Committee as secretary. The new party chief proceeded to display his authority by ordering Mao to undertake a military campaign in Southern Hunan. As a result, not only were Mao's forces defeated, but during their absence the base area itself was occupied by Kuomintang troops.[22]

Fortunately for Mao he was able, in the course of this expedition, to join forces with Chu Te, and also to recruit 8,000 additional peasant soldiers from Southern Hunan. Thus reinvigorated the Red forces returned to the Chingkangshan and reconquered the base area. On 20 May 1928 the 'First Party Congress of the Border Area' was held at Maop'ing, near Ningkang, and a

'Special Committee' of twenty-three members was elected, with Mao as secretary. Thus Mao reasserted control over the local party organization, while the military forces were organized into the 'Fourth Red Army', with Chu Te as commander-in-chief and Mao as party representative.* The army itself was composed of six regiments, numbered from twenty-eight to thirty-three, and totalled over 10,000 men. But only two of these regiments were fully equipped with rifles, the others being armed mainly with long spears. Moreover, the larger part of this apparently impressive force was composed of homesick peasants from Southern Hunan, and by the end of May the Thirtieth and Thirty-Third Regiments deserted and returned to their villages.[23]

In the ensuing weeks the remaining four regiments of the 'Chu–Mao Army', as it now came to be called, held off a succession of attacks from eight to eighteen enemy regiments. Already Mao and Chu Te had grasped one of the basic principles of partisan warfare, the concentration of forces 'so as to avoid being destroyed one by one'. But the higher party authorities were still thinking of the Red Army essentially as a catalyst to bring about a sudden revolutionary explosion, and not as a permanent feature of the Chinese political landscape over a relatively long period. They consequently sent instructions dictated by political and not by military considerations. Towards the end of June two emissaries of the Hunan Provincial Committee, Tu Hsiu-ching and Yang K'ai-ming, arrived in the border area and ordered Mao and Chu to advance immediately towards Southern Hunan. To justify their viewpoint, they produced a letter.[24]

Judging by the nature of the policies defended by Tu and Yang, this letter was probably a copy or paraphrase of the instructions sent by the Central Committee to the Hunan Provincial Committee towards the end of March. These instructions said that Southern Hunan was the centre of the uprisings in the three provinces of Hunan, Hupei, and Kiangsi, which were in turn the centre of the movement in the country as a whole. The Central Committee considered that the political situation, characterized by struggles among various cliques of militarists, offered a favour-

* The designation of 'Fourth Army' was deliberately chosen to emphasize the continuity with the Kuomintang Fourth Army, known as 'Old Ironsides', from which Chu Te's troops had come via the Nanchang Uprising.

able opportunity to liquidate the political and military power of the landlords, and to prepare for the general uprising in all of China. The letter denounced the 'objective abandonment and subjective slighting' of the cities, which could not only influence the composition of the party, but 'raise the fundamental question of the leadership of the agrarian revolution by the proletariat'. To correct these tendencies, the Central Committee demanded the 'immediate reorganization' of the party in Hunan. As for the border area, a 'Special Committee' should be appointed, with full powers in both political and organizational matters.[25]

Finding a 'Special Committee' already in existence, Yang K'aiming contented himself with ousting Mao as secretary and taking over himself. It is interesting to speculate whether Mao had got wind earlier of the instructions regarding a 'Special Committee' and had created his own version in May to forestall the initiatives of the party authorities in Hunan. It is certain, in any case, that despite what appeared to be the overwhelming support of higher party authority for the policies of Yang and Tu, Mao called a meeting of comrades in Yunghsin, a town about thirty miles from Ningkang, and secured the passage of a resolution rejecting the orders of the Provincial Committee to march south. But Yang and Tu took advantage of Mao's absence, and of the homesickness of the peasant soldiers from Southern Hunan, and succeeded in persuading them to undertake the campaign. Chu Te, willingly or not – it is impossible to say – went along as commander of the Red forces, which were composed of the Twenty-Eighth and Twenty-Ninth Regiments. The Thirty-Second Regiment served as rear guard, and only the Thirty-First Regiment, with a total of two hundred rifles, was left at Mao's disposal to defend the base area.

Taking advantage of this situation, five regiments of the Kuomintang Third Army and six regiments from the Kuomintang Sixth Army launched a joint assault on Yunghsin. With a single poorly-armed regiment Mao succeeded in holding them off for twenty-five days by guerrilla attacks, but eventually, in the middle of August, both Yunghsin and Ningkang were lost. At this point the Kuomintang Third and Sixth Armies began fighting among themselves, but the Red forces were too weak to take advantage of this. Mao therefore set out for Southern Hunan to persuade the main force under Chu Te to return. Chu had meanwhile lost virtually the whole of the Twenty-Ninth Regiment in a disastrous battle with his old friend Fan Shih-sheng on 27 July. But the stronger regiment, the Twenty-Eighth, was still intact, and occupied Kueitung on 18 August. There Chu Te was joined by Mao on 23 August, and the two decided to return to the Chingkangshan together.

When they got there they found a dramatic situation. The Red forces held only a few 'bandit-type mountain tops'.* The plains, and all the towns, were in the hands of the enemy. But by October a substantial portion – though not all – of the lost territory had been recovered. The situation having thus been stabilized, Mao and his comrades were able to hold the Second Party Congress of the Border Area on 14 to 16 October, and elect a new Special Committee with a secretary favourable to Mao.[26]

At this point Mao received the Central Committee's letter of 4 June, enclosing a copy of the February resolution of the Comintern condemning the theory of the 'uninterrupted revolution' and the tactics of 'uninterrupted uprising'. The letter also contained the welcome order to re-establish a Front Committee of five members with Mao Tse-tung as secretary. This was accordingly done on 6 November at a meeting which also decided to accept and apply the Central Committee's instructions as a whole, with the exception of one or two details regarding the organization of the army.† On 14 November the Sixth Party Congress of the Red Army named an Army Committee of twenty-three members, with Chu Te as secretary of the five-member standing committee. Thus Mao was firmly established in control of the whole party organization of the border area, for both the Special Committee and the Army Committee were subordinated to his Front

* *Hsüan-chi*, 1947 edition, Supplement, p. 85. This expression, which is applied by Mao to what might happen in case of new tactical blunders, but which might equally well be used to describe the situation which actually existed in September, has been removed from the current edition.

† *Hsüan-chi*, 1947 edition, Supplement, pp. 53–4, 80–81. In particular, Mao rejected the Central Committee's advice regarding the abolition of party representatives in the army. (For his arguments, see Schram, op. cit., pp. 196–7.) At the same time, he encouraged broad participation in political work by continuing to carry it out through soldiers' committees, rather than establishing a separate political department as recommended by the Central Committee. See John Rue, *Mao Tse-Tung in Opposition 1927–1935*, Stanford University Press, 1966, pp. 93, 96, 108.

Committee, and both were run by persons who shared his ideas.

Once again, as a result of the poor communications between Moscow and the Central Committee in Shanghai, and between Shanghai and Ningkang, Mao was reacting to decisions that were already out of date. In August and September 1928 the Sixth Comintern Congress and the Sixth Congress of the Chinese Communist Party had been held concurrently in Moscow. News of decisions taken by the latter reached Chingkangshan during the winter of 1928–9, and according to Mao, he and Chu Te were 'in complete agreement' with the line elaborated. 'From that time on,' he declared, 'the differences between the leaders of the Party and the leaders of the Soviet movement in the agrarian districts disappeared. Party harmony was re-established.'[27]

As everyone knows, and as we shall have many occasions to observe in the succeeding pages, this was an altogether misleading statement. To be sure, there was a kernel of truth in it, to the extent that the Sixth Congress did at least recognize the utility of establishing soviets in rural base areas and conducting partisan operations from them, instead of striking out wildly in all directions at once as Ch'ü Ch'iu-pai had recommended. But these tactics were regarded essentially as a holding operation until such time as a new revolutionary high tide in the whole country made possible a mass uprising and the overthrow of the Kuomintang Government. Moreover, although it was recognized that for the moment the peasantry in the countryside was more active than the working class in the cities, this was regarded as an unnatural state of affairs, the unfortunate consequence of the massacres in Shanghai and elsewhere, to be remedied as soon as possible. 'Proletarian hegemony' remained an unquestioned dogma. Particular attention in party work was to be accorded to the establishment of cells in industrial enterprises, and in general to the 'amelioration of the social composition of the party'. In the countryside party members were to be recruited essentially from the agricultural labourers and poor peasants. The party was thus not to be representative of the peasantry as a whole; still less was it to be the emanation of the *éléments déclassés.**

* All of the essential points of this policy had been concisely made in the February 1928 Comintern resolution, which was explicitly annexed to the political resolution of the Sixth Chinese Communist Party Congress and

All three of these points – the leading role of the urban proletariat, the narrow class basis of the Communist Party in the
countryside, and the tactics of planning an all-or-nothing struggle
in the near future rather than expanding the Red areas gradually
until the entire country was engulfed – were in clear contradiction
to the tactics which Mao Tse-tung was developing in practice.
Mao had, by this time, become convinced that base areas were
essential. In the autumn of 1928, while lecturing to his soldiers, he
explained that a base area bore the same relation to an army as
did the buttocks to a person. If deprived of either, one could
never rest and recuperate, but would be obliged to run about
until exhausted.[28] At the same time it should be emphasized that
Mao still had no very clear idea of how final victory might be
achieved. 'Wherever the Red Army goes', he wrote forlornly in
November 1928, 'the masses are cold and aloof, and only after
our propaganda do they slowly move into action. Whatever
enemy units we face, there are hardly any cases of mutiny or
desertion to our side, and we have to fight it out. . . . We have an
acute sense of our isolation which we keep hoping will end.'[29]

In the original version of his report Mao explained this isolation in a way which shows that he, too, was strongly influenced
by the ideas current at the time:

This cannot be called an insurrection, it is merely contending for the
country.* This method of contending for the country cannot succeed.
The reason for all this is that there is absolutely no revolutionary high
tide in the country as a whole. . . . The vast forces of the oppressed
classes have not yet been set in motion. So we are reduced to contending
for the country in this cold atmosphere.[30]

---

made a part of it. They were further developed both in the resolutions of the
Chinese Communist Party and in those of the Comintern, which were largely
identical in substance. See *Strategija i taktika, passim*; especially pp. 86–7,
207–11; Conrad Brandt, Benjamin Schwartz, and John K. Fairbank, *A
Documentary History of Chinese Communism* (Cambridge, Mass., Harvard
University Press, 1952), pp. 127–65, passim. For a lucid and precise analysis
of the line of the Sixth Congress and its relation to the pattern of revolution
ultimately developed by Mao, see Schwartz, *Chinese Communism and the
Rise of Mao*, pp. 115–26.

* The expression used (*ta chiang-shan*) implies that one is seeking to
conquer the country in an old-fashioned way, like the candidates for the
imperial throne who set out in the past from a limited territorial base.

Obviously the idea had not yet formed in his mind that it was possible to create a revolutionary climate in a limited area even if the Kuomintang régime remained firmly established in the country as a whole. One reason why he had not made this discovery lay in his doctrinaire and unrealistic agrarian policy. During the early period on the Chingkangshan the whole of the land was confiscated, and then divided up among the peasants, as Mao had proposed in his letter of 20 August 1927 to the Central Committee. But this policy failed to recognize that the peasant who had even a little bit of land was in general deeply attached to it, so that if it was taken away from him and put into a common pool, he would not be consoled for its loss even if the plot ultimately allocated to him was larger and more fertile. In April 1929 this mistake was corrected, and a policy adopted providing for the confiscation in the first stage only of estates from the landlords. This readjustment may have been inspired in part by the decisions of the Sixth Congress, received in the course of the winter, which contained a criticism of 'petty-bourgeois equalitarianism' in land distribution.*

The new agrarian policy developed in the spring of 1929 may also have resulted from the fact that by then Mao had transferred his base of operations to a larger and richer area with a more complex social structure. In November 1928 P'eng Te-huai had arrived on the Chingkangshan at the head of the Fifth Red Army. The mountainous border area was no longer adequate for the support of the combined forces, and the situation was made more severe by unusually bitter weather. Moreover, the Kuomintang armies had mounted a new offensive against the base. It was therefore decided to move south-east into Southern Kiangsi, leaving P'eng Te-huai and Wang Tso to hold the original base on the Chingkangshan.

P'eng Te-huai, who was henceforth to play a major role in the creation and direction of the Red Army until his disgrace in 1959, had been born about 1900 to a peasant family in Mao's own native Hsiangt'an *hsien*, in Hunan province. As a youth of ten,

---

* See, in particular, the resolution on the peasant movement, *Documentary History*, p. 158. This resolution may also have supplied Mao with the terms of 'middle' and 'rich' peasant, which he was to use in the future instead of 'peasant proprietor'.

he had run away from home and worked as a cowherd, a coal miner, and a shoemaker's apprentice. In 1916, he joined the army in Hunan, and gradually rose in rank, meanwhile abstaining, so far as is known, from any political activity. The Northern Expedition of 1926 found him in command of the First Regiment of the Fifth Division of Ho Chien's Thirty-Fifth Army, under T'ang Sheng'-chih. Following the split between the Communists and the Kuomintang, he succeeded in maintaining his position despite Ho's efforts to eliminate the leftists in his ranks, and in early 1928 he was in command of the First Brigade of Ho's army. In July 1928 he led his troops in an uprising at P'ingchiang, the survivors of which followed him up the Chingkangshan a few months later.*

Chu and Mao left the Chingkangshan base on 14 January 1929 and after a terrible battle at Tapoti, in which they lost half their forces,† established themselves in the area around Tungku and Hsingkuo. It was there that they began building up the base area which was eventually to become the Chinese Soviet Republic with its capital at Juichin. Meanwhile the forces on the Chingkangshan had been attacked from three sides by the Kuomintang troops. Wang Tso, with a small band of followers, remained in the mountain fastnesses and returned to his bandit ways. P'eng Te-huai led the remnants of the Fifth Army to join Mao Tse-tung and Chu Te in Southern Kiangsi. When the two armies were reunited in April 1929 Mao and Chu had 2,000 men left, and P'eng 800.[31]

It was in these circumstances that Mao Tse-tung received a message from the Central Committee, dated 9 February, expressing complete lack of confidence in his whole enterprise. He was advised, in order to preserve the Red Army and arouse the masses, to divide his forces into very small units and disperse them over the countryside. Chu Te and Mao himself were requested to withdraw from the army, so as to conceal the major targets. To this Mao replied, in a letter of 5 April, that 'the more adverse the circumstances, the greater the need for concentrating our forces and for the leaders to be resolute in struggle.'[32]

---

* For an account of the P'ingchiang Uprising, see *Hu-nan chin-pai-nien ta-shih chi-shu*, pp. 576–8.

† A classic description of the descent from the Chingkangshan and the battle of Tapoti is given in Smedley, *The Great Road*, pp. 236 et seq.

But the quarrel was not merely over tactics. The Central Committee (of which Li Li-san was the moving spirit, though he did not have the title of secretary*) was deeply concerned about the unorthodox and un-Marxist character of the goings-on in Kiangsi, and above all by the lack of proletarian direction. In November 1928 Li had complained, in a circular letter, of the 'peasant mentality' reflected in the party as a result of the 'particular development of the struggle in the countryside', and affirmed that only proletarian leadership could correct the dangers inherent in this peasant mentality, which might otherwise lead to 'a complete destruction of the revolution and of the party'.[33]

To this type of argument Mao replied in his letter of 5 April in highly characteristic fashion, accepting all the theoretical positions of the Central Committee provided that he were allowed to act as he liked in practice. He recognized that 'the laying of the party's proletarian basis . . . in key centres' was the greatest organizational task of the day, and added that, though the establishment of soviets in small areas in the countryside was useful for accelerating the revolutionary upsurge, it was a mistake to abandon the struggle in the cities and 'sink into peasant guerrilla-ism'. But he also added:

In our opinion it is also a mistake – if any party members hold such views – to fear the development of the power of the peasants, lest it overwhelm the leadership of the workers and hence become detrimental to the revolution. For the revolution in semi-colonial China will fail only if the peasant struggle is deprived of the leadership of the workers; it will never suffer just because the peasant struggle develops in such a way that the peasants become more powerful than the workers.[34]

The dubitative reference to 'any party members' who might hold such views as these can only have been deliberately ironical. In fact, not only Li Li-san and his colleagues in Shanghai, but Moscow itself, were having grave doubts about the utility of the peasant guerrilla tactics which were by now being employed in half a dozen bases scattered across China. In a letter of 7 June

* The secretary general from the Sixth Congress in 1928 until his arrest and execution in 1931 was a former boatman, Hsiang Chung-fa; he was largely a figure head, manipulated by other stronger personalities.

1929 on the peasant question, addressed to the Chinese Communist Party, the Comintern went even further than Li, and actually suggested the abandonment of the whole undertaking:

> The struggle of the peasant masses must be closely linked to the revolutionary struggle of the urban proletariat. Moreover, our tactics in the countryside should correspond to the work of the party in winning over the urban proletariat in the process of its day-by-day economic struggles. It is not at all necessary to begin the peasant movement immediately with calls for carrying out an agrarian revolution, with guerrilla warfare and uprisings. On the contrary, the current situation in China dictates to the party the task of exploiting particular and minor conflicts, of kindling and enlarging the day-by-day struggles of the fundamental masses of the peasantry against all forms of exploitation in order to lift them finally to a higher political level.[35]

The 'day-by-day' struggles in question included movements against taxes, and for the reduction or non-payment of rents. It was not explained how tactics of this type, based essentially on the conditions prevailing in a Western European parliamentary democracy, were to be applied in the Chinese countryside in the face of ruthless repression. The June letter of the Comintern also contains some curious references to agrarian policy, especially in the soviet areas. Citing a passage in a letter* from the Central Committee of the Chinese Communist Party to Mao Tse-tung calling on him to 'conclude an alliance with the rich peasants', the Comintern violently denounced this grave opportunist deviation. The struggle should be directed not only against the landlords, but against the rich peasants, without any regard for the attitude of the latter towards the anti-imperialist national liberation struggle.[36]

This document tends to confirm that, in these early years, Mao's line on the agrarian question was more radical than that of the Central Committee, instead of less so, as is usually supposed. Such an attitude would have been quite in harmony with Mao's continuing revolutionary excitement and impatience at the time. In his April letter to the Central Committee, he proposed a plan for conquering the whole of Kiangsi province, including the

---

* The citations may be from the Central Committee's letter of 9 February 1929, to which Mao was replying in his letter of 5 April; but since the former is known only from Mao's rebuttal of it, it is impossible to be sure.

capital of Nanchang, within a year – this at a time when he had a total of 2,800 men in his army. In January 1930 he still felt that, though the time-limit of one year had been 'mechanical', those who doubted the possibility of over-running the province very rapidly erred in 'over-estimating the importance of objective forces and under-estimating the importance of subjective forces'.[37]

These optimistic views were, as Mao himself pointed out afterwards, founded on the assumption that a 'revolutionary high tide' was on the way. At the time he formulated them the Comintern had adopted a relatively cautious attitude on this point, but in the autumn of 1929 Moscow's position suddenly changed. The explanation lay partly in the economic crisis that had seized the United States, encouraging the belief that capitalism was about to collapse, and partly in events within China itself, where Chiang Kai-shek was involved in a series of conflicts with rival generals. In a letter of 26 October 1929 Moscow therefore informed the Chinese Communist Party that the new upsurge was beginning, and that steps must be taken to 'overthrow the landlord-bourgeois régime and set up a dictatorship of peasants and workers of the soviet type' as soon as the tide had risen high enough.[38]

Henceforth both Mao and Li Li-san were on the look-out for a high tide in the near future, but their ways of preparing for it were categorically opposed. At the Kut'ien Conference of the Fourth Army Communists in December 1929 Mao admitted openly that a majority of the soldiers in the Red Army were *éléments déclassés*; he considered these elements responsible for the 'mentality of roving insurgents' that afflicted the troops, but he still thought it possible to transform them by 'intensified education'. It was this Red Army of dubious class composition which was to 'organize the masses, arm the masses, establish political power, annihilate the reactionary forces, carry forward the revolutionary high tide. . . .'*

For his part Li Li-san was unimpressed not only by an army of lumpenproletarians, but by anything at all that went on in the

---

* Only a brief extract from the Kut'ien resolution, in emasculated form, is given in the *Selected Works*. For brief passages from the original version, see Schram, op. cit., pp. 199–201. The last sentence cited above is to be found in *Ti-chiu-tz'u tai-piao ta-hui chüeh-i-an* (Hongkong, Hsin-min-chu Ch'u-pan-she, 1949), p. 26.

countryside. 'The villages are the limbs of the ruling class', he wrote in March 1930. 'The cities are their brains and heart. If we cut out their brains and heart, they cannot escape death; but if we simply cut off their appendages, it will not necessarily kill them.' In this he was merely echoing the opinion of the Comintern, which had written in its letter of the previous October: 'The soviet areas have been able to maintain themselves and even to extend and consolidate their activities of late. . . . All these peasant activities have become an important side-current in the revolutionary wave, which will grow and merge into the national revolutionary movement.'[39]

Certainly, as late as January 1930, Mao himself regarded his activities as a 'side-current' which would be lost in the sands unless there were a revolutionary high tide in the country as a whole, though he attributed to this same side-current an importance infinitely greater than did Moscow or Li Li-san. The general estimate of the importance to be attached to the soviet areas increased abruptly, however, in the wake of the remarkable successes scored by the Red Army in late 1929 and early 1930. By June 1930 Li Li-san himself had reached the conclusion that, although the workers' struggle in the cities remained the decisive factor in the coming revolutionary outbreak, 'powerful assaults by the Red Army' could make an indispensable contribution to victory. He therefore ordered a series of attacks on major cities in South-Central China.

The first of these led to the capture of Changsha on 28 July by the Third Army Corps under P'eng Te-huai, who held the city for about ten days. Meanwhile the First Army Corps under Chu Te, with Mao as political commissar, advanced on Nanchang, and the Second Army Corps under Ho Lung moved towards Wuhan. After a bloody battle lasting twenty-four hours Chu and Mao abandoned the attempt to take Nanchang and retreated towards the west. In the course of their march they encountered P'eng Te-huai, who had received from Li Li-san orders to make another attack on Changsha. The combined forces of the First and Third Army Corps, numbering some 20,000 men, therefore advanced once more towards the Hunanese capital; but after thirteen days' fighting against much more heavily armed Kuomintang forces, Mao 'persuaded' his comrades to abandon the

hopeless battle and return to the base in Southern Kiangsi.*
Once more, as with the decision two years earlier to reject the
instructions of the Hunan Provincial Committee to march to
Southern Hunan, Mao deliberately disobeyed the orders of higher
party authority when these orders appeared to him to put in
jeopardy the very existence of his army and of his revolutionary
base.

* On the military aspect of these events, see the clear and detailed account
of J. Ch'ên, *Mao and the Chinese Revolution*, pp. 156–9. *Hu-nan chin-pai-
nien ta-shih chi-shu* gives a somewhat different version (pp. 594–602),
stating in particular that the combined forces which staged the second
attack on Changsha totalled 32,000 rifles.

Notes

1. Eto Shinkichi, *China Quarterly*, No. 9, 1962, p. 162.
2. Snow, *Red Star Over China*, pp. 158–9.
3. See, for example, Ho Kan-chih, *A History of the Modern Chinese Revolution*, p. 158.
4. Brandt, *Stalin's Failure in China*, p. 158.
5. Cited by Roy in his article summarizing the work of the congress, North and Eudin, *M. N. Roy's Mission to China*, p. 281.
6. Full text of the resolution in *Strategija i taktika*, pp. 167–79; extracts in English in Eudin and North, *Soviet Russia and the East 1920–27* (O.U.P., 1957), pp. 369–76.
7. Eudin and North, op. cit., pp. 379–80; and also Stalin, *Works*, Vol. X, pp. 34–5.
8. The phrase was cited a few months later by Chang Kuo-t'ao. Translation of his testimony in C. Martin Wilbur, 'The Ashes of Defeat', *China Quarterly*, No. 18, April–June 1964, p. 47.
9. Directive of 13 June 1927, cited above.
10. Hu Hua, *Chung-kuo Hsin-min-chu-chu-i ko-ming shih*, revised edition (Shanghai, Hsinhua Shutien, 1950), p. 97.
11. *Hsiang-tao Chou-pao*, No. 197, 8 June 1927, pp. 2150–51.
12. Brandt, op. cit., pp. 138–9.
13. Snow, op cit., p. 161.
14. 'On the Question of Stalin', *Peking Review*, No. 38, 1963, p. 10.
15. Snow, op. cit., p. 166.
16. Schram, *Political Thought of Mao*, p. 196.
17. ibid., p. 176.
18. Letter to Chu Te, dated 21 December 1927 and published in *Chung-yang cheng-chih t'ung-hsün*, No. 16, pp. 81–8.
19. *Chung-yang cheng-chih t'ung-hsün*, No. 16, pp. 4–6.
20. *Mao Tse-tung hsüan-chi*, 1947 edition, Supplement, pp. 53–4.
21. *Selected Works* (Peking), Vol. I, p. 98.
22. ibid., p. 75.
23. ibid., pp. 79, 96; *Chung-yang chün-shih t'ung-hsün*, No. 1, 15 January 1930.
24. *Hsüan-chi*, 1947 edition, Supplement, p. 84.
25. *Chung-yang cheng-chih t'ung-hsün*, No. 30, 3 July 1928.
26. On this whole military campaign and the events from July to October 1928, see *Selected Works* (Peking), Vol. I, pp. 76–80 and 96.

27. Snow, op. cit., pp. 166–7.
28. *Hui-i Chingkangshan-ch'ü ti tou-cheng*, p. 11.
29. *Selected Works* (Peking), Vol. I, pp. 97–8.
30. *Hsüan-chi*, 1947 edition, Supplement, p. 82.
31. *Chung-yang chün-shih t'ung-hsün*, No. 1, 15 January 1930.
32. *Selected Works* (Peking), Vol. I, p. 123.
33. Schwartz, *Chinese Communism and the Rise of Mao*, p. 137.
34. Schram, op. cit., pp. 188–9.
35. *Strategija i taktika*, p. 242.
36. ibid., pp. 240–42.
37. *Selected Works* (Peking), Vol. I, pp. 125–8; *Hsüan-chi*, 1947 edition, Supplement, pp. 98–9.
38. Schwartz, op. cit., p. 134; Robert C. North, *Moscow and Chinese Communists* (O.U.P., 1954), p. 131.
39. Schwartz, op. cit., pp. 135 and 138.

Chapter 7
From Kiangsi to Yenan

Before advancing to elaborate a new policy for the Chinese revolution, the Comintern felt obliged, as usual, to attribute the responsibilities for past mistakes, and this task was entrusted in the first instance to a plenary session of the Chinese Communist Party's Central Committee, which met in September 1930. But despite the presence not only of the Comintern Representative Pavel Mif, but of Ch'ü Ch'iu-pai and Chou En-lai, both of whom had attended discussions of Li Li-san's policy in Moscow, the result was far from satisfactory. Ostensibly in order to safeguard the prestige of the Chinese leadership, and in fact because they agreed with a good many of Li's opinions, Ch'ü and Chou pushed through a highly indulgent verdict which found Li guilty only of minor errors in judging the objective situation, but not of major errors in his political line. The Comintern was therefore obliged to take matters into its own hands, and to conduct a formal inquest into Li's errors, for which both he and Ch'ü Ch'iu-pai were summoned to Moscow.

This episode is directly relevant to the story of Mao's own career, for two reasons. On the one hand, it has recently been claimed that Mao's line in 1930 was in fact that of the Comintern, which laid down the whole future course of the Chinese revolution in its resolution of June 1930. And on the other hand, the real root of the divergence between Li Li-san and Stalin lay in a Sinocentrism on Li's part which was curiously like that which characterizes Mao Tse-tung today.

The classical expression of the 'Li Li-san line' is to be found in a resolution of the Chinese Communist Party's Politburo dated 11 June 1930, which laid down the basic principles by which the Red Army was shortly ordered to attack Changsha, Nanchang, and Wuchang. This document reached Moscow towards the end of June, and after it had been discussed at Comintern headquarters a resolution was adopted which did not criticize Li's views expressly, but formulated a similar policy in more moderate terms.* Moscow and Li Li-san were agreed that the Red Army could play a most useful auxiliary role in the rapidly approaching crisis, but only on condition that the proletarian leadership of the army itself be strengthened, and that the urban proletariat intervene directly through general strikes and other such traditional methods of combat. Both spoke of a preliminary victory in one or several provinces. Both expected that the decisive combats would occur very soon. Li Li-san was unquestionably more sanguine; for him, 'very soon' meant 'immediately', while Moscow suggested that the Red Army would first have to be strengthened. More important, Li ruled out the existence of a regional or provincial soviet government over a significant period; victory in one or several provinces would immediately lead to uprisings, both urban and rural, in the whole country, and to the final decisive struggle against the Kuomintang régime.† Moreover, these events in China would provoke the intervention of the imperialists, and thus lead to a world-wide revolutionary war, with the final overthrow of capitalism.‡

In view of the Comintern's heavy emphasis on the workers' movement in the cities, with the call for 'general political strikes in all or a series of industrial centres', and in view of the prediction that the 'decisive battles' would take place in the very near future, it is impossible to claim that Moscow had already produced the blueprint for Mao's ultimate victory.§ It is true that Mao was closer to Moscow than to Li Li-san on tactical questions, for Li roundly denounced 'a stubborn adherence to the military

* A virtually complete translation of the resolution of 11 June is to be found in *Documentary History*, pp. 184–200. I take the date of the Comintern resolution to be late June, as indicated in the Russian text (*Strategija i taktika*, pp. 272–81), and not 23 July (or simply July) as found in Chinese-language sources.

† This did not prevent him from causing a proclamation to be issued, during P'eng Te-huai's occupation of Changsha, over his own signature as 'Chairman of the Hunan Provincial Soviet Government'. Text in *Hu-nan chin-pai-nien ta-shih chi-shu*, p. 599.

‡ Given these ideas, the charge of 'semi-Trotskyism', levelled against Li Li-san at the time and later repeated by Mao, was not altogether unjustified. See Schram, *Political Thought of Mao*, p. 159.

§ This is the main thesis of Hsiao Tso-liang's study *Power Relations within the Chinese Communist Movement, 1930–1934* (Seattle, University of Washington Press, 1961).

concept of guerrilla warfare' as 'a serious rightist mistake', and demanded 'a complete change from the guerrilla tactics of the past' to 'resolute attacks on the major forces of the enemy'. But on the other – and deeper – issue of China's role in the world revolution, and of the Russian inability to understand China, Li and Mao were strikingly alike.

The sinocentrism of Li Li-san was extreme. In the resolution of 11 June, he wrote:

China is the weakest link in the ruling chain of world imperialism; it is the place where the volcano of the world revolution is most likely to erupt. Therefore, with the present aggravation of the global revolutionary crisis, the Chinese revolution may possibly break out first, setting off the world revolution and the final decisive class war of the world.[1]

Li here placed himself in the direct line of descent from the Asia-centred world view of M. N. Roy at the Second Comintern Congress to the position defended by Peking since 1963. Not only did Li Li-san attach decisive weight to China in the world revolution; he did not try to conceal that his basic loyalty was to China rather than to Moscow. He is reported to have said at a Politburo meeting of 3 September that fidelity to the Comintern was one thing, and fidelity to the Chinese revolution another – adding that it would be possible to speak differently to Moscow after the capture of Hankow.[2]

These attitudes were naturally not appreciated in Moscow, but the Comintern was almost equally indignant at Li's claim to understand his own country better than did the Russians. This was, indeed, one of the main themes in the inquest on the errors of Li Li-san and Ch'ü Ch'iu-pai. Bela Kun attacked the two men for talking about China's 'national peculiarities', just as Ch'en Tu-hsiu and T'an P'ing-shan had done before them. The Russian China specialist L. Madiar emphasized that this was not merely a personal idiosyncrasy. 'A majority of the members of the Central Committee of the Chinese Communist Party who are not here is also profoundly convinced that Moscow does not know Chinese conditions', he declared, adding that this had been the habitual refrain of all Comintern rebels since Ruth Fischer. Another Comintern inquisitor quoted Li Li-san as saying: 'The Chinese revolution has so many peculiarities that the International has great difficulty in understanding it, and hardly understands it at all, and hence cannot in reality lead the Chinese Communist Party.' Summing up, Manuilsky declared that Li Li-san, who had dared complain of the Russians' 'narrow national prejudice', was himself guilty of 'extreme localism' (*hen li-hai ti ti-fang-chu-i*).*

To be sure, the Comintern authorities were far from minimizing the importance of China. Manuilsky even declared: 'In the future, the Chinese party should play the same role vis-à-vis the Communist parties of the colonies as the Soviet party vis-à-vis the International as a whole.'† But at the same time he emphasized that the Chinese Communists, who did not yet have 'their own Lenin and Stalin', could carry out their work successfully only if they accepted detailed Soviet guidance.

The Soviet claim to understand Chinese conditions better than did the Chinese themselves is the more remarkable since Moscow was so startlingly ignorant of what was going on in China. One celebrated example is the publication, in March 1930, of an obituary on a 'pioneer of the Chinese proletariat' and 'redoubtable enemy of the big landowners and of the bourgeoisie' by the name of . . . Mao Tse-tung.[3] Another is Stalin's statement, in June 1930: 'The Chinese workers and peasants have already . . . [formed] Soviets and a Red Army. It is said that a Soviet government has already been set up there. I think that if this is true, there is nothing surprising about it.'[4] Mao Tse-tung had in fact established a South-West Kiangsi Provincial Soviet Government in February 1930 (hitherto only local soviets had been created). Not only was Stalin poorly informed about this development, but he did not even seem greatly interested in the details.

If Stalin himself was far from China, however, he had sent his protégés there in the spring of 1930. Pavel Mif, the new Comintern emissary, had brought with him from Moscow the group of his former students at the Sun Yat-sen University there, later

* The principal documents regarding Li Li-san's trial, which took place in December 1930, were published in *Pu-erh-sai-wei-k'o* (*Bolshevik*), Vol. 4, No. 3, 10 May 1931. Summaries of them are contained in Hsiao Tso-liang, op. cit., pp. 77–92. Many of the details given above are not included in Hsiao's summaries, and are taken from my own notes on this periodical.

† Hsiao Tso-liang's summary of this point (op. cit., p. 86) is slightly deformed.

known ironically as the 'Twenty-Eight Bolsheviks'. These young men had no revolutionary experience, but they were thoroughly imbued with the view prevailing in Moscow that a good grounding in Leninist theory was far more important. Despite Mif's support they had not been able to capture control of the party organization, which Li Li-san had well in hand, and this was, in fact, one of the charges laid against Li during his interrogation in Moscow. 'We spent a lot of effort and money sending them back to China', complained one of the Comintern officials, 'and then they were prevented from doing leading work.'[5] With Li's disappearance from the scene (he was kept in Moscow until 1945), these returned students, headed by Ch'en Shao-yü (Wang Ming) and Ch'in Pang-hsien (Po Ku), were confronted only by a worker-oriented faction led by Ho Meng-hsiung, and by the supple Chou En-lai. The contest was utterly unequal, and at the fourth plenum in January 1931 the returned student faction captured firm control of the party.

The two leading figures among the 'Twenty-Eight Bolsheviks', Wang Ming and Ch'in Pang-hsien,* were strikingly alike in several respects. Born in the same year, 1907, they were only twenty-three when they returned to China, twenty-four when they captured control of the party. Both were representatives of the privileged strata in Chinese society, Wang being the son of a rich landlord and Ch'in the son of a *hsien* magistrate under the empire. Both had studied in Shanghai, and then gone to Moscow – Wang in 1925, Ch'in in 1926 – while still in their teens. Ch'in had joined the Communist Party in Shanghai in 1925; Wang joined it only in 1927 in Moscow. Thus neither of them had any first-hand knowledge of the peasantry, or any political experience except as adolescents in the great coastal city of Shanghai.

With these men, whose policy is now identified as the 'third "left" opportunist line', Mao was to carry on a bitter struggle that lasted four years. For the first time, he was to be a direct contender for supreme power in the party. Hitherto, since ascending the Chingkangshan in the autumn of 1927, he had simply

---

* Inasmuch as Ch'en Shao-yü, because of his role in the Comintern in the 1930s, became widely known even in the West under his pseudonym, I shall refer to him throughout as Wang Ming. Ch'in Pang-hsien on the other hand, whose pseudonym of Po Ku will probably not be familiar to most readers, I shall call by his real name.

observed the clash of factions within the Central Committee, endeavouring himself to avoid carrying out such directives as appeared to him catastrophic. Now the size of the soviet areas and the strength of the Red Army had reached such proportions that Mao and his 'real power faction', as they came to be called, appeared as the direct rivals of the party leadership.

The struggle between Mao and the returned students continued to have the same two basic themes that had characterized Mao's debates with the Central Committee ever since 1927: agrarian policy; and the relation of guerrilla tactics to revolution in the cities. Over land Mao's original policies on the Chingkangshan had been extremely radical. But by February 1930, when the conference which established the Kiangsi Provincial Soviet adopted a new land law, he had moved far to the right, until his 'rich-peasant deviation' surpassed even that of Li Li-san.* Not only did he allow peasant proprietors to keep all or part of their land, but he even permitted the rich peasants to retain their own good land, rather than requiring them to exchange it for the less fertile land of the poor peasants. He summed up his policy in the two principles of 'draw-on-the-plentiful-to-make-up-for-the-scarce' and 'draw-on-the-fat-to-make-up-for-the-lean'.[6]

All types of 'rich-peasant deviations' were sternly condemned during the Moscow trial of Li Li-san; it was even indicated that Stalin personally had denounced these errors. And yet, complained one of the interrogators, Moscow kept getting letters from the soviet areas declaring that the rich peasants were more powerful than ever.[7] In the directive of 1 September 1931, which marked the opening shot in the campaign of the returned student leadership to capture control of the soviet areas, this criticism was specifically turned against Mao's policies.[8]

Before it is possible properly to consider the manner in which problems of military strategy presented themselves to Mao at the end of 1930 and the beginning of 1931, when he began his confrontation with the returned students, it is necessary to deal with

---

* On the eve of his fall, Li Li-san had become guilty, in the eyes of the Comintern, of a simultaneous right and left deviation: right, because he was kind to the rich peasants; left, because he wanted to go forward to immediate collectivization. This double deviation was another of the charges against him during his trial in Moscow.

the obscure and unpleasant events known as the 'Fut'ien incident'. This affair revolved around a conflict between Mao Tse-tung and the leaders of the Twentieth Red Army in Southern Kiangsi, and on the basis of the numerous accounts available it is clear that considerations both of principle and of sheer, naked power were involved. The leaders of the Twentieth Army were linked to the South-West Kiangsi Special Committee, a bastion of Li Li-san's influence in the soviet areas, which had called a conference in July 1930 to oppose the agrarian policies adopted by Mao in the previous February. The policies that they themselves advocated when they rose in revolt against Mao in December 1930 were similar to those of Li Li-san both on agrarian questions and on military tactics. At the same time, it is clear that Mao was concerned not only to defend what he regarded as the correct policies, but also to eliminate a possible challenge to his own power.

Whatever his motive, he acted with decision. Early in December he had several leaders of the South-West Kiangsi Special Committee arrested on charges of being agents of the 'A-B [Anti-Bolshevik] League', an undercover organization created by the Kuomintang for struggle against the Communists. These men were imprisoned in the town of Fut'ien, in Southern Kiangsi, and on 8 or 9 December a battalion political commissar of the Twentieth Army in near-by Tungku rose in revolt and led his troops to Fut'ien to liberate the prisoners. The rebels fled to a town across the Kan River, where they were less exposed to Mao's counterblows, and continued to resist for more than two months, to the accompaniment of a complicated series of propaganda exchanges and clandestine manoeuvres. In the course of these polemics Mao's own General Front Committee revealed that more than 4,400 members of the A-B League in the ranks of the Red Army had been arrested before the leaders whose imprisonment in Fut'ien had led to the incident. In putting down the revolt Mao acted with extreme severity; some 2,000 or 3,000 officers and men lost their lives in the course of the repression.*

It is undoubtedly excessive to claim, as have some extremely

---

* The above account of the 'incident' is largely based on the two most recent summaries of all the available evidence, Hsiao Tso-liang, op. cit., pp. 98–113, and J. Ch'ên, *Mao and the Chinese Revolution*, pp. 164–5.

hostile critics, that the whole affair was simply a plot by Mao to kill off his rivals. But the affair does provide the first large-scale example of his ruthlessness. It also brought into action the very efficient secret police which he had created. There is no evidence whatever that Mao Tse-tung, like so many who have wielded absolute power (Stalin in particular), takes pleasure in killing or torturing the enemies of the revolution or of his own power. But he has never hesitated to employ violence whenever he believed it necessary. One must immediately add that he would never have survived otherwise. The tens of thousands who fell victim to the Kuomintang repression of 1927, and the few thousand victims of the Fut'ien incident and its sequel, were merely the prologue to the much vaster massacres which were to be perpetrated on both sides in the course of one of the greatest – and most pitiless – revolutionary upheavals of all time. Mao's wife and younger sister were both executed by the Kuomintang in Changsha in the wake of the Red Army's brief occupation of the city in July 1930. Deeply felt as they were, these private losses merely added two more names to the long list of friends and comrades who had already perished in the previous three years. No doubt Mao regarded it all as a natural part of revolutionary struggle. He gave no quarter, and he asked for none.

At the beginning of 1931 it seemed as though Mao Tse-tung was in an extremely favourable situation. The opposition to his person and/or his policies in the Kiangsi party organizations had been effectively destroyed in the Fut'ien incident. Moscow and the Central Committee of the Chinese Communist Party were attaching more and more importance to the type of work in the soviet areas of which he had been a pioneer. In fact, his very success was his undoing. For, to the extent that the higher authorities of the party and the International began to take the soviets seriously, they were naturally led to devote a larger part of their time and energy to devising policies for these areas, and to seeing that these policies were applied. And since the policies defended by the returned students, though different from those of Ch'ü Ch'iu-pai and Li Li-san, were quite as radically opposed to the policies of Mao, the latter was to find himself in a conflict with them which was even more bitter and violent because it was more direct and more central to the life of the party as a whole.

The conflict was all the more acute because in the summer or early autumn of 1931 the Central Committee decided to exchange its harried and clandestine existence in Shanghai for the relative security of the Central Soviet Area in Kiangsi.* Thus it was no longer a question of occasional written instructions from a distant body, which could be circumvented by delaying tactics, but of direct orders from higher party authorities who were present on the spot to make certain that their orders were obeyed.

The arrival of Chou En-lai and his comrades in Kiangsi was followed almost immediately by measures to curb Mao's authority in both party and government. In the latter domain a Central Soviet Government was formed in November. The need to form such a government had been mentioned in the Comintern directive of June 1930, and Li Li-san had also been a partisan of this move – though he wanted to establish the government in a large city rather than in the rural areas. Some preliminary steps had been taken towards the creation of a National Soviet Government during the summer and autumn of 1930, but they were halted by the fall of Li Li-san.[9] The International repeated its counsel in more precise and urgent form in a resolution of 26 August 1931 on the tasks of the Chinese Communist Party, in which it was categorically stated: 'A Central Soviet Government should be formed in the shortest possible time, in the most secure area.'[10]

The First All-China Soviet Congress, which opened at Juichin on the symbolic date of 7 November, proclaimed the establish-

---

* The date for the move to Kiangsi has been given in various sources as 1931, 1932, or even 1933. I am inclined to agree with Hsiao Tso-liang, who suggests, after a review of the available evidence, that it began about August 1931 (op. cit., pp. 161–2). But there is certainly room for doubt. Moreover, it is likely that the transfer did not take place all at once. Chou En-lai was almost certainly present in Kiangsi by the late summer or early autumn of 1931. Ch'in Pang-hsien, who had assumed the secretary generalship in September 1931, after the execution of Hsiang Chung-fa, followed probably in early 1932. For his part Wang Ming returned to Moscow, there to serve for several years as one of the Comintern's leading experts on colonial problems. According to Wang's autobiography as told to Nym Wales (cited by Shanti Swarup, *A Study of the Chinese Communist Movement*, Clarendon Press, 1966, p. 249) his return to Moscow took place in June 1931. Some of the leading figures in the Central Committee went, not to the Central Soviet Area in South-West Kiangsi, but to other bases. In particular, Chang Kuo-t'ao went to the Anhui-Hupei-Hunan soviet area.

ment of the Chinese Soviet Republic and elected a Central Executive Committee of sixty-three members. The Central Executive Committee in turn elected a Council of People's Commissars, with Mao as Chairman and Hsiang Ying and Chang Kuo-t'ao as vice-chairmen.[11] In this way Mao's authority over the governmental apparatus was diluted and circumscribed, but by no means eliminated.

The situation was even more unfavourable to Mao Tse-tung within the party hierarchy. The First Party Congress of the Soviet Area, in November 1931, contented itself with denouncing his policies in virtually every domain, with particular emphasis placed on shortcomings in military matters: the persistence of guerrilla tactics, and the stubborn refusal to enlarge and consolidate the Central Soviet Area by linking up with the other bases in neighbouring provinces. The following August, at the Ningtu conference of the party, Mao was not merely criticized but began to lose his authority over the Red Army, which passed to Chou En-lai. The process found its culmination in May 1933, when Chou, who strongly supported the military line of the 'Twenty-Eight Bolsheviks', was appointed political commissar of the entire Red Army.[12]

Mao's loss of control over the army was doubly grave because it occurred in the midst of the struggles to save the Kiangsi Soviet Area in the face of Chiang Kai-shek's campaigns of encirclement and annihilation; indeed it opened the door to the rigid tactics which ultimately led to defeat. The attacks on Changsha and Nanchang in the summer and autumn of 1930 had drawn the attention of the Kuomintang leaders to the growing strength of the Communists, and they had therefore called a temporary halt to the wars among themselves in order to unite against this single threat. The First Encirclement Campaign in December 1930 and January 1931, the Second Encirclement Campaign in the spring of 1931, and the Third Encirclement Campaign in the summer of 1931 were widely different in the number and nature of the forces involved, but the basic tactical principles followed by the Red Army were identical. They were the principles which Mao and Chu Te had been gradually evolving since their days on the Chingkangshan, and which now began to find their place in a coherent system.

In Mao's ideas on guerrilla warfare the military and political aspects are, of course, indissolubly linked, but it is possible to separate them for purposes of analysis. Both aspects appear as a mixture of the ancient and the modern, the traditional and the original. Not surprisingly, it is on the military side that the mark of tradition appears most clearly. The problems involved in manoeuvring troops on the ground have changed infinitely less over the centuries than the conditions under which the political support of the population can successfully be sought. Indeed, virtually all of Mao's tactical principles are to be found in the classical Chinese writings on the subject, and above all in the maxims of Sun Tzu, the famous military writer who flourished about 500 BC.

This is the case, first of all, with the fundamental idea that victory can be won only by concentrating in each successive battle a relatively large part of one's own forces against a relatively small part of the enemy's. Such a tactic is obviously indispensable when, as with the Red Army in the 1930s, one is fighting against an enemy who enjoys a great overall superiority in numbers, but Sun Tzu presented it long ago as a basic principle of all warfare:

> By discovering the enemy's dispositions and remaining invisible ourselves, we can keep our forces concentrated while the enemy must be divided. We can form a single united body, while the enemy must be split up into fractions. Hence there will be a whole pitted against separate parts of the whole, which means that we shall be many in collected mass to the enemy's separate few. . . . And if we are thus able to attack an inferior force with a superior one, our opponents will be in dire straits.[13]

Here is how Mao expressed the same basic idea, as applied to the specific conditions under which the Red Army fought during the 'Encirclement and Annihilation Campaigns' in Kiangsi:

> An army operating on strategically inferior lines suffers from many disadvantages, and this is especially so in the case of the Red Army, confronted as it is with 'encirclement and suppression'. But in campaigns and battles we can and absolutely must change this situation. We can turn a big 'encirclement and suppression' campaign waged by the enemy against us into a number of small, separate campaigns of 'encirclement and suppression' waged by us against the enemy. . . . We can change the enemy's strategic superiority over us into our super-

iority over him in campaigns and battles. . . . This is what we call exterior-line operations within interior-line operations, encirclement and suppression within 'encirclement and suppression', blockade within blockade, the offensive within the defensive, superiority within inferiority, strength within weakness, advantage within disadvantage, and initiative within passivity. The winning of victory in the strategic defensive depends basically on this measure – concentration of troops.

. . . The Chinese Red Army, which entered the arena of the civil war as a small and weak force, has since repeatedly defeated its powerful antagonist and won victories that have astonished the world, and it has done so by relying largely on the employment of concentrated strength. When we say, 'Pit one against ten, pit ten against a hundred', we are speaking of strategy, of the whole war and the overall balance of forces, and in the strategic sense that is just what we have been doing. However, we are not speaking of campaigns and tactics, in which we must never do so. Whether in counter-offensives or offensives, we should always concentrate a big force to strike at one part of the enemy forces. . . . Our strategy is 'pit one against ten', and our tactics are 'pit ten against one'; these contrary and yet complementary propositions constitute one of our principles for gaining mastery over the enemy.

We use the few to defeat the many – this we say to the rulers of China as a whole. We use the many to defeat the few – this we say to the enemy on the battlefield. That is no longer a secret, and in general the enemy is by now well acquainted with our method. But, he can neither prevent our victories nor avoid his own losses, because he does not know when and where we shall act. This we keep secret. The Red Army generally operates by surprise attacks.*

---

* 'Strategy in China's Revolutionary War', *Selected Works* (Peking), Vol. I, pp. 235–9, passim. (I have modified the translation slightly to bring it into conformity with the original Chinese text.) This work, written in 1936, sums up what Mao had learned in the course of the encirclement campaigns in Kiangsi. Mao himself affirms (op. cit., p. 213) that his military thought transcended its 'originally simple nature' to take form as a 'complete set of operational principles' in the course of the third campaign, i.e. in the summer of 1931. To the extent that this is true – and, contrary to the opinion of Jerome Ch'ên (op. cit., pp. 163–4), I believe that it is, though naturally Mao continued to learn and to develop his ideas in subsequent years – it is not inappropriate to cite this text of 1936 to characterize the essential traits of Mao's military thinking five years earlier. For a further discussion of Mao's guerrilla tactics as they developed in the course of the anti-Japanese war, see Chapter 8. On the relations between Mao's tactics and those of Sun Tzu, see also General Samuel B. Griffith's introduction to his translation of Sun Tzu, *The Art of War* (Oxford University Press, 1963), pp. 45–56.

As the first sentence in the above quotation from Sun Tzu and the last sentences in the quotation from Mao clearly indicate, an indispensable condition for 'pitting ten against one' is to maintain secrecy on the movements of one's own troops. This idea is fundamental to Sun Tzu's whole doctrine:

All warfare is based upon deception. Hence, when able to attack, we must seem unable; when using our forces, we must seem inactive; when we are near, we must make the enemy believe that we are far away; when far away, we must make him believe we are near. Hold out baits to entice the enemy. Feign disorder, and crush him. If he is secure at all points, be prepared for him. If he is superior in strength, evade him. If your opponent is of choleric temper, seek to irritate him. Pretend to be weak, that he may grow arrogant. If he is inactive, give him no rest. If his forces are united, separate them. Attack him where he is un-prepared; appear where you are not expected. These military devices, leading to victory, must not be divulged beforehand. . . .

The general who is skilled in defence, in effect, hides in the most secret recesses of the earth; he who is skilled in attack flashes forth from the topmost heights of heaven. . . .

Hence the general is skilled in attack whose opponent does not know what to defend; and he is skilful in defence whose opponent does not know what to attack. O divine art of subtlety and secrecy! Through you we learn to be invisible, through you inaudible; and hence hold the enemy's fate in our hands. You may advance and be absolutely irresist-ible if you make for the enemy's weak points; you may retire and be safe from pursuit if your movements are more rapid than those of the enemy.[14]

Mao Tse-tung does not develop this point at quite such length, essentially because he considers the problem of secrecy regarding the Red Army's movements (and the complementary problem of intelligence regarding the enemy) as essentially political in nature. He quotes the Kuomintang general Ch'en Ming-shu as saying: 'Everywhere the National Army gropes in the dark, while the Red Army walks in broad daylight.'[15] Mao's explanation for this is naturally that the peasant masses are favourable to the Red Army and hostile to the Nationalists. But at the same time he is far from disdaining deception on the purely military level. Not surprisingly, he cites Sun Tzu directly on this point: 'We can in-duce the enemy to make mistakes by our own actions, for instance by "counterfeiting an appearance", as Sun Tzu called it, that is, by making a feint to the east but attacking in the west.'[16] Mao also learned a great deal about the role of deception in warfare from his favourite novels, the *Romance of the Three Kingdoms* and *Water Margin*.

These were Mao's essential tactical principles, and while they by no means earned the approval of the returned students, they were undoubtedly a great deal less repugnant to the party leadership than his highly flexible strategic concepts. These grew out of the celebrated four-line slogan invented on the Chingkangshan:

> The enemy advances, we retreat;
> The enemy camps, we harass;
> The enemy tires, we attack;
> The enemy retreats, we pursue.

From these basically defensive principles for the guidance of a weak force in a hostile environment, Mao and Chu Te developed, in the course of the First Encirclement Campaign, the principle of 'luring the enemy in deep' into soviet territory, for the purpose of attacking his forces piecemeal in the most favourable circumstances. Needless to say, Mao did not envisage remaining indefinitely in a position of inferiority; but increasingly, in contrast to his own earlier optimism, he began to feel that the struggle would be a long one, and that the Red Army would not be in a position to seize the strategic offensive in the foreseeable future. Beginning in September 1931, Ch'in Pang-hsien and Chou En-lai, who supported the returned student leadership, began to denounce these ideas of Mao's as 'guerrilla-ism', quite out of date in a situation where the Red Army had grown to a strength of several hundred thousand and the conflict between the Communists and the Kuomintang appeared as a war between two states. They therefore insisted on replacing the old slogans by new ones such as 'Attack on all fronts', 'Seize key cities', and 'Don't let our pots and pans be smashed'.[17]

The meaning and consequences of this shift in tactics can best be understood from a comparison of the five Encirclement Campaigns. During the First Campaign, in December 1930 and January 1931, the situation was complicated by the consequences of the Fut'ien affair, which had led to suspicion and hostility amongst the population of Fut'ien and Tungku toward the Red Army.

But this difficulty was more than balanced by the fact that the Kuomintang forces were not crack units, and enjoyed only a relatively limited numerical superiority over the Red Army – 100,000 to 40,000. The tactics of attacking the enemy piecemeal, and of waiting patiently for a favourable opportunity before undertaking the first engagement, were fully employed. In Mao's own words:

> ... we originally planned to strike at the troops of T'an Tao-yüan (the commander of one of the three enemy divisions); we advanced twice, but each time had to restrain ourselves and pull back, because they would not budge from their commanding position on the Yüant'ou heights. A few days later we sought out Chang Hui-tsan's troops, which were more vulnerable to our attack.

In the first assault the Red forces struck at two of Chang's brigades as well as his divisional headquarters, capturing the entire force of 9,000 men and Chang Hui-tsan himself. This victory put the other two divisions to flight; the Communists pursued one of them and wiped out half of it. Thus the First 'Encirclement and Annihilation' Campaign ended in total failure. General Chang was later executed, after being tried before the local peasantry.[18]

The Second Campaign was both more difficult and more interesting as an example of Mao's tactics. The relation between the opposing forces had become more than twice as unfavourable to the Red Army; the attacking force numbered nearly 200,000 men, whereas the Red Army had only 30,000. This made it all the more important to strike first against the weakest units in the enemy forces. For twenty-five days in late April and early May the Red Army encamped at Tungku,* despite the danger of discovery, waiting for Wang Chin-yü's division to leave its strongpoint at Fut'ien. At last the moment came; the Red Army fell on elements of both Wang's division and Kung Ping-fan's, totalling eleven regiments, and routed them in a battle near Fut'ien. This engagement took place only three or four miles from another Kuomintang division to the north-east, and only a dozen miles from Ts'ai T'ing-k'ai's crack Nineteenth Route Army to the

---

* By this time the echoes of the Fut'ien affair had largely died away and the Red Army was no longer obliged to avoid this area as in the first campaign.

south-west; but no alarm reached these forces to incite them to intervene. Truly, the National Army 'groped in the dark', while the Red Army 'walked in daylight'. After this first successful engagement, the Red Army marched 700 *li** in fifteen days, fought five battles against the weaker units of enemy forces, and captured 20,000 rifles. This was the end of the Second Campaign.[19]

Mao expressed his elation in a poem written shortly afterwards:

> About the peak of White Cloud Mountain clouds are pressing upward,
> Beneath White Cloud Mountain the cries grow more desperate,
> Dry wood and hollow trees gather together for the struggle.
> A forest of rifles press forward,
> The flying generals sweep down from the void into battle.
>
> In 15 days we have driven 700 *li*,
> Green and majestic are the waters of Kiangsi,
> And the mountains of Fukien like jade.
> We have swept through armies thousands strong like rolling up a mat.
> Someone is weeping,
> How bitterly he regrets the strategy of step-by-step advance!

White Cloud Mountain is on the border of Fukien, near the end of the Red Army's seven-hundred-*li* march. According to the commentaries published in Peking, the clouds pressing upwards represent the anger of the Red soldiers at the atrocities perpetrated by the Kuomintang armies in the base areas temporarily abandoned during the campaign; the 'dry wood and hollow trees' represent the enemy, who is already fated to perish though he still stands upright. The last two lines reflect the bitterness of Ho Ying-ch'in, the commander-in-chief of the Second Campaign, at the unsuccessful tactics of slow advance by several isolated columns.†

The Third Campaign was to witness significant changes in the scale and tactics of the Kuomintang effort. Now thoroughly aroused, Chiang Kai-shek arrived in Nanchang to take personal command of the operations; he also threw into the battle 100,000

---

* One *li* is approximately one third of a mile.

† The above translation is my own. The interpretation is taken from *Mao Chu-hsi Shih-tz'u chien-i* (Peking, 1962).

of his own crack troops. In these new circumstances the Red Army had a much more difficult time, and did not succeed in foreseeing and shaping the whole course of the campaign as it had done in the two previous ones. Here is Mao's own account of the campaign, illustrated by his own sketch map.

The disposition of our forces and those of the enemy in the third 'encirclement and suppression campaign' was as shown in [the] Figure.

The situtation at the time was as follows:

1. Chiang Kai-shek personally took the field as commander-in-chief. Under him there were three subordinate commanders, each in charge of a column – the left, the right, and the centre. The central column was commanded by Ho Ying-ch'in, who like Chiang Kai-shek had his headquarters at Nanchang, the right was commanded by Ch'en Ming-shu, with headquarters at Kian, and the left by Chu Shao-liang with headquarters at Nanfeng.

2. The 'suppression' forces numbered 300,000. The main forces, totalling about 100,000 men, were Chiang Kai-shek's own troops and consisted of five divisions (of nine regiments each), commanded by Ch'en Ch'eng, Lo Cho-ying, Chao Kuan-t'ao, Wei Li-huang and Chiang Ting-wen respectively. Besides these, there were three divisions (totalling forty thousand men) under Chiang Kuang-nai, Ts'ai T'ing-k'ai, and Han Te-ch'in. Then there was Sun Lien-chung's army of twenty thousand. In addition, there were other, weaker forces that were likewise not Chiang's own troops.

3. The enemy's strategy in this 'suppression' campaign was to 'drive straight in', which was vastly different from the strategy of 'consolidating at every step' that he used in the second campaign. The aim was to press the Red Army back against the Kan River and annihilate it there.

4. There was an interval of only one month between the end of the second enemy campaign and the beginning of the third. The Red Army (then about thirty thousand strong) had had neither rest nor replenishments after much hard fighting and had just made a detour of a thousand *li* to concentrate at Hsingkuo in the western part of the Soviet area, when the enemy pressed it hard from several directions.

In this situation the plan we first decided on was to move from Hsingkuo by way of Wanan, make a breakthrough at Fut'ien, and then sweep from west to east across the enemy's rear communication lines, thus letting the enemy's main forces make a deep but useless penetration into our base area in southern Kiangsi; this was to be the first phase of our operation. Then when the enemy turned back northward, inevitably very fatigued, we were to seize the opportunity to strike at his vulnerable units; that was to be the second phase of our operations.

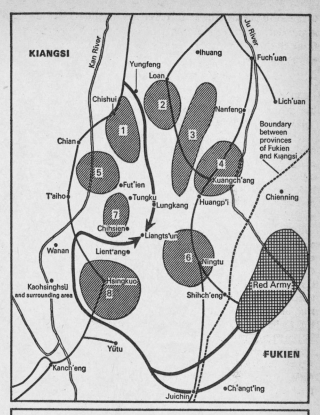

KIANGSI

Kan River

Ju River

Fuch'uan

Yungfeng
Loan

●Ihuang

Lich'uan

Chishui

2

Nanfeng

Boundary
between
provinces
of Fukien
and Kiangsi

Chian

1

3

5

4

Kuangch'ang

Chienning

Fut'ien

T'aiho

Tungku

Huangp'i

Lungkang

7

Chihsien

Liangts'un

Wanan

Lient'ang

6

Ningtu

Kaohsinghsü
and surrounding area

8

Hsingkuo

Red Army

Shihch'eng

Yütu

FUKIEN

Kanch'eng

●Ch'angt'ing

Juichin

KUOMINTANG
MILITARY FORCES

1 Hao Meng-ling's and
  Shang-kuan Yün-hsiang's divisions

2 Five weaker divisions

3 Mao Ping-wen's division

4 Hsü K'o-hsiang's division

5 Ch'en Ch'eng, Lo Cho-ying

6 Sun Lien-chung's army

7 Ts'ai T'ing-k'ai and
  Chiang Kuang-nai

8 Chiang Ting-wen's division
  and other, weaker forces

The Third Encirclement Campaign

The heart of this plan was to avoid the enemy's main forces and strike at his weak spots. But when our forces were advancing on Fut'ien, we were detected by the enemy, who rushed the two divisions under Ch'en Ch'eng and Lo Cho-ying to the scene. We had to change our plan and fall back to Kaohsinghsü in the western part of Hsingkuo county, which, together with its environs,* was then the only place for our troops to concentrate in. After concentrating there for a day, we decided to make a thrust eastwards towards Lient'ang in eastern Hsingkuo county, Liangts'un in southern Yungfeng county, and Huangp'i in northern Ningtu county. That same night, under cover of darkness, we passed through the forty-*li* gap between Chiang Ting-wen's division and the forces of Chiang Kuang-nai, Ts'ai T'ing-k'ai and Han Te-ch'in, and swung to Lient'ang. On the second day we skirmished with the forward units under Shang-kuan Yün-hsiang (who was in command of Hao Meng-ling's division as well as his own). The first battle was fought on the third day with Shang-kuan Yün-hsiang's division and the second battle on the fourth day with Hao Meng-ling's division; after a three-day march we reached Huangp'i and fought our third battle against Mao Ping-wen's division. We won all three battles and captured over ten thousand rifles. At this point all the main enemy forces, which had been advancing westward and southward, turned eastward. Focusing on Huangp'i, they converged at furious speed to seek battle and closed in on us in a major compact encirclement. We slipped through in the high mountains that lay in the twenty-*li* gap between the forces of Chiang Kuang-nai, Ts'ai T'ing-k'ai and Han Te-ch'in on the one side, and Ch'en Ch'eng and Lo Cho-ying on the other, and thus, returning from the east to the west, reassembled within the borders of Hsingkuo county. By the time the enemy discovered this fact and began advancing west again, our forces had already had a fortnight's rest, whereas the enemy forces, hungry, exhausted, and demoralized, were no good for fighting and so decided to retreat. Taking advantage of their retreat, we attacked the forces of Chiang Kuang-nai, Ts'ai T'ing-k'ai, Chiang Ting-wen and Han Te-ch'in, wiping out one of Chiang Ting-wen's brigades and Han Te-ch'in's entire division. As for the divisions under Chiang Kuang-nai and Ts'ai T'ing-k'ai, the fight resulted in a stalemate and they got away.[20]

Mao's account of this campaign is extremely interesting as illustrating the application of his military ideas – the emphasis on concealment and on striking the enemy at his weakest point, and the strategy of 'luring the enemy deep' into the base area. Much

---

* In the 1951 edition, Mao added that the area involved was only 'some tens of square *li*', i.e. less than ten square miles.

of the factual information it contains is no doubt also accurate, but it cannot be considered as a complete or objective account of the campaign. Despite the skilful manoeuvring of the Red Army described by Mao, the strongest units of Chiang's army remained virtually intact, and might very well have advanced on the capital of the 'Soviet Republic' in Juichin if news had not arrived, precisely at the crucial moment, of the 'Mukden Incident' on 18 September. In the face of the Japanese menace, Chiang Kai-shek was unable to give his undivided attention to the 'suppression' campaign against the Communists, which was therefore postponed. In fact, Chiang found it expedient to resign the presidency in October 1931, only to return to power the following January in another guise.[21]

In the long run the continuing Japanese aggression was to prove perhaps the most important single factor in Mao's rise to power. For the moment, however, its impact in the soviet area was relatively limited. The Communist leaders, like the International, adhered to the view that nothing could be done to resist Japanese aggression until the 'traitorous and reactionary' Kuomintang Government had been overthrown. For the moment, therefore, there was no thought of a 'united front' of any kind even with dissident forces in the Nationalist camp, let alone with Chiang himself.

As indicated above, it was in the months immediately following the end of the Third Encirclement Campaign and the beginning of the Sino-Japanese conflict that Mao was progressively deprived of such authority as he still possessed in the party and the army.

At the same time, he retained his position as Chairman of the Central Executive Committee of the Chinese Soviet Republic. In this capacity, he continued to be one of the principal spokesmen for the régime, but his freedom of expression was strictly limited. As a result, in trying to trace Mao's policies in the years 1932–4, we find ourselves in a situation unlike that prevailing as regards any other period of his existence. Normally, when we read a contemporary text of one of Mao's speeches or writings, we can assume that this accurately represents his views at the time. For the period of the Kiangsi Soviet Republic, on the other hand, the piecing together of what Mao actually said or wrote is only the first step; we must then endeavour to decide, on the basis of the scanty evidence available, whether he meant it or not.

Inasmuch as the years 1932–4 do not, on the whole, constitute one of the most interesting and characteristic periods in Mao's life, there is no point in speculating at great length about what his position may conceivably have been on every possible issue. In general, it appears that he enjoyed considerably more influence and independence in the realm of economic policy than in matters of military strategy. Ch'in Pang-hsien, Chou En-lai, and the others probably thought that, while Mao's guerrilla tactics were out of date, he could still be useful as a man close to the peasants and able to mobilize their support.

One aspect of the economic and social policy of the Chinese Soviet Republic merits particular attention, for it corresponds so closely to a fundamental bent of Mao's personality that it seems likely he not only approved of it, but actually originated it. This is the so-called 'land verification movement' (*ch'a-t'ien yün-tung*). The land reform law adopted in November 1931 was a relatively moderate one. To be sure, the rich peasants were to receive inferior land* – and even this only provided they did not adopt a counter-revolutionary attitude – and the landlords were to receive no land at all. But the middle peasants were to be allowed to keep their own land and remain completely outside the whole process of land reform unless they agreed of their own free will to participate in it.[22] The purpose of the 'land verification movement', launched by Mao on a large scale in June 1933, was, as the name suggests, to verify whether the redistribution had been properly carried out. But this economic objective was less important in Mao's mind than the political aim of promoting class struggle.

In an article published at the time, Mao explained that there were three stages in the revolutionary transformation of the countryside: land reform, the verification movement, and agricultural development. In other words, he held that the

---

* Mao's policy of February 1930 was thus reversed, but he may well have changed his own view on this point, and not simply yielded to pressure. As we have seen, he had applied far more radical agrarian policies in Chingkangshan days.

development of farm production (which was an urgent matter in the face of the Kuomintang blockade) would be possible only after a thoroughgoing transformation of social relationships.

His reasons for thinking in this way are probably to be found less in an abstract commitment to revolution than in his view of the nature of Chinese rural society. Mao explained that, because of their inherited prestige, and the cultural advantages they enjoyed ('it is they who know how to speak, it is they who know how to write'), the former landlords and rich peasants were frequently able to capture the leadership of the mass organizations such as the peasant associations, and in this way regain a great deal of their ancient privileges in a disguised form. In order to counteract this tendency, it was necessary to promote class struggle by the poor peasants against their former exploiters, thus awakening them to a consciousness of their true interests. 'The revolutionary masses', wrote Mao, 'are engaged in a serious struggle with the landlords and the rich peasants, but this struggle is not, like that which took place during the first stage (i.e., the stage of land reform), an open struggle between the white flag and the red flag; it is a struggle of the revolutionary peasant masses against elements of the landlords and rich peasants wearing false masks'.[23]

The land verification movement, in Mao's view, would play a key role in opening the peasants' eyes to the true class nature of these hidden enemies. What this meant in practice Mao explained a few months later, in describing what had happened in a model district near the Soviet capital of Juichin:

For 55 days the masses of the whole district were set in motion. The feudal remnants were radically destroyed. In the course of the verification more than three hundred families of landlords and rich peasants were discovered, twelve counter-revolutionary elements, called 'big tigers' by the peasants, were shot, and counter-revolutionary activities were repressed.[24]

It is in this context that one must interpret the directives for determining class membership in the countryside, penned by Mao in June 1933, which affirmed: 'The landlord class is the principal enemy of the revolution. The policy of the Soviets toward the landlords consists in confiscating the whole of their property, and in annihilating the landlord class'.* As can be seen from the fact that in the example cited by Mao only twelve 'big tigers' were executed out of 300 families of landlords and rich peasants, the 'annihilation' of the landlord class advocated by Mao in 1933–4 did not mean the physical extermination of all its members. The total confiscation of the landlords' property destroyed the economic basis for their existence as a class; the mobilization of the peasant masses to 'struggle' against them was intended to destroy the roots of their position in tradition and habit.

The 'land verification movement' is worthy of note not merely as an important aspect of the policy of the Kiangsi Soviet Government, but as a precursor of similar policies and preoccupations in the years since 1949. On the one hand, it reflects Mao's conviction that land reform is a political quite as much as an economic measure, and at the same time a process of education of the peasantry. As in the early years of the Chinese People's Republic, meetings were held in Kiangsi to associate the masses with the decisions of the authorities, and to approve the treatment meted out to hostile classes. 'To conduct these tasks solely through the activities of a few Soviet functionaries', declared Mao, summing up the work of the 'land verification movement' in January 1934, 'involves the danger of lowering the fighting spirit of the masses'.[25]

The concern with mass mobilization as the key to the transformation of the Chinese countryside, and the idea that the existing situation must be constantly reviewed and called into question to prevent the re-emergence of the former exploiting classes in positions of influence, have survived to this very day, in the form of repeated 'class struggle' movements. More generally, Mao's conception of revolution as above all a psychological process has inspired a whole series of policies of 'thought reform' and 'rectification' over the past quarter of a century, culminating in the current 'Great Proletarian Cultural Revolution'.

---

* 'How to analyse class membership', *Hung-se Chung-hua*, No. 89, 29 June 1933, p. 8. This item was not signed by Mao at the time, but has since been attributed to him; see *Selected Works* (Peking), Vol. I, pp. 137–9. The current version has been doctored even more than that of most of his other writings.

If Mao was thus able to place his stamp to a considerable extent on the economic and social policy of the Chinese Soviet Republic, his influence on the conduct of military matters seems to have been virtually nil. And it was this aspect of things which was to prove decisive, for the inflexible strategy imposed by the majority of the Central Committee ultimately led to a catastrophic defeat which put an end to the very existence of the Kiangsi Republic.

This so-called 'forward and offensive line', which had been developed by Wang Ming, Ch'in Pang-hsien, and Chou En-lai beginning in the autumn of 1931, gained momentum in late 1932 and early 1933 as a result of the position adopted by the Twelfth Comintern Plenum in September 1932. On this occasion, it was proclaimed that the 'revolutionary upsurge' growing out of 'the sharpening of the general crisis of capitalism' set the Communists the task of preparing the masses for 'the impending fight for power' in the little time remaining before the revolutionary crisis matured. The immediate tasks of the Chinese Communist Party, as itemized by the Comintern, included not only mobilizing the masses for 'the national revolutionary struggle against the Japanese and other imperialists', developing and uniting the Soviet territories, and strengthening the Red Army, but also fighting directly 'for the overthrow of the Kuomintang régime'.[26]

Thus encouraged in its belief that the victory of the revolution throughout the world was near at hand, the majority of the Central Committee was further emboldened in its advocacy of a head-on struggle against Chiang Kai-shek's army, and hence even less inclined than previously to tolerate Mao's flexible guerrilla tactics. It must be recognized that more was involved than contempt by these city-dwellers for Mao's 'peasant methods'. The tactics of 'luring the enemy deep' into soviet territory, so that the battles could be fought in an environment favourable to the Red Army, had proved militarily successful against Chiang's first three campaigns, but politically they had the decided disadvantage of constantly subjecting the population of the soviet areas to the sufferings and dangers of war. What sort of political entity was the Chinese Soviet Republic, which claimed to be the only legitimate government of the country, but could not protect its own citizens any better than this?

Even afterwards, in analysing the reasons for the collapse of the Kiangsi base, Mao was able to produce only a negative answer to these objections. He pointed out that the effort to maintain the stability of the soviet areas had led to a catastrophe; he did not claim that his own tactics, if pursued, would have led to victory on a national scale:

> Fluidity of battle lines leads to fluidity in the size of our Soviet areas. Our Soviet areas are constantly expanding and contracting, and often as one area falls another rises. . . .
>
> Fluidity in the war and in our Soviet territory produces fluidity in our Soviet construction. Construction plans covering several years are out of the question. Frequent changes of plan are all in the day's work.
>
> It is to our advantage to recognize this characteristic. We must base our planning on it, and must not have illusions about a war of advance without any retreats, or take alarm at any temporary fluidity of our Soviet territory. . . . It is only out of today's fluid way of life that tomorrow we can secure relative stability, and eventually full stability.
>
> The strategy of 'regular warfare' . . . denied this fluidity. Its adherents opposed what they called 'guerrilla-ism', and managed affairs as though they were the rulers of a big state. The result was an extraordinary and immense fluidity – the 25,000-*li* Long March.[27]

If Mao did not have a blueprint for rapid victory in 1936, when he wrote these comments, it is even less likely that he had one to offer when the decisive strategic controversy took place in 1933–4. His attitude appears to have been that the best the Communists could do was to maintain their existence, and wait for the emergence of some new circumstance which would permit them to expand their territory and influence. But by this period he was no longer in a position to defend his own views openly, and was even obliged to countenance a thinly veiled attack on himself in the guise of a campaign against the 'Lo Ming Line'.

Lo Ming, the acting secretary of the Fukien Provincial Committee of the Party, had followed, in the course of the Fourth Encirclement Campaign, the flexible guerrilla tactics which had always been associated with Mao. Today, in the official historiography, the success of the Communist defence against the Fourth

Campaign, which took place during the winter of 1932–3,* is attributed entirely to Mao's tactics.† In fact, it is evident from Mao's own highly brief and incomplete account of the campaign that the tactics applied by the Red Army were different from those adopted previously, and involved, as Chou En-lai, Ch'in Pang-hsien, and the others advocated at the time, meeting the enemy as far as possible from the centre of the soviet area.‡ In any case, the unfortunate Lo Ming, who questioned the wisdom of direct attacks on the Kuomintang armies by the relatively weak guerrilla forces at his own disposal, was condemned and vilified by the 'returned students', beginning in February 1933, for his 'pessimism', 'liquidationism', 'flight-ism', and other 'right opportunist' errors. Though Mao himself did not explicitly condemn Lo, he obediently denounced all the deviations associated with the 'Lo Ming Line', and proclaimed that the victory against the Fourth Campaign was entirely due to the 'forward and offensive line' of the Central Committee. And yet his own brother, Mao Tse-t'an, was among those adherents of the 'Lo Ming Line' whose pernicious influence was singled out by the Party leadership as an obstacle to the liquidation of this erroneous tendency.§

In the 1945 resolution on Party history, Lo Ming is presented as a wise commander whose basically correct policy was unjustifiably attacked by the 'leftists' in control of the Party. While,

* There is some disagreement regarding the dates of this campaign. Edgar Snow, on the basis of Mao's oral account, indicated that the campaign took place from April to October 1933 (*Red Star Over China*, p. 178). Jerome Ch'ên (op. cit., p. 176) opts for June 1932–March 1933. Hsiao Tso-liang sets some kind of record by giving both dates (June 1932 and spring 1933) in his chronology for the beginning of the campaign, without any further explanation (op. cit., pp. 310–11). In reality, the former date marks the beginning of the first preparatory moves against some of the other less-important soviet areas, and the second the major battles for the central soviet area itself.

† See the 1945 'Resolution on Certain Questions in the History of our Party', *Selected Works* (Peking), Vol. III, p. 190.

‡ It is perhaps because this campaign was victorious despite the fact that his counsels were not heeded that Mao accords so little space to it. See *Selected Works* (Peking), Vol. I, pp. 230–31.

§ The principal documents regarding the 'Lo Ming Line' are summarized in Hsiao Tso-liang, op. cit., pp. 230–47.

as we have already seen in various contexts, the Maoist school of historiography does not always show a scrupulous respect for the facts, there is every reason to believe that in this particular instance the 1945 resolution accurately reflects Mao's views at the time. His position was so far weakened, however, that he could not even defend his brother, but was obliged to praise instead the 'Bolshevik Line' of the Central Committee.

In the course of the Fourth Campaign, this 'forward and offensive line' led, as already indicated, to a victorious result. This was certainly gratifying in the short run, but in the long run it had the unfortunate effect of appearing to justify the idea that it was possible for the Red Army to fight with the Kuomintang forces on a basis of equality, keeping them at all times out of the base area. In the next campaign, when Chiang Kai-shek adopted the new strategy of building blockhouses and gradually tightening the stranglehold on the Kiangsi base area, this optimistic and offensive mentality was to lead to quite different results.

But before this fifth and decisive campaign got well under way, the problem of the united front against Japan was at last significantly injected into the struggle by the revolt of the Nineteenth Route Army under Ts'ai T'ing-k'ai. The first reaction of the Chinese Communists to the Japanese aggression of September 1931 had been strictly in accordance with the Comintern line of the time, which envisaged only a 'united front from below'. It was recognized that there were those among the masses of the people, and even among the Kuomintang armies, who genuinely desired to resist Japan – but such resistance could be effective only under Communist leadership. The complete overthrow of the Kuomintang régime was thus a necessary pre-condition to an effective anti-Japanese policy. This was the viewpoint expressed by the Comintern at its twelfth plenum in September 1932, and again at the thirteenth plenum in December 1933.[28] It was also the position adopted by the Chinese Soviet Republic in its declaration of war against Japan in April 1932.[29] But in January 1933 the Chinese Communists set forth a notably more flexible line. Under three conditions (cessation of the attacks against the soviet areas, the granting of democratic rights, and the arming of the masses against Japan), they were prepared to conclude an agree-

ment with 'any armed force' – i.e., with any dissident commander.[30] Although there was no hint of a willingness to deal
with the top leadership of the Kuomintang, still less with Chiang
Kai-shek himself, this was further than the Comintern was inclined to go at the time. There was apparently no outright disapproval in Moscow, for this appeal of January 1933 was published
in the Comintern press.[31] But even in December 1933 Wang Ming,
at the Thirteenth Comintern Plenum, proposed only that the
Communists should seek to gain the support of the non-commis-
sioned officers, and perhaps of some officers of lesser rank, in the
Kuomintang Army, for the struggle against Japan. He did not
advocate collaboration with organized units of the Nationalist
Army; the offer of January 1933 to unite with 'any armed force'
he treated as a pure propaganda manoeuvre.[32]

And yet, a month before he spoke, on 26 October 1933, the
Chinese Soviet Government had actually concluded a preliminary 'anti-Japanese, anti-Chiang' agreement with the
'People's Revolutionary Government' which had been established in Fukien with the military backing of Ts'ai T'ing-k'ai.[33]
In fact, Ts'ai, whose Nineteenth Route Army had fought hero-
ically in defence of Shanghai in January 1932, and who had
never accepted Chiang Kai-shek's policy of concessions to Japan,
had gone ahead with his rebellion against Nanking only after
the conclusion of the accord with the Communists. And yet the
Red Army stood by idly while Chiang crushed the Fukien
Government.

What was Mao's role in these events? As early as 1936, in an
interview with Edgar Snow, he denounced the failure to unite
with the Fukien rebels as an extremely grave one,[34] and he has
continued to denounce it as such ever since. His avowed position
at the time was completely different. In January 1934, shortly
after Chiang's decisive offensive against the Fukien rebels, Mao
made extremely hostile remarks about them in his speeches to the
Second All-China Congress of Soviets. 'Regarding the so-called
People's Revolutionary Government in Fukien', he stated in his
concluding remarks, 'one comrade has said that it had a certain
revolutionary character and was not entirely reactionary. This
viewpoint is erroneous'. The 'People's Revolutionary Government', Mao declared, was merely a trick by a part of the ruling

class to deceive the people with the false slogan of a 'third way' between Communism and reaction.[35]

It is not quite so easy to conclude in this case that in condemning the Fukien rebels Mao was merely saying what was required of him in a situation where he had lost all real power. It is a fact that, while Mao early evolved a relatively tolerant attitude towards the bourgeoisie as a social category, he often showed extreme reticence when it came to sharing power with bourgeois nationalists, however patriotic they might seem. Kung Ch'u, the former chief of staff of the Red Army in Kiangsi – a witness who is sometimes fallible, but not systematically hostile to Mao – declares that, in the debate within the Central Committee regarding Ts'ai T'ing-k'ai's offer of alliance, Mao was the leading spokesman for a 'guerrilla faction' which did not wish to risk the forces of the Red Army in such an adventure.* (For this reason he calls Mao 'conservative', in the sense of one who was primarily interested in conserving his own strength.) Mao himself, in his report of December 1935, though he judges that Ts'ai T'ing-k'ai's revolt was 'beneficial to the revolution', criticizes the Fukien Government (as he had done at the time) for its 'adherence to the old practices in failing to arouse the people to struggle'.†

It is thus entirely possible that Mao was far more reticent in 1933–4 regarding collaboration with Ts'ai T'ing-k'ai than he would now have us believe. On the other hand, his influence was so limited at the time that he could not singlehandedly have prevented action in support of Ts'ai if the other members of the Central Committee had insisted on it. Moreover, Wang Ming's remarks, cited above, at the December 1933 Comintern Plenum do not suggest that the 'Twenty-Eight Bolsheviks' were very

---

* Kung Ch'u, *Wo yü Hung-chün (The Red Army and I)* (Hongkong, Southwind Publishing Co., 1954), pp. 362–7, 395–400. Kung even goes so far as to affirm that as a result of his negative attitude Mao was put on probation as a party member.

† *Selected Works* (Peking), Vol. I, p. 156. In December 1933 Mao had addressed similar but much sharper criticisms to the Fukien régime and called for action on the part of the latter to remedy these defects as a condition of Communist support. For a summary of his telegram to the 'People's Revolutionary Government', also signed by Chu Te, see Hsiao Tso-liang, op. cit., p. 253.

enthusiastic about the Fukien rebels either. Perhaps Ch'in Pang-hsien, fighting for his life in China, adopted a more realistic attitude than Wang in Moscow. Or perhaps, as Kung Ch'u tells us, Chou En-lai was the leader of those who took a positive attitude toward Ts'ai T'ing-k'ai. But if so, why were Ch'in and Chou unable to carry through their policy of alliance with the Fukien Government? It must also be pointed out that a hostile attitude toward Ts'ai T'ing-k'ai as the exponent of a 'third way' between Communism and reaction would have been altogether in keeping with the line of Moscow at the time, which held that the main blows should be aimed at the 'intermediate forces'. (One manifestation of this was the short-sighted policy designating social democracy as a more dangerous enemy than Fascism, which greatly contributed to bringing Hitler to power.) On the whole, the majority of the Central Committee in 1933–4 espoused this line of hostility to the 'intermediate forces'. Why would they have violated it in the case of Ts'ai T'ing-k'ai?

The problem of the policy of the Chinese Soviet Republic towards the Fukien rebels thus remains one of the most obscure ones in the whole history of the Chinese Communist Party. In any case, whoever was responsible for the failure to support the People's Revolutionary Government, the fact is that as a result of this error the Communists soon found themselves fighting alone in increasingly difficult conditions.

These difficulties were in large part the result of the new and devastatingly effective tactics employed by Chiang's armies, under the guidance of his German military advisers, von Seeckt and von Falkenhausen. These involved, as already indicated, the building of a ring of blockhouses linked by barbed wire and other defences around the periphery of the Central Soviet Area in Kiangsi. On the one hand, this positional warfare made it infinitely more difficult for the Red Army to shatter isolated Kuomintang units, as it had done in the mobile warfare employed previously by Chiang. On the other hand, it made it possible for the Kuomintang assailants to enforce a strict economic blockade, which ultimately had very grave effects on the situation within the soviet areas. By the summer of 1934 there was not even an adequate supply of salt for the population, and food prices had risen enormously.

On the military front, despite one initial success, the Red Armies, according to Mao, 'milled around between the enemy's main forces and his blockhouses, and were reduced to complete passivity. This,' he added, 'is really the worst and most stupid way to fight.'* It remains open to question, however, exactly how Mao himself wanted to fight at the time. In 1936 he explained very clearly what should have been done. Taking advantage of the Fukien rebellion, the Red Army should have seized the initiative and undertaken a sweeping offensive toward the north and the east, in the direction of Nanking and Hangchow, thus drawing the Kuomintang armies away from the encirclement of the soviet area and undoing the effects of Chiang's blockhouse tactics. On the one hand, such a strategy would have been hardly compatible with the prudence which has always characterized Mao's behaviour – at least until very recent years. On the other hand, in late 1933 and early 1934 Mao continued to put forward views virtually identical with those of the majority of the Central Committee, according to which the Kuomintang was on the verge of military and political collapse, and the Red Army, which had become an 'invincible force', was perfectly capable of meeting Chiang Kai-shek's armies head on.† In other terms, he was too optimistic in words, and probably too cautious in acts, to have espoused the tactics he later said should have been employed.

Even with the very best strategy, it is not certain that the Communists could have withstood von Seeckt's blockhouse tactics, but they would have had a better chance in any event. By the early summer of 1934, the basic decision to withdraw from Kiangsi had already been taken.

The first concrete manifestation of the new policy was the announcement, on 15 July 1934, that an 'Anti-Japanese Vanguard' was being sent north to fight the invaders. In this document, signed by Mao and Chu Te, it was stated that if this detachment

---

* *Selected Works* (Peking), Vol. 1, pp. 231–2.

† For Mao's *ex post facto* analysis in December 1936, see *Selected Works* (Peking), Vol. I, pp. 221, 247–8. For his intransigent utterances at the time of the Fifth Campaign, see his speech of 12 August 1933 (*Hung-ch'i*, No. 62, 20 November 1933), and his report to the Second All-China Congress of Soviets, *Chih yu Su-wei-ai neng-kou chiu Chung-kuo*, pp. 80–89.

should meet with any armed force prepared to accept the three conditions for concluding an alliance set forth in 1933, 'the main units of our Workers' and Peasants' Red Army will follow the advance guard to unite with all armed forces in China for a common struggle'. This vanguard detachment, led by Fang Chih-min, was soon cut to pieces by the Kuomintang forces, and only a small part of it survived to wage guerrilla warfare in the Chekiang-Fukien border.*

In an interview on 1 August 1934 Mao Tse-tung declared that all the units of the Red Army throughout the country, in the other bases in Central China as well as in Kiangsi, had been ordered to make ready for departure northwards,[35] and from this time on, active preparations were made for the start of what was later to be celebrated as the Long March. Mao's role in these events continues to be obscure. His influence was certainly circumscribed, and possibly nil. It has even been claimed that he was expelled from the Central Committee and held under virtual house arrest in Yütu, a town some fifty miles west of Juichin. He was also gravely ill with malaria in August and September.†

Early in August Hsiao K'e, who was in command of the old Chingkangshan base, received orders to join Ho Lung in his base on the Hunan-Kweichow border. (See the map of the Long March on pp. 178–9.) This mission was successfully carried out, and in October their combined forces were reorganized as the

---

* For extracts from the proclamation of 15 July 1934, see Schram, op. cit., pp. 152–4. Apparently the decision to send the 'Anti-Japanese Vanguard' northwards was taken in April, on the eve of the fall of the important strategic centre of Kuangch'ang to the Kuomintang forces under Ch'en Ch'eng. On this, and on the significance of the vanguard in general, see Hsiao Tso-liang, op. cit., pp. 286, 293–5; also J. Ch'ên, op. cit., p. 182.

† The strongest affirmations to the effect that Mao was in disgrace are to be found in Kung Ch'u, op. cit., pp. 395–400. In an interview with Hsiao Tso-liang (op. cit., pp. 296–7), Chang Kuo-t'ao, though he could not confirm all the details of Kung Ch'u's account, likewise stated that by late 1933 Mao had been deprived of virtually all power of decision. His illness is referred to by Kung Ch'u in the passage cited. It has also been described by his physician Dr Nelson Fu, whose account is to be found in *Hung-ch'i P'iao-p'iao*, X, pp. 3–12; summary in J. Ch'ên, op. cit., p. 183.

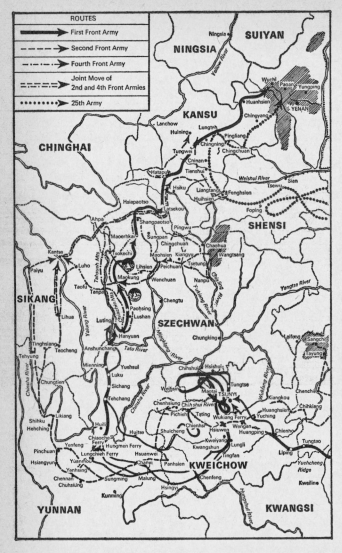

Routes of the Long March,

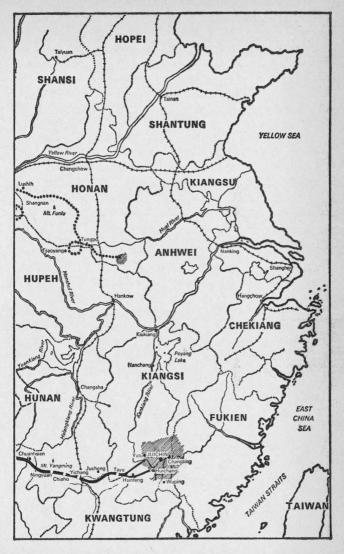

October 1934–October 1936

Second Front Army, with Ho Lung as commander and Jen Pi-shih as political commissar.[36]

At last, on 15 or 16 October, the main body of the Red Army began moving towards the south-west. A brief and successful battle at Hsinfeng enabled the troops to break the encirclement, and they moved through the northernmost tip of Kwangtung and the southern tip of Hunan, before crossing the Hsiang River in Northern Kwangsi.

There were 100,000 of them altogether, 85,000 soldiers and 15,000 cadres of the party and the government. Neither their destination nor their purpose was altogether clear. The two points were linked. If they really intended to fight Japan, they would have to go relatively far northwards. If, on the other hand, their purpose was simply to find a safer haven after the destruction of the base in Kiangsi, they might well stop in Szechuan, where Chang Kuo-t'ao had established himself, after being driven out of the Hupei–Hunan–Anhui base during the Fourth Encirclement. We may assume that Mao Tse-tung himself was sincerely interested in fighting Japan, both in order to gain a new theme for appealing to the masses, and because of his intense concern for the fate of the Chinese nation. But Mao was still not in control of events. The army was under the leadership of Chu Te as commander in chief and of Chou En-lai as political commissar, with the assistance of a German military expert, sent by the Comintern, by the name of Otto Braun, who went under the Chinese pseudonym of Li T'e.* Braun had been in large part responsible for the tactics of positional warfare which had cost the Red Army so dearly during the Fifth Encirclement. Now he once more advocated taking the most direct path to the goal,

---

* Braun recently emerged into the light of day when he published a violently hostile article entitled 'In Whose Name Does Mao Tse-tung Speak?' in the East Berlin press (*Neues Deutschland*, 27 May 1964). Born near Munich at the turn of the century, he escaped from prison in Germany in 1928 (he had been condemned for treason as a result of his Communist activities), and lived in Moscow until he was sent to China. Contrary to what has usually been assumed, he was not (according to the biography published with his article in *Neues Deutschland*) a professional soldier, but a schoolteacher and later a full-time Communist Party functionary. Whatever military training he had he apparently picked up in the Soviet Union between 1928 and his arrival in China.

even though this meant a head-on clash with much stronger forces.*

A logical first step seemed to be a joining of forces with Ho Lung's Second and Sixth Army Groups operating in the north-west corner of Hunan. The Hsiang River was therefore forced in Northern Kwangsi under heavy enemy fire; the crossing was effected after a week-long battle, and when the Red Army regrouped on the other side, it had lost half of its strength. In the face of this disaster Mao succeeded in rallying support for his proposal to move westwards into Kweichow province, where the enemy forces were relatively weak, rather than to stake all on a desperate battle in Central Hunan.†

At the end of the year the Red Army crossed the Wu River in Central Kweichow, an operation that succeeded only through the heroism of a few picked men, who crossed the broad river on rafts under heavy enemy fire, and then scaled a sheer rocky cliff to take by storm the Kuomintang strong-point guarding the ferry. The city of Tsunyi was captured virtually without firing a shot, thanks to a stratagem straight out of the *Romance of the Three Kingdoms*, involving the use of captured Kuomintang uniforms and banners. There, during the first week in January 1935, the Politburo held the celebrated Tsunyi Conference, which was at last to place Mao Tse-tung in control of the Chinese Communist Party.

* The first, and one of the most eloquent, accounts of the Long March is that of Edgar Snow (op. cit., pp. 189–208). A recent attempt at a synthetic account, 'The Long March', by Anthony Garavente (*China Quarterly*, No. 22, 1965, pp. 89–124), is useful, but its documentary basis is sadly weak, being composed mostly of western-language newspaper reports. See also J. Ch'ên, op. cit., pp. 185–200, and several of the contributions to *Red Dust*. Of the very voluminous output of memoirs and historical studies on this theme published in China, a selection has been issued in English. See, in particular, Ch'en Chang-feng, *On the Long March with Chairman Mao* (Peking, Foreign Languages Press, 1959 – by Mao's bodyguard), and *The Long March. Eyewitness Accounts* (Peking, Foreign Languages Press, 1963). The figure of 85,000 soldiers and 15,000 civilians is taken from Ch'en Chang-feng, op. cit., p. 88.

† According to the then chief of staff of the Red Army, Liu Po-ch'eng, in his article 'Looking Back on the Long March', *The Long March*, pp. 207–8. Liu, a veteran of the Nanchang Uprising, was known as the 'one-eyed dragon' because he had lost one of his eyes in combat.

Apart from its tactics of striking directly at strong enemy concentrations, the Red Army had been hindered during the first part of its march by the attempt to carry along all the panoply of a fully-organized state: machines, bullion, printing presses, etc. At Tsunyi this whole series of policies was finally abandoned as a failure. Ch'in Pang-hsien, who had been responsible as secretary-general for imposing this line, was forced out of his post, and replaced by another 'returned student', Chang Wen-t'ien. But henceforth the real power belonged to Mao, who was given the newly created post of chairman of the Politburo.*

Like another great military and political leader in different but equally dramatic circumstances, Mao was clearly exhilarated at having at last 'the authority to give directions over the whole scene', though this enthusiasm, which never left him throughout the Long March, despite the perpetual fatigues and dangers encountered, was tempered by the thought of those left behind to fight a delaying action. Among them were Mao's own brother Mao Tse-t'an, and Ch'ü Ch'iu-pai, who was suffering from tuberculosis. In March 1935 the former was killed in action, and the latter captured and subsequently executed.

Now that the Red Army had reached a point close to the border of Szechuan, the natural course was to effect a junction with Chang Kuo-t'ao's forces in the northern reaches of that province, as it had earlier tried to do with Ho Lung's in Hunan. Unfortunately, like the attempt to join Ho Lung, this objective was all too obvious to Chiang Kai-shek as well, who took quick steps to block its accomplishment. His task was somewhat more difficult than in the areas where the five 'Encirclement and Annihilation campaigns' had been carried out, for Szechuan was not under the direct control of the National Government, and Chiang was obliged to enlist the cooperation of the local war-lords. But by the time Mao and his comrades began marching north from Tsunyi in the middle of January, both they and Chang Kuo-t'ao's Fourth Front Army were under attack by a combination of Kuomintang and local forces.†

* Mao's title of Chairman may have been *de facto* rather than official until 1938. Cf. Rue, op. cit., pp. 3–4 and 270–2.

† For details regarding the political and military situation in Szechuan and Chiang's moves against the Red Armies, see Garavente, op. cit., pp. 108–19.

In these circumstances Mao Tse-tung did not hesitate to apply the tactics he had espoused long ago, in Li Li-san's day: 'When pursued by a powerful enemy, employ the policy of circling around in a sinuous motion.'[37] The result can be seen on the map. In the course of its sinuous movement about Northern Kweichow the First Front Red Army won a major battle at Loushan Pass against the military governor of the province, Wang Chia-lien. The natural grandeur of the scene, and the challenge of his new power and responsibilities, inspired Mao to write the first of several poems composed during the Long March:

> Keen is the west wind;
> In the endless void the wild geese cry at the frosty morning moon,
> The frosty morning moon.
> The clatter of horses' hoofs rings sharp,
> And the bugle's note is muted.
> They say that the strong pass is iron hard,
> And yet this very day with a mighty step we shall cross its summit,
> We shall cross its summit!
> The hills are blue like the sea,
> And the dying sun like blood.[38]

Despite this success, Mao prudently decided to re-cross the Wu River at Tsunyi and make a long detour through Yünnan and Sikang provinces to Western Szechuan, rather than attempt to force a passage directly through the strong enemy positions in Central Szechuan. By this time Chiang Kai-shek was himself established in Kueiyang, to direct operations against the Red Army. After crossing the Wu River Mao made straight for Kueiyang as though to attack it. 'We shall win the battle', he remarked while planning operations, 'if we can lure the enemy troops out of Yünnan.'[39] As a result of this feint Chiang did indeed call for reinforcements from the neighbouring province, whereupon the Red Army abruptly turned west and marched straight across Yünnan. After another feint towards the provincial capital of Kunming, the Red Army marched northwards towards the Chinsha (Golden Sands) River, which marked the border between Yünnan and Szechuan. Once more, a few men crossed the turbulent river in small boats – this time disguising themselves as Kuomintang scouts, policemen, and tax-payers, in order to make their way into various key points in the village on the opposite shore which served as the terminus of the ferry. Then for nine days the boats plied back and forth taking the entire Red Army across.

As the Reds headed north through Sikang province towards Szechuan, Chiang Kai-shek was convinced that fate had delivered his enemies into his hands. For they would now be obliged either to turn west once again into the impossibly difficult terrain on the Sikang-Szechuan frontier, or to make their way across the sheer rock gorges and rushing torrents of the Tatu River. It was in this very area that the Taiping general Shih Ta-k'ai had been trapped and his army wiped out by the imperial forces in 1865. The parallel was clearly present in the mind of Chiang Kai-shek, who had come to regard the Taipings as peasant riff-raff like the Communists, though they had been praised by Sun Yat-sen as a great revolutionary movement. The Kuomintang air force even dropped leaflets declaring that Mao Tse-tung would be another Shih Ta-k'ai,[40] and Chiang himself flew to Chungking to direct operations.

The crossing of the Tatu River is one of the legendary exploits in the history of the Long March. It began with an episode similar to that which had marked the crossing of the Wu River – a surprise attack on the small enemy force guarding the ferry terminus at Anshunch'ang, the capture of a small boat, and an assault by eighteen picked men against the enemy position on the other side, which was defended by blockhouses at the top of a sheer cliff. But this time the securing of the ferry was only the beginning and not the end of the affair, for it would have required many precious days for the whole army to cross in the few small boats available, and meanwhile strong enemy reinforcements would have had time to arrive. Moreover, the army was exposed to bombardment from the air. As the current was too rapid to permit building a temporary bridge, the only solution was to attempt to capture the suspension bridge at Luting, some 110 miles up the river to the north-west. Meanwhile the First Division, which had crossed in the small boats before the decision was taken to drive on to Luting Bridge, would advance along the other bank of the river to support the action of the main force.

The armies on both banks marched over narrow, precipitous trails above the sheer cliffs, continually accompanied by the roaring of the waters. They slithered in the mud, built temporary bridges over small tributaries of the Tatu which blocked their path, fought skirmishes with occasional enemy units. At the outset the Army Group Headquarters had given the vanguard on the left bank three days to cover the 110 miles to Luting Bridge; but on the morning of the second day, with 240 *li* or 80 miles still to go, the soldiers were suddenly informed that they must complete the march within twenty-four hours. There was not even time for the political commissars to call a meeting and exhort the men; explanations were given during the march. Towards dusk they found themselves opposite an enemy column on the right bank, making a forced march for the bridge, and as night fell the column on the other side lit torches. The Red Army needed torches too, if it was to make its way along the treacherous paths. In order to avoid drawing the fire of the enemy across the river, the commanders ordered some Szechuan men to shout replies to any questions, so as to give the impression that they were the three enemy battalions defeated earlier in the day. Early in the morning of 25 May the Reds reached the bridge on schedule, and easily captured the enemy post.

Before them, on the opposite shore, stood the city of Luting, heavily fortified and defended by machine guns and mortars. Between the two banks stretched thirteen heavy iron chains sealed into the rocks. Two of them served as hand railings on either side; the other nine had originally supported a floor of planks, but this had been removed by the enemy. (See the picture of the bridge plate, 11.) Once more volunteers were wanted. At four in the afternoon twenty-two men, carrying machine guns, broad-swords, and hand-grenades, began climbing across the swaying chains, followed by another company carrying planks to re-build the bridge. Some of the assault party were killed and dropped into the river; but the others continued to inch forward. Near the opposite shore part of the bridge was still covered with planks, and at the last moment the enemy set fire to these in order to block the Red Army's progress; but the remnants of the vanguard dashed through the flames throwing hand-grenades, followed by the company of plank-layers and the remainder of the regiment. By dusk the two enemy regiments in Luting were defeated and scattered, and the Red Army was in complete control of the city. Soon afterwards the First Division arrived along the right bank, having defeated the enemy troops encountered on the way. Thus the entire Red Army crossed the Tatu River and escaped the fate of Shih Ta-k'ai.*

Henceforth the troops under Mao Tse-tung would do relatively little fighting until they arrived in Shensi, but they would be confronted with tremendous natural difficulties that made their march one of the great feats in military history. Mao also had to contend with political difficulties in the shape of a conflict with Chang Kuo-t'ao. After struggling over the high snow-covered passes of the Chiachin Mountains the First Front Army met Chang's Fourth Front Army at Maokung, and together they marched northwards as far as Maoerhkai. Here the differences between the two men finally came to a head.

Mao Tse-tung continued to advocate moving northwards to the Shensi-Kansu border area, where Kao Kang and Liu Chih-tan had established a soviet, with the ultimate purpose of exploiting the anti-Japanese sentiment of the Chinese people. Chang Kuo-t'ao believed this policy impractical, and proposed withdrawing to Sikang or even to Tibet. This is now stigmatized as having been 'flightism' and 'war-lordism'. 'Flightism' it certainly was, for it would have taken the Red Army as far as possible from its Nationalist pursuers. 'War-lordism' it also was in a certain sense, for it would have involved establishing the army in an area inhabited by national minorities where the political mobilization of the masses would have been difficult. But the principal meaning of this latter term is no doubt simply that Chang set his own prestige and his own authority against that of Mao. He had been Mao's senior in the party organization a dozen years earlier; he had no desire to be his subordinate now.

This trial of strength ultimately ended in a stalemate. The army was divided into two, with Chang Kuo-t'ao taking charge of one column and Mao of the other. Chang led his half of the army where he had always wanted to go, towards Sikang. Mao and his force headed northwards.

The only real mystery about this whole affair is why Chu Te

---

* There are innumerable accounts of this celebrated episode. The above description is taken primarily from the two chapters devoted to the subject in *The Long March*, pp. 82–109.

accompanied Chang Kuo-t'ao in his westward move, instead of casting in his lot with Mao. Early explanations, according to which the separation of the two armies and/or that of Chu Te from Mao was due to natural hazards or to the action of the Kuomintang armies, are altogether unconvincing. The only explicit reference to the motives of Chu Te's action is to be found in his own story as told to Agnes Smedley, in which he claims that Chang Kuo-t'ao forced him at gunpoint to go along to Sikang.* Compulsion or pressure there may have been, but it is highly unlikely that this is the whole story. Regional loyalties were probably also a factor. Chang Kuo-t'ao had recruited a new army since his move to Szechuan, composed largely of natives of the province. (It was for this reason, too, that he was reluctant to go northward as Mao desired.) Chu Te was a Szechuanese, as was Liu Po-ch'eng, who also cast in his lot with Chang. They may have shared Chang's desire to remain near their home province. In any case, there seems to be reason to believe that Chu was not entirely in agreement with Mao at this time. (After all, he had already cooperated with Ch'in Pang-hsien, Chou En-lai, and Otto Braun in 1933 and 1934 in pursuing a line to which Mao was strongly opposed.) In any event, the two men, who had been physically if not morally united without pause since early 1928, were now separated, and Mao led his column northwards alone.

Mao Tse-tung and his companions still had before them the most harrowing experience of the whole Long March, the crossing of the terrible grasslands. This was a vast expanse of swampy ground where anyone taking the wrong step was sucked down into the morass and rapidly disappeared altogether from sight, unless he was lucky enough to have a comrade near by who could pull him out with a pole. To make matters worse, the non-Chinese minority peoples here proved more hostile to the Hans, red or white, than in any of the other territories through which the Red Army had come. In Yünnan Liu Po-ch'eng had sealed an alliance with one of the I tribes by drinking cock's blood and swearing brotherhood with the chief.[41] There and elsewhere, the Red Army had succeeded in convincing the minorities that it was different from other Chinese armies and authorities they had known.

* Smedley, *The Great Road*, p. 331. For an analysis of the Mao–Chang conflict at this time, see Garavente, op. cit., pp. 120–22.

But the Man tribes in the grasslands simply drove away all their cattle, hid or removed their stocks of food, and disappeared into the mists that shrouded the marshes on every side. They revealed their presence to the Red Army only when they rolled boulders down on the troops in the rocky passes which occasionally broke the monotony of the swamps, or sniped at stragglers from ambush. Thus deprived of all possibility of obtaining food from the inhabitants, even in exchange for payment in the silver dollars which they had always given for everything throughout the march, the Red Army soldiers soon exhausted the grain rations that they had brought with them. After that they ate the horses, and then they boiled or broiled their leather shoes and belts, or any other object they could come upon which appeared to offer some promise of nourishment. Many of the wild vegetables they plucked or dug up proved to be poisonous, and led to nausea, diarrhoea, or even sudden death. In addition to all this, it rained almost constantly and then at night turned deathly cold. The poisonous mud in which their legs were continually plunged produced agonizing sores, and there were no medical supplies available for treating them except boiling water.

Only 7,000 or 8,000 men emerged from this ordeal to head eastwards to Latsek'ou pass, and then northwards over the Minshan Mountains. In Southern Kansu a strong concentration of enemy troops was once more encountered, but after a feint in the direction of T'ienshui, Mao and his comrades marched rapidly north-eastwards, passed through areas inhabited by Muslim minorities, and broke through the enemy blockade between Huining and Chingning.

As they made their way over the Liup'an Mountains, the last major natural barrier before their arrival in Shensi, Mao expressed his feelings in another poem:

> The sky is high, the clouds are pale,
> We watch the wild geese flying south till they vanish;
> If we reach not the Great Wall, we are no true men!
> Already we have come two thousand leagues.
>
> High on the crest of Liup'an Mountain
> Our banners idly wave in the west wind.
> Today we hold the long cord in our hands;
> When shall we bind fast the grey dragon? [42]

The 'grey dragon' is the name of an eastern constellation. Here it refers to the Japanese invaders from the East. If by 'binding' them Mao meant achieving final victory over Japan, then the event was still ten years in the future. But if by his question he meant merely, 'When will the struggle against Japan begin to play a tangible part in our fate?' then the answer was: 'Very soon indeed.'*

* As this chapter concludes, Mao was not yet, as the title would suggest, installed in Yenan, which was taken only at the end of 1936; but the 'Yenan Period' in the history of Chinese Communism is usually considered to begin with the arrival of the Red Army in the north.

Notes

1. *A Documentary History of Chinese Communism*, p. 185.
2. *Strategija i taktika*, p. 290.
3. *Inprecorr*, Vol. X, No. 14, 20 March 1930.
4. Stalin, *Works*, Vol. 12, p. 258.
5. 'Remarks of Comrade P'i' (apparently a Russian), *Pu-erh-sai-wei-k'o*, Vol. 4, No. 3, p. 61.
6. Hsiao Tso-liang, *Power Relations within the Chinese Communist Movement*, p. 7.
7. *Pu-erh-sai-wei-k'o*, Vol. 4, No. 3, passim, especially the remarks of Madiar, pp. 21–7.
8. Hsiao Tso-liang, op. cit., pp. 159–62.
9. ibid., pp. 39–49.
10. *Strategija i taktika*, p. 306.
11. Hsiao Tso-liang, op. cit., p. 173.
12. ibid., pp. 165, 210–11, 220–21.
13. Sun Tzu, *The Art of War*, Lionel Giles's translation, Ch. VI, pars. 13–15.
14. ibid., Ch. I, pars. 21–7; Ch. IV, par. 7; Ch. VI, pars. 8–10.
15. *Selected Works* (Peking), Vol. I, p. 217.
16. ibid., p. 218. The quotation is actually from the commentaries to Sun Tzu rather than from Sun Tzu himself. See Griffith's translation of *The Art of War*, p. 80.
17. ibid., p. 214.
18. *Selected Works* (Peking), Vol. I, pp. 226–7, 231; Smedley, *The Great Road*, pp. 287–90; J. Ch'ên, *Mao and the Chinese Revolution*, pp. 166–7.
19. *Selected Works* (Peking), Vol. I, pp. 227–8; J. Ch'ên, op. cit., pp. 167–8.
20. *Selected Works* (Peking), Vol. I, pp. 228–30. With the exception of the added detail regarding the area around Kaohsinghsü, the elimination of the term 'soviet' which I have reinstated, and the disappearance of the map illustrating this (and the other) campaigns, there are no significant differences between the original and revised texts of this passage.
21. J. Ch'ên, op. cit., pp. 168–70.
22. *A Documentary History of Chinese Communism*, pp. 224–6; Hsiao Tso-liang, op. cit., pp. 178–9.

23. *Hung-se Chung-hua*, No. 86, 10 June 1933, p. 3; also in *Hung-ch'i*, No. 59, 31 August 1933, pp. 30–34.
24. *Tou-cheng*, No. 24, 29 August 1933, pp. 4–12; also in *Hung-ch'i*, No. 61, 30 October 1933, pp. 45–61.
25. Report to the Second All-China Soviet Congress, 24 January 1934, from the abridged translation in *A Documentary History of Chinese Communism*, p. 235.
26. Jane Degras (ed.), *The Communist International 1919–1943*. Documents, Vol. III (Oxford University Press, 1965), pp. 221–30, passim.
27. *Selected Works* (Peking), Vol. I, pp. 240–1; translation corrected on the basis of the original Chinese text.
28. *Strategija i taktika*, pp. 316–70, passim.
29. *Su-wei-ai Chung-kuo*, pp. 71–4.
30. ibid., pp. 91–4.
31. *Inprecorr*, 26 January 1933, pp. 91–2.
32. *Strategija i taktika*, p. 362.
33. *Hung-se Chung-hua*, No. 149, 14 February 1934, p. 4; summary in Hsiao Tso-liang, op. cit., pp. 248–9.
34. Snow, *Red Star Over China*, p. 179.
35. *Hung-se Chung-hua*, No. 221, 1 August 1934; summary in Hsiao Tso-liang, op. cit., p. 296.
36. J. Ch'ên, op. cit., p. 183; Nym Wales, *Red Dust*, p. 139.
37. From his letter of 5 April 1929 to the Central Committee, *Selected Works* (Peking), Vol. I, p. 124.
38. Schram, *Political Thought of Mao*, p. 206.
39. Liu Po-ch'eng, in *The Long March*, pp. 212–13.
40. ibid., p. 83.
41. ibid., pp. 77–80.
42. Schram, op. cit., p. 206.

Chapter 8
The Struggle on Two Fronts

In the autumn of 1935, as the remnants of his armies neared their goal in Shensi province, Mao Tse-tung had at last captured control of his own party – though his authority would by no means go unchallenged for several years to come. Geographically and politically placed as he was, he could now present himself as the staunchest partisan of resistance to the Japanese, and so seize and shape the enormous forces of social mobilization for nationalist ends released in China by the continuing Japanese aggression. But in order to achieve his goal he would be constantly obliged to struggle on two fronts – against the Kuomintang, and against the domination of Moscow.

No explanation is required for the first of these quarrels. Though the Communists and the Kuomintang would shortly find it expedient to collaborate in fighting the Japanese rather than go on fighting each other, neither of the two parties regarded this as a permanent solution. Mao Tse-tung and Chiang Kai-shek were each equally convinced that his was the mission of leading the Chinese nation as a whole, and neither had the slightest intention of sharing power indefinitely. Even at its period of greatest cordiality and effectiveness their collaboration was therefore marked by continuous manoeuvring, with each of the allies attempting to strengthen his own position before the inevitable split.

The problem of relations with the Soviet Union was more complex; for it involved alongside rivalry a genuine feeling of solidarity between the Communists of the two countries. At the same time Stalin had two distinct but related goals, both of which were totally unacceptable to Mao Tse-tung. The first was to avoid pushing the revolution in China too hard if this was likely to endanger Moscow's diplomatic position. The second was to make sure that the Communist movement in China remained under Soviet guidance and control. Although the form in which these issues presented themselves varied substantially over the years, the two basic tensions persisted during the entire period from 1935 to 1949. To be sure, Mao was at no time in a position to flout Stalin's authority openly, though on occasion he ventured to suggest that Moscow's rights and influence in China had their limits. But there is reason to believe that the problem of who was ultimately to control the destinies of the Chinese revolution was never totally absent from the mind of either of the two men.

In September 1935, shortly before the arrival of the 'Central Red Army' under Mao, the Twenty-Fifth Army Corps under Hsü Hai-tung, which had been operating in the south-east corner of Hunan province, completed its march and joined Liu Chih-tan in the Shensi-Kansu base area. The two armies were fused together to create a new unit, the Fifteenth Army Corps, with a total of about 7,000 men, and this force successfully resisted one Kuomintang attack, then under way. In October Mao and his 8,000 veterans of the Long March joined it, and the following month a further Kuomintang 'Encirclement and Suppression Campaign' was smashed. Thus the total force under Mao's command numbered about 15,000. It would be further increased in the course of 1936 by the arrival of Ho Lung's men from the base in Hunan, and the army of Chang Kuo-t'ao and Chu Te, which finally returned from its expedition to the borders of Tibet; but the total number remained small, in comparison either with the size of the Nationalist forces round about, or with the 'Red Army of one million' which had been the Kiangsi objective in 1934.*

Under the circumstances, it was not easy to exploit the theme of united resistance by all patriotic Chinese against Japanese aggression, without at the same time capitulating to the Kuomintang and losing all semblance of initiative or independence. Mao was exceptionally well qualified for this task, because he continued to be – as he has been throughout his career – both a convinced revolutionary and a passionate Chinese nationalist.

* On the circumstances of the arrival of the various fragments of the Red Army in the north in 1935 and 1936 and their respective strengths, see J. Ch'ên, *Mao and the Chinese Revolution*, pp. 195–202; Ch'en Chang-feng, *On the Long March with Chairman Mao*, pp. 65–70; *The Long March*, pp. 219–24; *Selected Works* (Peking), Vol. I, pp. 175–6. Mao's fellow Hunanese Jen Pi-shih, a graduate of the First Normal School who was later to become one of his most faithful lieutenants, apparently played an important role in persuading Chang to return eastward.

As a revolutionary, he was at last assimilating the basic principles of Marxism-Leninism and learning to cast his policies in the proper ideological shapes. This is apparent from the report on the tactics of fighting Japanese imperialism in which Mao summarized the results of the important Waiyaopu Conference of the Politburo held at the end of December 1935. Here Mao, instead of merely mentioning an alliance with 'any armed force', as he had been doing since 1933, produced a relatively sophisticated class analysis of the forces existing in China and their political attitudes. Not only the petty and 'national' bourgeoisie, but even a part of the *compradores* and landlords, were recognized as potential allies in an anti-Japanese united front, to the extent that their interests were linked with those of countries such as England and America which were in conflict with Japan. Moreover, the creation of a single 'government of national defence', enjoying the support of all these forces, was proposed.*

At the same time it should be pointed out that this proposal – for a formal coalition government – represented only a rather limited concession to the Nationalists. For though it was indicated (as in the earlier appeal of 1 August) that the 'government of national defence' should emerge from discussions among representatives of various parties and social groups, it was clear that the Communists themselves would take the lead in these discussions, and that the existing soviet government in the north would constitute the nucleus of the new anti-Japanese régime. In any event, the central role was denied to Chiang and the Kuomintang. This position was entirely in conformity with the Comintern line of the time, adopted at the Seventh Congress in

---

* The only available text of Mao's speech of 27 December 1935 is that published in the official edition, *Selected Works* (Peking), Vol. I, pp. 153–78; extracts in Schram, *Political Thought of Mao*, pp. 154–8. This was undoubtedly rewritten for the occasion, like nearly everything in the current canon, but on the whole I believe it corresponds to the substance of what Mao actually said at the time. (See the discussion of this question in Schram, op. cit., pp. 137–8.) I have, however, replaced 'people's republic' by 'government of national defence', which was the term used at the time. (See the contemporary version of the resolution adopted by the Politburo on 25 December 1935, in *Mu-ch'ien hsing-shih ti fen-hsi*, n.p., Li-lun yü shih-chien She, 1936, pp. 32–47.) The slogan of a 'government of national defence' had first been put forward in an appeal to the country adopted on 1 August 1935, at Maoerhkai.

the summer of 1935, which had called for the creation, in the colonies and semi-colonies, of an 'anti-imperialist united front', but which had emphasized as well that in China 'the Soviets must become the centre around which the whole Chinese people is united in its struggle for liberation.'*

An image of Mao's overall view at this time, a view both more striking and more balanced than that furnished by an ideological text such as his speech of December 1935, can be found in his poem 'Snow', written in February 1936:

> This is the scene in that northern land;
> A hundred leagues are sealed with ice,
> A thousand leagues of whirling snow.
> On either side of the Great Wall
> One vastness is all you see.
> From end to end of the great river
> The rushing torrent is frozen and lost.
> The mountains dance like silver snakes,
> The highlands roll like waxen elephants,
> As if they sought to vie in height with the lord of heaven,
> And on a sunny day
> See how the white-robed beauty is adorned with rouge,
> Enchantment beyond compare.
>
> Lured by such great beauty in our landscape
> Innumerable heroes have rivalled one another to bow in homage.
> But alas, Ch'in Shih Huang and Han Wu Ti
> Were rather lacking in culture,
> T'ang T'ai Tsung and Sung T'ai Tsu
> Had little taste for poetry,
> And Genghis Khan
> The favourite son of heaven for a day
> Knew only how to bend his bow to shoot great vultures.
> Now they are all past and gone.
> To find heroes in the grand manner
> We must look rather in the present.†

---

* *Rezoljutsii VII Vsemirnogo Kongressa Kommunisticheskogo Internatsionala* (Moscow, Partizdat, 1935), pp. 25–6. At this congress, Wang Ming had already called for a 'government of national defence'.

† Schram, op. cit., pp. 107–8. This poem has recently been re-dated, from 1945 to 1936; see next footnote.

Here we have the affirmation that the past is not enough; that the greatest heroes in Chinese history are the revolutionaries of today. But we also have a tremendous glorification of the past. The majesty of the Chinese landscape, and the heroic exploits of those who, for over two thousand years, have struggled for mastery of the country, find their culmination in the present, but they are also worthy of admiration in themselves.

Particularly striking is the evocation of Genghis Khan, for the only other known reference to the Mongol conqueror in Mao's writings occurs almost simultaneously, in an appeal of December 1935 to the people of Inner Mongolia:

> We are convinced that it is only through a common struggle by ourselves and the people of Inner Mongolia that we can overthrow our common enemies – the Japanese imperialists and their running dog, Chiang Kai-shek; at the same time, we are persuaded that it is only by fighting together with us that the people of Inner Mongolia can preserve the glory of the epoch of Genghis Khan, prevent the extermination of their nation, embark on the path of national revival, and obtain the freedom and independence enjoyed by peoples such as those of Turkey, Poland, the Ukraine, and the Caucasus.*

The reference to Chiang Kai-shek in this text as the 'running dog' of the imperialists is typical of the transitional period in which Mao was moving towards proposals for collaboration with the Kuomintang, but had not yet resigned himself to the necessity of embracing his old rival. Denunciations of the Kuomintang's reactionary character have virtually disappeared, giving place to the single, overriding accusation of failing to defend the existence of the Chinese nation. The most common insult directed at Chiang throughout the first half of 1936 is that of 'traitor to the nation' (*mai-kuo-tsei*: literally a rebel or traitor who sells out his country).

Such attacks on Chiang were, of course, the logical negative corollary of the policy which set out to mobilize every conceivable group, national or social, around the soviet government as

---

* *Mu-ch'ien hsing-shih ti fen-hsi*, pp. 52–3. In view of the general conformity between the poem 'Snow' and Mao's mentality at the time, and in particular in view of this contemporary reference to Genghis Khan, I see no reason to reject (as does Jerome Ch'ên, op. cit., p. 341) the re-dating of this poem in the latest Chinese edition of Mao's poetry.

the principal centre of resistance to the invaders. Mao's efforts in this direction were varied and picturesque. In addition to the Inner Mongolians he also appealed, in May 1936, to the Muslim minorities of North-West China, recommending to them the 'glorious fame' of the national revival by Muslim Turkey as a 'compass' which should guide all Muslims.[1] Within the country he appealed to the Ko Lao Hui, the famous secret society which he had so often encountered in Hunan and elsewhere, and to which Liu Chih-tan, the founder of the Shensi base, belonged. He even went so far as to try to persuade the members of this society that their ideology and that of the Chinese Communist Party were virtually identical, for both favoured the poor against the rich, and the Chinese against their foreign oppressors.[2]

Appeals such as these are extremely significant as reflections of Mao's mentality, and of the situation at the time; but politically the decisive problem was obviously that of relations with the Kuomintang, and with its leader Chiang Kai-shek. Little by little both men were obliged, by force of circumstances, to overcome their instinctive repugnance at dealing with an old enemy. In May 1936 – in the face of attacks by the forces of Yen Hsi-shan, supported by two of Chiang's divisions under Ch'en Ch'eng, and by the armies of Chang Hsüeh-liang, the 'Young Marshal' who had been driven from Manchuria by the Japanese* – the Communists at last decided to address themselves directly to the Military Council of the Nanking National Government. Their telegram of 5 May was later referred to by Mao as marking the 'abandonment of the anti-Chiang Kai-shek slogan',[3] and while there were still a few scattered hostile references to Chiang after this date, the insults that had been hurled at him for nine years in the Communist press rapidly died away.

In the summer of 1936, in the course of his lengthy interviews with Edgar Snow, Mao explicitly indicated his willingness to co-operate with his old enemy:

There must be a day of decision, a day when he must either oppose Japan or be overthrown by his own subordinates, who are not all ready to become the slaves of Japan. Many of his own generals are now very restive, and demanding a policy of resistance. . . .

* These attacks were provoked by the 'anti-Japanese' expedition described on p. 204.

This increasing pressure from his own generals and from the anti-Japanese mass movement may yet compel Chiang to realize his mistakes and grant the demands of the people. If he stops civil war, begins to fight Japan, re-establishes the union of the Kuomintang and the Chinese Communist Party, we will welcome this change and cooperate whole-heartedly. But this only Chiang Kai-shek can determine for himself. The decision cannot much longer be delayed.[4]

For the moment Chiang himself showed no disposition to co-operate with the Communists. But there were those who listened more willingly to the appeals of Mao and his comrades for national unity against Japan. Chief among these was Chang Hsüeh-liang, whose troops, captured in the battles with the Com-munists and then indoctrinated before being released, were filled with nostalgia for their homeland in Manchuria. Chang therefore entered into contact with the Communists through a variety of channels.[5] By October, indeed, Mao was ready to proclaim that the Red Army would no longer engage in hostilities with units of Chang Hsüeh-liang's 'People's Revolutionary Army' unless attacked.

Commenting on this order in a declaration to the press, Mao affirmed that he and his comrades were prepared to 'join hands' not only with the People's Revolutionary Army, but with the Nanking Government to resist Japan. 'We are extremely desirous of cooperating with the Nanking Government', he declared.[6]

Disturbed by these developments, and anxious to promote a new campaign against the Communists which would finish them off completely, Chiang sped to Chang Hsüeh-liang's headquarters in Sian. But there he found himself arrested and held prisoner by his host, who set out to persuade him of the need to fight the Japanese first and the Communists later, rather than the other way round. This celebrated 'Sian Incident' ended, as everyone knows, in a compromise which apparently favoured Chiang Kai-shek, but which in fact involved major concessions on his part. Outwardly, Chang Hsüeh-liang professed himself to have been completely wrong, and humbly begged Chiang for suitable punish-ment. (In this the generalissimo gladly complied; nearly thirty years after the 'incident', Chang Hsüeh-liang is still under house arrest in Taiwan.) But at the same time, theoretically of his own free will, Chiang modified his own earlier attitude and rapidly moved towards the acceptance of cooperation with the Communists and all other patriotic forces against Japan.*

The role of the Communists, and of Mao Tse-tung in particular, during this crisis is the subject of considerable dispute. There is no doubt whatever that Mao and his comrades were much more reticent about cooperation with Chiang than was Moscow. Indeed, so preoccupied was Stalin with maintaining some semblance of political cohesion in China under Chiang Kai-shek, who appeared to him the only possible national leader, that the Soviet press violently denounced the Sian coup, on first hearing of it, as a Japanese plot.† The Chinese Communists, for their part, who were soon brought by Chang Hsüeh-liang into the discussions over Chiang's fate, were themselves reluctant to forgo this opportunity of squaring accounts with their old enemy. According to Edgar Snow, voices were raised in the Red capital in Shensi in favour of staging a public trial of the generalissimo.‡ But in the end Mao and his comrades apparently concluded that, since they were not yet strong enough to take the initiative in forming a coalition government themselves, Chiang was a lesser evil than the even more reactionary leaders who were likely to gain the upper hand in Nanking if he were sacrificed.

This is, in any case, the explanation which Mao himself gave in a statement at the time:

Throughout the incident, the Communist Party stood for a peaceful settlement and made every effort to that end, acting solely in the interests of national survival. Had the civil war spread and had Chang and Yang kept Chiang Kai-shek in custody for long, the incident could only have developed in favour of the Japanese imperialists and the Chinese 'punitive' group.[7]

* Of the innumerable accounts of the Sian Incident, see Chiang's own story, published as an appendix to his wife's account, *Sian, a Coup d'État* (Shanghai, 1937); and J. M. Bertram, *Crisis in China: The Only Story of the Sian Mutiny* (Macmillan, 1937).

† This is abundantly documented by Charles B. MacLane in his book *Soviet Policy and the Chinese Communists 1931–1936* (New York, Columbia University Press, 1958), pp. 79–91.

‡ E. Snow, *Random Notes on Red China* (Cambridge, Mass., Harvard University Press, 1957), pp. 1–11. This is perhaps going too far; in any case, it is out of harmony with Mao's subsequent attitude in the years 1937–8, as will be shown below.

During the first half of 1937 rapid progress was made towards an agreement on a new period of Kuomintang–Communist co-operation. The basis for a compromise was laid in a telegram of 10 February 1937 from the Central Committee of the Chinese Communist Party to the Third Plenary Session of the Central Executive Committee of the Kuomintang. In this proposal the Communists offered four concessions:

1. Abandonment of the policy of armed insurrection to overthrow the National Government.

2. Reorganization of the soviet régime in Shensi as the government of a 'Special Region' of the Republic of China, and rebaptism of the Red Army as a unit of the national army, under the direction of the Nanking Government and its Military Council respectively.

3. Establishing in the 'Special Region' a democratic system based on universal suffrage.

4. A stop to the confiscation of landlords' land.

These concessions were offered on condition that the Kuomintang accept the national policy in five points proposed in the same telegram:

1. An end to civil war.

2. Democratic freedoms and the release of all political prisoners.

3. A conference of representatives from all parties to concentrate on saving the country.

4. Speedy preparations for resisting Japan.

5. An improvement in the people's livelihood.[8]

On 22 September 1937, two months after the Marco Polo Bridge Incident initiated a full-scale Japanese invasion of China, final agreement was reached between the Communists and the Kuomintang on terms similar to those proposed by the former in February. The principal difference lay in the addition of a paragraph pledging the Communists to work in favour of Sun Yat-sen's Three People's Principles. The idea that these principles constituted a suitable programme for the current stage in China had, in fact, been put forward by the Communists themselves in a letter of 25 August 1936 to the Kuomintang, calling on the latter to 'revive the spirit of Dr Sun's revolutionary Three People's Principles', and 'reaffirm his three great policies of alliance with

Russia, cooperation with the Communist Party, and assistance to the workers and peasants.'[9]

This type of appeal to the principles of the Nationalists themselves led to speculation on whether the Chinese Communists had abandoned their ideas and ceased to be 'real' Communists. Such illusions, which eventually became relatively widespread (though not so widespread as was later claimed), were quite without justification, for from the very beginning Mao himself took pains to make matters perfectly clear. In an interview of 1 March 1937 with the American journalist Agnes Smedley, who had reached Yenan in January,* he dealt in considerable detail with precisely such questions as whether the Communists had capitulated to the Kuomintang, and the relations between short-term and long-term objectives.

In the course of this interview Miss Smedley asked Mao point-blank if the policy of a united front implied that the Chinese Communists had abandoned the class struggle, and turned into simple nationalists. To this Mao replied:

> The Communists absolutely do not tie their viewpoint to the interests of a single class at a single time, but are most passionately concerned with the fate of the Chinese nation, and moreover with its fate throughout all eternity. . . . The Chinese Communists are internationalists; they are in favour of the world Communist movement.† But at the same time they are patriots who defend their native land. . . . This patriotism and internationalism are by no means in conflict, for only China's independence and liberation will make it possible to participate in the world Communist movement.

Thus the ultimate goal remained Communism, both in China and throughout the world; but a social revolution could be promoted only if the nation itself was first saved from destruction. In conclusion Mao recalled that he and a large number of his comrades had worked within the Kuomintang and its military forces during the period 1925–7, and that they had been able to do so because they were believers in the Three People's Prin-

* For Miss Smedley's own account of her adventures in Sian at the time of the incident, and her journey to Yenan, see her book *Battle Hymn of China* (London, Gollancz, 1944), pp. 96–143.

† Here the Chinese text (the only one available) has the ancient term of *ta t'ung*, or 'great harmony', to express the idea of Communism.

ciples, which were likewise in conformity with China's present need.[10]

In May 1937 a national conference of the Chinese Communist Party was held in Yenan to discuss the new policy of collaboration with the Kuomintang, which had aroused some anxiety among many members of the party.* In his report on this occasion Mao affirmed once more that, at the present stage, the Communist Party's programme, though more radical than the Three People's Principles, was not in fundamental conflict with them. Communists and Kuomintang could therefore find a basis of collaboration in a common struggle for 'the three great objectives of national independence, democracy and freedom, and the people's livelihood and happiness.'[11]

Although the temporary subordination of the social revolution to the national revolution against imperialism, alongside the corresponding policy of alliance with the 'bourgeois nationalists', was entirely in conformity with Lenin's theory and Stalin's practice, there continued to be a decisive difference between Mao's view of the problem and that of the Soviet leaders. For Mao himself, the alliance of all Chinese for the salvation of their country was not merely skilful tactics; it was a value in itself. How much this was so is shown by the enthusiasm with which he threw himself into the execution of the policy. Not only did he predict a 'brilliant future' for the Kuomintang, and praise its 'great leader' Chiang Kai-shek,[12] but over the organizational form of the alliance he came out explicitly and energetically, in his report to the Sixth Plenum of the Central Committee of the Chinese Communist Party, for the 'bloc from within' which had prevailed in 1924–7. He even accepted in advance two limitations similar to those which Chiang had imposed in 1926 on the activity of the Communists in the Kuomintang: a complete list of Communist Party members who joined the Kuomintang would be handed over to the latter, and Kuomintang members would not be recruited into the Communist Party.[13]

* In his interview, cited above, with Agnes Smedley, Mao specifically referred to a certain group of people (manifestly within the Communist Party) who were afflicted with 'left-wing infantilism' and hence had opposed the release of Chiang Kai-shek after the Sian Incident and the whole policy of collaboration with the Nationalists.

When these proposals were made to Chiang Kai-shek he not unnaturally saw in them merely an attempt to 'infiltrate the Kuomintang on a large scale', as in 1924–7.[14] Moreover, all Mao's words and actions at the time, including a paragraph in the very same report of October 1938,[15] show that he had no intention whatsoever of sacrificing the essential basis of his own independent power. But at the same time it seems likely that during this 'second united front', as during the first one, Mao was more optimistic than many of his comrades about the possibilities of long-term (though not indefinite) cooperation with the Kuomintang.

Paradoxically, Mao's very sincerity (or at least partial sincerity) in proposing collaboration to Chiang Kai-shek made him more dangerous as a rival. For it grew out of an intense concern with the fate of China, and this greatly increased his prestige as a leader in the eyes of the now thoroughly aroused peasant masses, who were ultimately to sweep him into power on a wave that was as much a wave of nationalism as a wave of social revolt.

That Mao's conquest of power owed a great deal to his success, far greater than Chiang Kai-shek's, in mobilizing the patriotic sentiments of the Chinese masses in the face of the Japanese invasion, has long been assumed by most students of the question. Recently a work based on a wealth of evidence drawn from the archives of the Japanese Army has substantiated this impression by solid historical evidence.*

The decisive role in the emergence of the Chinese Communist

* See Chalmers A. Johnson, *Peasant Nationalism and Communist Power: The Emergence of Revolutionary China 1937–1945* (Stanford, Stanford University Press, 1962). In the process, the author also overstates his point and encumbers it with a largely unconvincing theoretical framework. He would have us believe that the acute consciousness of China as an historical and political reality, which has characterized the Chinese for the past two thousand years, and the preoccupation with saving China from the incursions of the barbarians, which made itself felt in all strata of Chinese society beginning with the Opium War, have little or nothing to do with modern Chinese nationalism, which was somehow born *ex nihilo* in 1937. On the other hand, he maintains that resentment by the peasants at the economic injustices perpetrated on them by the landlords and the tax-collectors played only a marginal and negligible role in their decision to support the Communists. But this one-sided interpretation does not deprive the book of its value as a piece of historical research.

Party as the incarnation of the country's resistance to Japan was played by the guerrilla base areas. It will be recalled that the Long March had been explained and justified in large measure as a move northwards to fight Japan. In January and February 1936 Mao and his comrades actually embarked on what they called an 'anti-Japanese expedition'. A substantial force of the Red Army, under the command of Mao, Liu Chih-tan, and Hsü Hai-tung, crossed the Yellow River, and soon occupied eighteen *hsien* in Shansi province against the feeble opposition of Yen Hsi-shan. At this point Chiang's own army intervened; by May 1936 most of the Red forces had been driven back across the river, and Liu Chih-tan had died of his wounds.[16] The Communists thereupon embarked on the campaign for a united front with Nanking already described, which finally led to the agreement of 22 September 1937. The paragraph of this accord dealing with military matters read in full:

> The name of the Red Army will be abolished, and its designation changed; it will be reorganized as the National Revolutionary Army and will be subject to the orders of the Military Council of the National Government. Moreover, it will await orders to march forth and take up its duties on the anti-Japanese front.*

In conformity with this stipulation the Red Army in the Yenan area, which had been renamed the Eighth Route Army, crossed the Yellow River once more in September 1937, and entered into action against the Japanese almost immediately.[17] Thus the Communists were at last able, with the authorization of Chiang Kai-shek, to do what they had proclaimed they wished to do since 1934.

But the struggle of the Communist armies against the Japanese, and the reinforcement of their position in any future conflict with the Nationalists, was not to be achieved by positional warfare between regular armies. In any case the newly rebaptized Eighth Route Army, which was limited by Nanking to 45,000 men, was in no position to meet the powerful Japanese forces head-on. They could therefore most effectively fight the invader by the

---

* *Chung-kuo Kung-ch'an-tang k'ang-chan wen-hsien*, Vol. I (Hongkong, Hung-mien Chu-pan-she, 1946), p. 8. This declaration was proposed by the Central Committee of the Chinese Communist Party to the Kuomintang on 15 July 1937, but published by the latter only on 22 September. The full text is translated in *A Documentary History of Chinese Communism*, pp. 245–7.

guerrilla tactics for which their training and doctrine prepared them, especially as the Japanese were incapable of occupying the vast countryside, with the exception of the lines of communication along which their armies advanced. At the same time this type of military activity was the most difficult for the National Government to control. It was virtually impossible even to keep track of the growth in the Communist-controlled forces when they were split up into units of 1,000 or 2,000 men behind the Japanese lines – so much so that they had multiplied tenfold by 1945, from 50,000 to 500,000 men.*

Militarily, the Communist forces expanded both by direct recruitment and by supporting, re-shaping, and eventually incorporating into their own ranks a variety of self-defence and anti-Japanese guerrilla units that had been formed spontaneously in the wake of the invasion and the ill-treatment of the population by the Japanese. Politically, the Communists applied the 'three thirds' system, under which local government in the guerrilla areas was composed of one third Communists, one third from other organizations, and one third non-party people. The result was of course, not democracy in the sense of majority rule, which in any case would have been impossible under these conditions, but it was, on the whole, honest government, which earned the respect of a large part of the population – as the Japanese ruefully discovered.[18] The appeal to all classes in the name of unity against Japan was buttressed by the moderate economic policies, involving a limitation of rents but no confiscation of land, which had been adopted in 1937. The landlords got less, but they were at least certain of collecting it, for the government in Yenan and its various emanations guaranteed their right to do so. The peasants, on the other hand, paid no more than one third of their crop in rent – no small advantage, since hitherto rents of well over one half the crop had not been uncommon. The term 'soviet' was abandoned, and the base area in the north-west was renamed the 'Border Region'.

In Central China tactics similar to those employed in the north by the Eighth Route Army were pursued by the 'New Fourth

* Johnson, op. cit., p. 73. Johnson advances these figures as the most likely, but suggests that the actual total both at the beginning and at the end of the process may have been considerably higher.

Army', created in September 1937 from the survivors of the various guerrilla bases who had remained in that area after the Long March. Yeh T'ing, the leader of the Nanchang Uprising, was called back from exile in Hongkong to assume command of this army. (As his Communist Party membership had lapsed in the interval, he was regarded as a relatively less alarming figure than some of the other Communist generals.) His second in command was Hsiang Ying, who had been (together with Chang Kuo-t'ao) one of the two vice-chairmen of the Chinese Soviet Republic in the Kiangsi days. Perhaps the ablest of his subordinates was Ch'en I, the present foreign minister. By the spring of 1938 the reorganization of this army had been completed, and it began emulating the behaviour of the Eighth Route Army in the north by carrying out guerrilla attacks on Japanese communication lines.[19]

The obvious sincerity of Mao's anti-Japanese sentiments, and the effectiveness of the guerrilla warfare waged by the Communists against the invader, increasingly drew large numbers of students and other intellectuals from the coastal cities to Yenan, which became one of the liveliest centres in China. The participation of the students in the anti-Japanese struggle had first sounded on 9 December 1935, when several thousand students had demonstrated in Peking under the slogan: 'Resist Japan and save the country!' Such demonstrations had soon spread to other parts of China, and aroused the sympathies not only of students and teachers, but of other articulate elements in the population. From the outset the demonstrations had been under strong Communist influence,* and the Kuomintang leadership had naturally disapproved of them, both because it did not like mass movements in general, and because the impulse behind them had menaced the official policy of avoiding open conflict with the Japanese. Increasing numbers of students had therefore begun streaming to the Red base in Shensi for a friendlier environment.

---

* The official Peking historiography today claims that the whole movement took place 'under the leadership of the Communist Party'. Ho Kanchih, *A History of the Modern Chinese Revolution*, pp. 285–6. As John Israel shows (*Student Nationalism in China, 1927–1937*, Stanford University Press, 1966), the Communists did not launch the movement, though they ultimately came to control it.

Mao and his colleagues had soon taken measures to accommodate these visitors. In February 1936, only a few months after the end of the Long March, a declaration signed by Mao himself, together with Chou En-lai and P'eng Te-huai, had announced the establishment of the 'North-West Anti-Japanese Red Army University', which would begin its work on 1 April. The purpose of the university, it had been explained, was to train cadres for the 'anti-Japanese national revolutionary war', in order to save the Chinese nation from annihilation which menaced it as a result of the actions of the Japanese imperialists and 'the chief traitor (*mai-kuo-tsei t'ou-tzu*) Chiang Kai-shek'. The programme had been divided into four sections, designed to train officers (1,000 students), political workers (1,200 students), guerrilla leaders (300 students), and specialists such as cavalry, artillery, or engineer officers (200 students). The course in the first three sections would last six months, and in the fourth section one year. Students would be welcomed without consideration of their political background, class origin, or sex, provided only that they were resolved to fight Japan. The minimum academic preparation required would be a diploma from a higher primary school or the equivalent, but middle school students and university students would naturally be welcomed.[20]

Edgar Snow had found this university already in operation when he had visited the soviet area in the summer of 1936.[21] The formal agreement with the Kuomintang in 1937, and the beginning of the anti-Japanese war, made it easier for students to make their way to Yenan and also increased the motivation for going there. The resulting mass exodus to the hinterland provided Mao and his comrades with an exceedingly large number of valuable new cadres, many of whom are still to be found in the state and party apparatus today.

The Commissioner of Education in the Government of the Border Region was Mao's old teacher from Changsha, Hsü T'e-li. Hsü was singularly well equipped to understand the problems of teaching people of varying ages and backgrounds, for in 1919, when he was over forty, he had gone to Paris as a work-and-study student, and in 1928 he had enrolled in the Sun Yat-sen University in Moscow. On his return to China in 1930, he had become Assistant Commissioner of Education in the

Kiangsi soviet régime under Ch'ü Ch'iu-pai, succeeding Ch'ü when the latter was executed by the Kuomintang in 1935.*

This movement of the students, intellectuals, and artists to Yenan had one important consequence for Mao's personal life.

In 1928, when he was thirty-five, Mao had begun to live with a girl of eighteen, Ho Tzu-chen. Although she may not have been officially his wife until after the execution of Yang K'ai-hui in Changsha in 1930, she bore him five children in the course of the ensuing nine years – one of them during the Long March, which she had been one of the few women to make, despite the fact that she was pregnant at the time. In 1937 she had been sent to the Soviet Union for medical treatment. While she was away Lanp'ing, a well-known film actress from Shanghai, arrived in Yenan and began work in the theatre there. It appears to have been love at first sight. Mao divorced Ho Tzu-chen and married the glamorous newcomer, with whom he has lived happily ever since.†

The fact that he found consolation in the present did not mean that Mao had forgotten the companion of his youth, Yang K'ai-hui. In the summer of 1937, before he met Lan-p'ing, he once asked Agnes Smedley whether she had ever loved any man, and why, and what love meant to her; and on another occasion he recited for her a poem in memory of his first wife. This poem, like the majority of those Mao has written in the course of his life, has never been published.‡ But twenty years later he dedicated another poem to K'ai-hui's memory, and this time allowed it to be printed for the general public. (See inside back cover and last page of book.)

---

* See his own account in Nym Wales, *Red Dust*, pp. 44–5.

† According to the biography of Ho Tzu-chen, published as an appendix to the Chinese translation of Mao's autobiography as told to Edgar Snow and issued by the Li-ming Shu-chü in Shanghai in November 1937, with a preface in Mao's own calligraphy, she was the daughter of a landlord; she had joined the Communist Party in 1927, and led a women's regiment during the Nanchang Uprising. Whether or not this last detail is accurate, she was clearly a forceful character. See also J. Ch'ên, op. cit., p. 150; Meng Ch'iu, *Sui-chün hsi-hsing chien-wen-lu* (Shanghai, 1938), pp. 94–6.

‡ A. Smedley, *Battle Hymn of China*, pp. 122–3; Robert Payne, *Portrait of a Revolutionary: Mao Tse-tung* (London, Abelard-Schuman, 1961), p. 233. According to Payne, Mao collected about seventy of his poems while he was in Yenan and had them printed in a very small number of copies for his intimate friends, under the title *Wind Sand Poems*.

1. (*right*) Mao Tse-tung about 1919
2. (*below*) Mao Tse-tung at the time of his election to the Central Committee of the CCP in 1923

3 and 4. The First Congress of the CCP, July 1921: (*right*) the house in Shanghai where the Congress began; (*below*) the lake where the concluding session was held

5. (*opposite left*) Mao's intimate friend from Changsha, Ts'ai Ho-sen

6. (*opposite right*) Mao's mentor in Communism, Li Ta-chao

7. (*opposite below*) Peasant Movement Training Institute, 1926

8 and 9. Early collaboration
between the CCP and the
KMT: (*right*) a list of names
of Communists, including that
of Mao, proposed by
Sun Yat-sen for alternate
membership of the KMT
Central Executive Committee
and written in his own hand;
(*below*) Sun Yat-sen and
Li Ta-chao leaving the
First Congress of the KMT
side by side

大本營公用殘

中央執行委員候補十七人

邵元冲 鄧家彦 沈定一 林祖涵

茅祖權 李宗黃 白雲梯 張知本

彭素民 毛澤東 傅汝霖 于方舟

張葦村 瞿秋白

張國燾

張秋白 韓麟符

中華民國 年 月 日

10 and 11. The Long March: (*top*) elements of the Red Army on the march near the end of their journey in the autumn of 1935; (*bottom*) the bridge over the Tatu River crossed by the Red Army after a famous battle

12 and 13. (*opposite*) From
one war to another: (*top*)
Mao outside the Anti-
Japanese Military and
Political University of
Yenan in 1937; (*bottom*)
Mao and Chu Te during
the civil war against the
Kuomintang in 1948

14. (*right*) Cover of a book
entitled *Songs to the Glory
of Mao Tse-tung*, published
in 1951

15. (*below*) Mao proclaiming
the establishment of the
Chinese People's Republic
on 1 October 1949

毛澤東頌歌

工農兵歌曲集之五

中央音樂學院研究部編·萬葉書店刊

中國圖書發行公司總經售

16. (*left*) Mao and Voroshilov during the latter's visit to China in 1957
17. (*left below*) Mao with Krushchev during his visit to Moscow in 1957
18. (*right*) Mao talking to a peasant during his tour of the country in the summer of 1958
19. (*below*) Mao and other party and government leaders receiving agroscientific, technical, and medical workers attending their national conference in Peking, 1963

20. (*left*) Some of the 50,000 workers involved in the construction of the Ming Tombs Dam
21. (*below*) Women of the Shiu Shin Commune park their guns while they hoe

培养自己成为有社会主义觉悟、有文化的劳动者。

22. (*left*) Children in a kindergarten are exhorted: 'Educate yourselves to become a cultured worker endowed with socialist consciousness'
23. (*top*) Mao discards his formal attire to meet some of the student delegates to the Ninth Congress of the Communist Youth League of China in Peking, 1964
24. (*bottom*) Mao exchanges greetings with Asian and African writers in Peking, 1964

25. (*below*) Mao enjoys food and conversation simultaneously with Voroshilov's Russian delegation, 1957
26 and 27. (*opposite*) Solidarity with North Vietnam. Mao with Ho Chi Minh: (*top*) in life, 1959; (*bottom*) in art, 1964

友谊万岁

TINH HUU NGHI MUON NAM

28 and 29. Old friends and comrades: (*top*) Mao's third wife, Lan-p'ing (Chiang Ch'ing),·making an unprecedented public appearance at a Peking reception in honour of President Sukarno of Indonesia, 1962. Mao is shaking hands with Mme Sukarno; (*bottom*) Mao welcomes Chou En-lai at Peking airport in 1964 on the latter's return from the celebration of the 47th anniversary of the October revolution in Moscow

At the same time that he gained a new wife Mao was separated from his one remaining brother, Mao Tse-min, who had followed his political fortunes faithfully ever since, at Mao's urging, he had attended the Kuomintang's Peasant Movement Training Institute in 1925. In 1938 Tse-min gave up his post in Yenan as one of those responsible for the economy of the Border Region to work in Sinkiang as head of the finance department under General Sheng Shih-ts'ai, the local war-lord, who was then collaborating with the Soviet Union. When General Sheng shifted his allegiance from Moscow to Chungking Mao Tse-min was arrested in September 1942, and executed in 1943. Mao's sister, Tse-hung, had been executed in 1930 at the same time as his first wife, and, his brother Tse-t'an having perished in 1935, he thus remained alone of all his siblings.*

The years in the North-West saw Mao's emergence as a public figure, both in China and abroad. Though often mentioned in the press during the previous decade, he had remained little more than a name: a bandit chieftain for some, a revolutionary hero for others, but in either case an abstraction. Now, following Edgar Snow's pioneering journey in the summer of 1936, a succession of visitors contrived to reach the 'Border Region', and to bring back first-hand reports on its inhabitants.

The portraits they painted of Mao were varied, except in one respect: he left no one indifferent. Edgar Snow, the first foreign journalist to interview him, described Mao justly as characterized by a combination of intellectual depth and peasant shrewdness. Though the cult of the leader was not to develop seriously for another five years, Snow noted Mao's 'deep sense of personal dignity', as well as his 'power of ruthless decision'. One of the most perceptive accounts of Mao's personality has been left by an intense and highly sensitive woman, Agnes Smedley. She found him, at their first meeting, effeminate and vaguely repellent. 'Whatever else he might be,' she wrote, 'he was an aesthete.' Later she discovered the strength behind the mobile features and graceful gestures, and the human warmth and simplicity behind the dignified reserve. But though he could communicate intensely with a

* See Allen Whiting and General Sheng Shih-ts'ai, *Sinkiang: Pawn or Pivot?* (East Lansing, Michigan, Michigan State University Press, 1958); also J. Ch'ên, op. cit., pp. 204, 210.

few intimate friends, he remained on the whole reserved and aloof. 'The sinister quality I had at first felt so strongly in him', continued Agnes Smedley, 'proved to be a spiritual isolation. As Chu Teh was loved, Mao Tse-tung was respected. The few who came to know him best had affection for him, but his spirit dwelt within itself, isolating him.'

The Westerners who met Mao in these years were impressed by his earnestness, and by his willingness to sacrifice personal comfort in the pursuit of an ideal, which contrasted sharply with the venality and self-indulgence of most politicians in Kuomintang China. They were, on the whole, disconcerted by the peasant simplicity of Mao's manners, and his lack of concern with personal appearance. Snow was particularly struck by seeing Mao 'absent-mindedly turn down the belt of his trousers and search for some guests', or take off his trousers on a hot day in the presence of visitors. (Some of this shock, as Jerome Ch'ên justly points out, was due to unfamiliarity with Chinese customs.) Another observer has recorded Mao's curious habit of sucking in the smoke of his cigarette with an unpleasant noise characteristic of the peasants in certain parts of China.*

As it was far easier for foreigners to pass through the Kuomintang blockade, and to publish their impressions of the Communist-controlled area once they had been there, it was in large part through the eyes of Western journalists that the Chinese first learned about a major political force in their own country. (*Red Star Over China* was immediately translated into Chinese, as was the full text of Edgar Snow's interviews with Mao.) The image projected in these writings of Mao and his comrades as patriots animated by a serious moral purpose produced a considerable impact on Chinese public opinion. (A certain disillusionment was caused, however, especially among the students, who had seen in Mao the incarnation of an austere revolutionary puritanism, by the news that he had abandoned his faithful companion of the Long March for the seductive Lan-p'ing.)

By the summer of 1938, despite strong Chinese resistance at Shanghai and one major Chinese success at T'aierhchuang, the

---

* Snow, *Red Star Over China*, pp. 79–88; A. Smedley, *Battle Hymn of China*, pp. 121–2; G. Stein, *The Challenge of Red China* (London, Pilot Press, 1943), p. 82; J. Ch'ên, op. cit., p. 211.

Japanese Army had over-run the great coastal cities of China, and Wuhan fell in October. The sack of Nanking by the Japanese Army following the city's capture in December 1937 was one of the most savage acts of mass terror in modern times and further mobilized Chinese sentiment against the invader. The Nationalist headquarters moved to Chungking, and behind the Japanese lines the Communist-led guerrillas remained virtually alone as an effective political force. In these circumstances, Mao Tse-tung began to develop his theory of the anti-Japanese war in a series of important writings.

The Yenan period, especially the two years from the beginning of 1938 to the beginning of 1940, is an altogether exceptional one in Mao's literary career. Not only was his total output very high, but it consisted more of relatively long and systematic writings and less of reports, speeches, and directives than at any other time in his life. It is not difficult to understand the reasons for this. On the one hand, the stable situation in the base area resulting from the accord with the Kuomintang, and the fact that he was not personally involved in combat against the Japanese, gave him more leisure than he had enjoyed since his student days, except perhaps for a year or two during the heyday of the Kiangsi Soviet Republic. And on the other hand, he was at last achieving the grasp of Marxist theory, the self-confidence, and the breadth of vision necessary to deal globally with the problems of the Chinese revolution.

His writings in the years 1938–40 fall into two groups, the first concerned primarily with the military aspects of the anti-Japanese war, and the second with political questions. The military writings, all produced during the first half of 1938, constitute a trilogy analysing the war at various levels. Characteristically, Mao began with a small volume called *Basic Tactics*,* dealing in concrete terms with the day-to-day conduct of guerrilla warfare. He then proceeded to discuss the place of guerrilla warfare in the anti-

---

* *Chi-ch'u Chan-shu* (Hankow, Tzu-ch'iang Ch'u-pan-she, March 1938). This work is made up of lectures given at the Anti-Japanese Military-Political University; it is not included in the *Selected Works*. An English translation with an introduction by myself and a foreword by General Samuel B. Griffith was published by the Pall Mall Press early in 1967. References to *Basic Tactics* below are to this translation.

Japanese struggle,* and only afterwards the strategy of the war as a whole.†

The guerrilla tactics recommended by Mao for the anti-Japanese war were basically identical with those he had employed in the civil war against the Kuomintang, and expounded systematically in 1936. They consisted in 'avoiding strength and striking at weakness' (*pi-shih kung-hsü*),‡ in concentrating one's own forces to annihilate enemy units one by one, and in emphasizing the killing of enemy soldiers rather than the conquest of territory or strong points. 'It is not worthwhile to kill 1,000 of the enemy and lose 800 killed ourselves,' Mao declared.[22]

The principal differences between Mao's view of the anti-Japanese war and his view of the civil war lay in the themes to be employed for the mobilization of the masses, and in the fact that the strategic problems of the anti-Japanese war were envisaged in a world setting. The main content of political work both within the army and among the population was to preach national revival, to stimulate national consciousness, and to publicize Japanese atrocities, in order to strengthen 'the resolve to fight to the death to kill the enemy'.[23]

Particularly characteristic of Mao's personality, and of the warrior's temperament which was ultimately to carry him to final victory, was his insistence on utter fearlessness as the key to success:

When we see the enemy, we must not be frightened to death like a rat who sees a cat, simply because he has a weapon in his hands. We must not be afraid of approaching him or infiltrating into his midst in

---

* In 'Questions of Strategy in the Anti-Japanese Guerrilla War', dated May 1938 in the current canon of his works, and originally written as Chapter VII of *K'ang-Jih yu-chi chan-cheng ti i-pan wen-t'i* (*On All the Problems of the Anti-Japanese Guerrilla War*), published on 7 July 1938 on the first anniversary of the war. On the relation between this volume and that translated by General Samuel B. Griffith, *Mao Tse-tung on Guerrilla Warfare* (New York, Frederick A. Praeger, 1961), see Schram, op. cit., p. 315.

† 'On the Protracted War', consisting of lectures delivered between 26 May and 3 June 1938 before the Society for the Study of the Anti-Japanese War, and published in July 1938 in *Chieh-fang*, No. 43–4.

‡ This expression of Mao's (*Basic Tactics*, p. 54) clearly derives from Sun Tzu (see *The Art of War*, Ch. V, par. 4, and Ch. VI, especially pars. 6 and 10), and in fact sums up the essence of the ancient strategist's teaching.

order to carry out sabotage. We are men, the enemy is also composed of men, we are all men, so what should we fear? The fact that he has weapons? We can find a way to seize his weapons. All we are afraid of is getting killed by the enemy. But when we undergo the oppression of the enemy to such a point as this, how can anyone still fear death? And if we do not fear death, then what is there to fear about the enemy? So when we see the enemy, whether he is many or few, we must act as though he is bread which can satisfy our hunger, and immediately swallow him.*

As already emphasized repeatedly, this nationalist appeal, if it by no means summed up the whole of Mao's aims, was nonetheless profoundly sincere, and this sincerity multiplied his influence tenfold at a time of supreme national trial. China's war of resistance, Mao declared, was 'an extraordinary spectacle in the history of warfare, an heroic exploit of the Chinese people, a tremendous and earth-shaking† enterprise.' It would in the future, he added, influence not only the fate of China and Japan; it would also 'urge forward the progress of all countries, and in particular of oppressed peoples such as India'.[24]

That Mao's pride in the heroism of his people was altogether justified, the barest catalogue of Chinese courage and sufferings suffices to prove.‡ He was also not wrong in predicting that the Chinese revolution would ultimately have a great impact on all colonial and dependent countries, though the precise nature and scope of that impact are still not altogether clear. In 1938 a more immediate problem was that of what impact the world situation would have on China's struggle.

In his analysis of the anti-Japanese war as a whole Mao envisaged three phases. During the first, which was nearly com-

* *Basic Tactics*, p. 53. A comparison immediately suggests itself between this passage and one which Mao wrote nearly twenty years earlier, in the course of the May 4th Movement of 1919: 'What is the greatest force? The greatest force is that of the union of the popular masses. What should we not fear? We should not fear heaven. We should not fear ghosts. We should not fear the dead. We should not fear the bureaucrats. We should not fear the militarists. We should not fear the capitalists.' (Schram, op. cit., p. 105.)

† Literally, 'which startles heaven and shakes the earth', a Chinese cliché of which Mao is particularly fond.

‡ A passionate and partisan but highly moving account is that of Agnes Smedley, *The Battle Hymn of China*.

pleted, the Japanese would be on the offensive and the Chinese on the defensive. During the second, which would be the longest, there would be a certain equilibrium, and the Chinese would carry on guerrilla warfare behind the Japanese lines. During the third and last phase the guerrilla units which had been organized and forged in the course of the preceding phases would swing over to the offensive and abandon their guerrilla tactics for conventional mobile warfare on a large scale. But final victory would be possible only if their efforts were supported both by the conventional forces of the Nationalist army, and by other democratic powers opposed to Japanese imperialism.[25]

In the uncertain and rapidly changing international situation on the eve of the Second World War it was difficult for Mao, or for anyone else, to know exactly on whose support China could count. As late as the beginning of 1939 he professed to believe that, despite Chamberlain's policy of appeasement and his 'cowardly attitude towards Japan', the 'broad popular masses, including all progressive people from the various social strata who sympathize with China', would in the end succeed in convincing the governments of England, France, and America that it was in their own interests to oppose Japanese aggression.[26] In September 1939, following the Nazi-Soviet pact, he condemned all the 'imperialist powers' with equal violence as 'mad dogs' who could not do otherwise than 'hurl themselves pell mell against their enemies and against the walls of the world'. He even suggested that Chamberlain, the leader of 'the most reactionary country in the world', was worse than Hitler.[27]

The problem of what could be expected from the Soviet Union was a more complex one. As a Leninist and a sincere believer in 'proletarian internationalism', Mao could hardly doubt that the only Socialist great power would come to the assistance of an oppressed people in their struggle against imperialist invasion. Moreover, by momentarily relieving Stalin of anxieties over the security of his western frontier, the pact of August 1939 freed his hands in the East – as the Japanese, who protested violently in Berlin against the pact, were well aware. This was the aspect stressed by Mao in his statements at the time. But the pact was also such a flagrant example of subordinating principle to Russian national interest that, whatever its immediate practical repercussions, it could not fail to arouse misgivings over Soviet intentions in China as well. Apart from this, tensions were already beginning to develop (or, rather, had never entirely disappeared) between Mao and Stalin on the role of the Chinese Communists and the Kuomintang in the anti-Japanese united front.

For the moment this controversy took place indirectly, through the person of Wang Ming. After half a dozen years in Moscow, where he had risen to the status of Stalin's leading expert on colonial problems as well as membership in the Executive Committee of the Comintern, Wang returned to China late in 1937. Shortly after his arrival he announced to an American correspondent that he had come to China for discussions with Chiang Kai-shek, which might lead to Communist representation in the Nanking Government.[28] During 1938 and 1939 he devoted himself to united front work, and appears to have adopted the line that leadership in the war belonged to the Kuomintang.

This is, in any case, the deviation of which he was accused in 1951.[29] Like all denunciations of rivals (and at the time of his return to China he appeared as one of Mao's two chief rivals, together with Chang Kuo-t'ao), this judgement no doubt represents an oversimplification. We have seen that Mao himself, in his speech of October 1938, predicted a 'glorious future' for the Kuomintang, although he also emphasized that the Communists had an independent role to play. But in the course of the following year, and especially after the change in the world situation brought about by the Nazi-Soviet pact, Mao moved towards a much firmer line. In 'The Chinese Revolution and the Chinese Communist Party', written in December 1939 largely as a textbook for party members, he affirmed clearly that the Chinese revolution must be led by the proletariat and the Communist Party.[30] In 'On New Democracy', written a month later for a wider audience, he was more discreet, and even included a sentence recognizing that leadership would continue to belong to the Kuomintang – provided it proved itself capable – which he has removed from the current edition. But even in this second text he made it plain that the Communists already required a share in the leadership of the revolution, and that if (as they rather expected) the Kuomintang proved itself unworthy of the tasks before the country, they were prepared to shoulder the whole burden themselves.[31]

Building on the well-established Leninist principle that revolution in a pre-capitalist country must have two stages, a 'bourgeois-democratic' and a socialist one, Mao proceeded to declare that in China the first of these would take the form of what he called 'new democracy'. This new democracy would be distinguished from the 'old' bourgeois-democratic revolution by the fact that it would be led, not by the bourgeoisie alone, but by the 'joint revolutionary-democratic dictatorship of several revolutionary classes'.[32]

This affirmation represented a shaping of orthodox Communist doctrine in a direction adapted to China's particular needs. Lenin had spoken of the 'revolutionary-democratic dictatorship of the workers and peasants'. He had also recommended a united front with the bourgeois nationalists under certain conditions. But in extending the limits of the dictatorship itself to include 'several revolutionary classes', one of which was the bourgeoisie as represented by the Kuomintang, Mao was already groping towards the 'people's dictatorship' he was to proclaim in 1949. This innovation tended to emphasize the revolutionary character of the Chinese people as a whole. On the other hand, though Mao followed Stalin in considering China's anti-imperialist revolution as 'a part of the proletarian-socialist world revolution' – and moreover as a 'mighty part' – he did not claim for Asia, as he does today, the central role in the world revolution. In more orthodox fashion he declared that the proletariat of the capitalist countries constituted the main force, of which the oppressed colonial and semi-colonial peoples were merely the allies.

By the time he published these two very clear and important writings during the winter of 1939–40, Mao was at last in complete control of the policy and ideology of the Chinese Communist Party. His two chief rivals had been eliminated. Chang Kuo-t'ao, who had, like Wang Ming, favoured a more conciliatory line towards Chiang Kai-shek, had fled from Yenan to Kuomintang China in early 1938, and was read out of the party in April. Wang Ming was relegated to subordinate functions and ceased to play any real role by the end of 1939.[33]

Mao's declarations on the major, and probably dominant, role to be played henceforth by the Communists in the leadership of China, and on the socialist revolution which would inevitably follow, cannot have given any pleasure to Chiang Kai-shek. But

Chiang was probably less disturbed by Mao's words than by his actions. By the spring of 1940 the Eighth Route Army counted some 400,000 regular troops and full-time guerrillas. The New Fourth Army lagged behind, and had not attained the goal of 100,000 men set for it a few months earlier by the Central Committee. One reason for this was no doubt the exclusive concentration on military affairs to the detriment of political work for which the leaders of the new Fourth Army were criticized at the time.* On 4 May 1940 Mao penned a directive, in the name of the Central Committee, ordering Hsiang Ying and Ch'en I to expand their forces speedily to this level, and to 'gain control of as many districts as we can in the area extending from Nanking in the west to the sea coast in the east'.[34]

To be sure, the development of the guerrilla base areas was the most effective manner in which Mao and his comrades could fight the Japanese. But Chiang Kai-shek could not be blind to the enormous increase in the Communists' power which resulted from these activities, and which would fortify their bargaining position with the Nationalists once the invaders were expelled. As Chalmers Johnson has put it: 'Although the Communist Army was actually continuing to fight the Japanese, its methods left no room for more than one Chinese victor over Japan.'[35]

The tension generated by this Communist challenge to Kuomintang predominance was further aggravated by the international situation in 1940. The military disasters suffered in Europe by the democratic powers on which Chiang had relied for foreign support greatly increased the pessimism in Chunking, and rumours began to circulate of a compromise peace. Wang Ching-wei's pro-Japanese puppet government, established in Nanking in March, gave concrete expression to even more sinister tendencies. In these circumstances Mao and his comrades launched their famous 'Hundred Regiments' Offensive' against the Japanese on 20 August, demonstrating that the war was very much alive and reducing the likehood of a negotiated peace.†

No less than 400,000 troops of the Eighth Route Army attacked the Japanese simultaneously in five provinces of Northern China in this offensive. The damage inflicted on the enemy was severe.

* Johnson, op. cit., p. 136.
† On the Hundred Regiments Offensive, see Johnson, op. cit., pp. 56–9.

Railroads and other communications were blasted, and an important coal-mine destroyed. Fighting continued until early December, and disrupted the plans of the Japanese High Command.

Meanwhile friction was developing in Central China between the New Fourth Army and various Kuomintang forces in the area. In the face of hostility shown by Chungking's representatives in Southern Kiangsu, Ch'en I in the summer of 1940 sent a significant portion of the units under his command across the Yangtze into Northern Kiangsu. There they attacked and routed Han Te-ch'in's Kuomintang troops, chasing the remnants towards the north-west[36]; this led to a further hardening of Chungking's attitude. An understanding had been reached in June 1940 between representatives of the Communists and the Kuomintang: the Communists would be accorded a free hand in the area north of the Yellow River (except for Southern Shansi), in exchange for the evacuation of Central China by the New Fourth Army. But as Chalmers Johnson points out, the fact that Chungking wanted the Communist troops to move north did not mean that they were authorized to destroy all Kuomintang troops in their path.[37] As a result of the fighting in Northern Kiangsu the government in Chungking, on 9 December 1940, ordered all New Fourth Army troops to move north of the Yangtze before 31 December. (They were given another month to join the Eighth Route Army in the area north of the Yellow River.) The bulk of the army had crossed by the prescribed date, but for reasons which remain obscure and controversial the headquarters force of approximately 9,000 men was still south of the river. On 4 January 1941 it was surrounded by Nationalist troops and virtually wiped out in a ten-day battle. Hsiang Ying was killed, and the commander-in-chief, Yeh T'ing, taken prisoner, remained in custody until 1946.*

This incident marked the end of any real collaboration between

---

* The above account of the New Fourth Army incident is largely based on Johnson, op. cit., pp. 136–40. A Japanese account cited by Johnson indicates that the Kuomintang commander in the area ordered Yeh T'ing to cross the Yangtze at a point in Anhui which would have placed him in the hands of a strongly anti-Communist general, whereupon he moved south-west instead. A contemporary Communist account affirms that the New Fourth Army was prepared to cross at the point designated, but was prevented from doing so by Japanese forces and by lack of ammunition. ibid., note 40, p. 229.

Communists and Nationalists during the anti-Japanese war.* The Chungking Government loudly denounced the insubordination of the New Fourth Army and ordered its dissolution. The Communists not only refused to comply, but promptly set about rebuilding and strengthening it. Ch'en I was appointed acting commander, and no less a personage than Liu Shao-ch'i himself was sent as political commissar. The army's guerrilla operations soon extended over seven military areas, defined by Liu in the spring of 1941, all of which had been transformed into base areas with functioning local governments by the end of the war.[38]

As Chalmers Johnson points out, the New Fourth Army Incident was in the long run extremely useful to the Communists, for it enabled them to supplement the reputation of heroic fighters for the national cause, which they had already won, with the even more valuable title of national martyrs. 'The Communists could violate Nationalist military directives with impunity so long as they appeared more anti-Japanese than the Government, and they knew it.'[39] This judgement in fact sums up the whole subsequent course of the anti-Japanese war, during which Mao and his comrades ceased to make even a pretence of obeying Kuomintang directives, yet continuously increased their prestige as the staunchest defenders of China's national interests.

At the same time that their relations with the Kuomintang thus deteriorated the Communists found themselves under increasingly fierce attack by the Japanese. The latter, who had already been building blockhouses to separate and isolate the guerrilla areas, intensified this policy in order to prevent any recurrence of the Hundred Regiments' Offensive. In July 1941 the notorious 'three-all policy' (burn all, kill all, loot all) was adopted in North China. The result was to reduce the size of the Eighth Route Army from 400,000 to 300,000 men, and to diminish the population of the Communist areas in North China from 44 millions to 25 millions.[40]

The mass burning and killing associated with the 'three-all' policy on the whole intensified the peasants' will to resist, and thus aided the Communist cause. But at the same time the Japanese construction of blockhouses, walls, and barbed-wire

* It is surely no accident that Mao's long study of the civil war in the early 1930s, written in 1936, was published in February 1941.

enclosures, with the virtually total economic blockade of the Communist areas henceforth imposed by the Kuomintang, produced increasingly difficult economic and political problems for Mao and his comrades. The economic challenge was met by a successful drive to increase production in the base areas themselves, involving the encouragement of mutual-aid teams among the peasantry, and the participation of the army in industrial and agricultural work.* The political challenge was met by the 'Cheng Feng' or rectification campaign, which was formally launched in February 1942.

The rectification campaign had a dual purpose, neither aspect of which should be ignored. On the one hand, it was designed to strengthen the unity and discipline of the Chinese Communist Party in difficult circumstances, imparting a minimum knowledge of Marxism-Leninism and the political methods developed in the Soviet Union to the large number of new recruits who had been taken into the party in the course of its wartime expansion. But it was also designed to give to the ideological consciousness of party members a special and characteristic quality directly inspired by Mao Tse-tung himself. In short, Mao's aim was 'sinification' of Marxism. He had launched this slogan at the sixth plenum in October 1938, where he defined it as follows:

The history of our great people over several millennia exhibits national peculiarities and many precious qualities. . . . We are Marxist historicists; we must not mutilate history. From Confucius to Sun Yat-sen we must sum it up critically, and we must constitute ourselves the heirs of all that is precious in this past. . . . A Communist is a Marxist internationalist, but Marxism must take on a national form before it can be applied. There is no such thing as abstract Marxism, but only concrete Marxism. What we call concrete Marxism is Marxism that has taken on a national form. . . . If a Chinese Communist, who is a part of the great Chinese people, bound to his people by his very flesh and blood, talks of Marxism apart from Chinese peculiarities, this Marxism is merely an empty abstraction. Consequently, the sinification of Marxism – that is to say, making certain that in all of its manifestations it is imbued with Chinese peculiarities – becomes a problem that must

* Mao dealt very concretely with the material difficulties encountered in the border region in a 200-page report of December 1942, entitled 'Economic and Financial Problems', of which only the first half-dozen pages have been retained in the current edition of the *Selected Works*.

be understood and solved by the whole Party without delay. . . . We must put an end to writing eight-legged essays on foreign models . . . we must discard our dogmatism and replace it by a new and vital Chinese style and manner, pleasing to the eye and to the ear of the Chinese common people.[41]

In a speech to a meeting of cadres on 5 May 1941 Mao stressed a similar theme:

When many scholars of Marxism-Leninism speak, they must talk about Greece; they can only repeat quotes from Marx, Engels, Lenin and Stalin from memory, but about their own ancestors they have to apologize and say they've forgotten.

They are not ashamed but proud when they understand very little or nothing about their own history. . . . Many are ignorant of anything which is their own, yet hold on to Greek and foreign tales (they are nothing more than tales) which are pathetically abstracted and presented from a pile of old foreign papers. For the past few decades, many returned students have been making this mistake. They return from Europe, America or Japan, and all they know is to recite a stock of undigested foreign phrases. They function as phonographs but forget their own responsibility to create something new.*

Launching the rectification campaign in February 1942, Mao once more energetically denounced party formalism in general, and 'foreign formalism' in particular. In the second of his two long speeches he complained that his appeal of 1938 for the adaptation of Marxism to Chinese conditions had not been heard, and demanded that it be put into practice immediately.[42]

The previous May Mao had criticized only the students returned from Europe, America, or Japan; but now it was clear that the 'foreign formalism' he desired to eliminate consisted in the

* *Mao's China. Party Reform Documents, 1942–1944* (Seattle, University of Washington Press, 1952), translation and introduction by Boyd Compton, pp. 61–3. This volume contains translations from the original texts of the rectification campaign, before Mao rewrote and edulcorated them for the current official edition. (In the above passage, he removed the criticism of those who 'repeat quotes from Marx, Engels, Lenin and Stalin from memory', and also an ironic reference, later in the same speech, to 'seventeen- and eighteen-year-old babies' who are taught to 'nibble on *Das Kapital* and *Anti-Dühring*' instead of learning about their own country.) Mr Compton's introduction to this volume, though written nearly fifteen years ago, stands up very well in the light of more recent research as a global evaluation of the rectification movement. The book is now available as a paperback.

imitation of Soviet models. The rectification campaign was thus directed in part against the predominance of Soviet influence in the party. The most recent official history of the Chinese Communist Party makes this quite clear when it explains that the campaign was aimed at 'doctrinaires as represented by Comrade Wang Ming' who were 'ignorant of the Party's historical experience', and could 'only quote words or phrases from Marxist writings'.[43]

To be sure, the fact that the rectification campaign was designed to make the Chinese Communists more aware of their own history and traditions, and teach them to deal with the problems of the Chinese revolution in an independent and original manner, does not mean that it was, properly speaking, anti-Soviet. Approximately one fourth of the materials eventually put together in a volume for study by party members consisted of translations from Lenin, Stalin, and Dimitrov.[44] But if the Chinese Communists were to be instructed in the ideas and methods of their Soviet comrades, they were instructed as well to regard all such with detachment, to choose and assimilate only what was useful to them. A fully autonomous Communist movement in China, with its own ideology specifically adapted to Chinese conditions, was Mao's goal.

Mao's attempt to provide the Chinese Communist Party with an independent *philosophical* basis, as distinguished from an independent policy orientation, merits account. During the Yenan period, beginning in 1936, several important Soviet writings on Marxist philosophy were translated into Chinese, and thus made available to Mao for the first time. In addition, trained philosophers among the Communist intellectuals in Yenan began to produce writings of their own, popularizing the current tendencies in Soviet thought on various levels. Since excellence as a Marxist philosopher (and not merely as a Marxist political thinker) had traditionally been regarded as indispensable for consecration as a leading figure in the world Communist movement, Mao tried his hand at producing essays of this kind. In July and August 1937 he delivered two speeches, published for general distribution only in 1950 and 1952, entitled 'On Practice' and 'On Contradiction'. These two pieces are of interest as reflecting a mind which sees the world in ceaseless and perpetual flux, with no final and definitive harmony attainable even under Communism – a view which has found more striking expression subsequently in the theory of the 'permanent revolution'. But as philosophic writings in the strict sense they are the work of a beginner. Three years later, in early 1940, Mao produced another theoretical essay, entitled 'On Dialectical Materialism'. This was not only a pale and lifeless effort, but so closely modelled on translations from Soviet philosophical writings that Mao decided to stop its publication after only two instalments had appeared.* These three pieces represent the totality of Mao's attempts to qualify as a Marxist philosopher. Perhaps he was better aware of his own limitations than those in China who now vaunt his merits as a theoretician. In any event, it is not here that his greatness lies. He has given the Chinese Communist Party its own strategy and tactics, rooted in Chinese conditions, and he has made Communism infinitely more familiar and acceptable to his countrymen by writing about it in a style heavily coloured by allusions to classical writings and precedents from Chinese history. He has not – despite the extravagant claims made for his 'Thought' today – contributed substantially to Marxist theory.

Whatever the merits of Mao's contribution as a philosopher, there is no doubt that, by the end of the rectification campaign of 1942–4, he had succeeded in inculcating in his comrades the habit of viewing political problems in their specific Chinese context. He had also affirmed the independence of his own leadership from Moscow. In May 1943 he used the dissolution of the Communist International as an occasion to hail the independent achievements of his own party. The Chinese Communist Party, he declared, had lived through revolutionary movements which were 'continuous and uninterrupted and extraordinarily complex,

---

* The most meticulous and detailed study of Mao's three philosophical essays is that by Karl A. Wittfogel, 'Some Remarks on Mao's Handling of Concepts and Problems of Dialectics', *Studies in Soviet Thought*, III, 4 December 1963, pp. 251–77 (including a complete translation of 'On Dialectical Materialism'). Although I differ profoundly in many respects with Professor Wittfogel regarding the interpretation of Mao's life and thought, his analysis here seems to me wholly convincing. For my own views of Mao as a dialectician, and comments on other writings on the subject, see my article in *China Quarterly* review No. 29, pp. 155–65.

even more complex than the Russian Revolution'. In the course of this experience it had already acquired 'its own excellent cadres'. Furthermore, since the Seventh Comintern Congress in 1935, the International had not 'meddled' in the internal affairs of the Chinese Communist Party. And yet, he added, 'the Chinese Communist Party has done its work very well, throughout the whole anti-Japanese War of National Liberation.'[45]

Meanwhile the Second World War was drawing to its climax. Although Mao was bent on establishing the independence and originality of his own leadership in China, his views of the world scene as a whole were, as already indicated, largely identical with those of Moscow. In January 1940, in 'On New Democracy', he even went so far as to state that China would soon be obliged to take sides in the increasingly acute struggle between the Soviet Union and Anglo-American imperialism (not German imperialism).* And in May 1941 the organ of the Chinese Communist Party denounced 'Roosevelt's plot to enlarge the imperialist war' by dragging America into war with Germany.†

Hitler's invasion of the Soviet Union brought an abrupt change, and Mao was soon emphasizing the urgency of American military and economic aid both to the Soviet Union and to China, and appealing for an immediate declaration of war against Germany.[46] Two days after Pearl Harbour the Central Committee of the Chinese Communist Party issued a directive emphasizing the importance of British and American participation in the anti-Japanese united front in the Pacific.[47]

In 1942 Mao, like the Soviet leaders, grew impatient at the lack of offensive action by Britain and America. The two countries, he

* See the paragraphs translated in Carrère d'Encausse and Schram, *Le Marxisme et l'Asie*, pp. 357–8. This passage has now been cut and rewritten so that it refers only to the intensified conflict between the Soviet Union and the imperialist powers in general. cf. *Selected Works* (Peking), Vol. II, pp. 364–5.

† *Chieh-fang Jih-pao* (*Liberation Daily*), 29 May 1941 (editorial). Mao took a close interest in this newspaper, and frequently contributed anonymous editorials to it, which have now been attributed to him and included in the current edition of his *Selected Works*. The editorial in question was further directed against the possibility of an 'Eastern Munich', which Roosevelt was supposed to desire in order to free his hands in Europe. Mao himself had written a party directive on this latter theme a few days earlier. *Selected Works* (Peking), Vol. III, pp. 27–8.

remarked bitterly in an article hailing victory at Stalingrad, would finally resolve to open a second front only when they could attack an already dead tiger.* In the course of the following year, as the Allied effort unfolded in North Africa and Italy, Mao, like Stalin, adopted a much more amicable tone towards London and Washington, and several times spoke favourably of their contribution to the war.†

This evolution, parallel to that of Moscow, in Mao's view of the world situation, moved a step further in late 1943 and early 1944 as a result of developments in the situation within China itself. On the one hand, relations between the Communists and the Kuomintang, which had been stabilized since the New Fourth Army Incident of 1941 at a level of grudging tolerance, underwent a further sharp deterioration. On the other hand, Chiang and his ghost writers began propagating opinions on the superiority of traditional Chinese civilization and the vanity of modern innovations from the West that must have made Mao wonder whether the battles in which he had ardently supported Ch'en Tu-hsiu during the May 4th period, in favour of 'Mr Democracy' and 'Mr Science', had really been won after all. The most striking manifestation of these tendencies on Chiang's part was to be found in his book *China's Destiny*, published late in 1943, in which he denounced Communism and liberalism as equally pernicious and affirmed that the last word on nearly everything had been said by Confucius.‡

In these circumstances Mao and his comrades began to manifest a more and more pro-American attitude. The most curious example is no doubt the *Liberation Daily* editorial of 4 July 1944, which heaps lavish praise on the American democratic tradition, and establishes an explicit parallel between America's struggle for

---

* *Selected Works* (Peking), Vol. III, p. 107. This scornful sentence was included in the original version of the editorial, as published in the *Chieh-fang Jih-pao* on 12 October 1942.

† See, for example, his message of 20 February 1943, *Ch'ün-chung*, 8 (4), 1943, p. 81.

‡ Some of these passages were eliminated or attenuated in the official English translation (New York, Macmillan, 1947); they were restored in an unauthorized version by Philip Jaffe, the editor of *Amerasia* (Chiang Kai-shek, *China's Destiny and Chinese Economic Theory*, New York, Roy Publishers, 1947).

democracy and national independence in the eighteenth century, and China's struggle in the twentieth:

Democratic America has already found a companion, and the cause of Sun Yat-sen a successor, in the Chinese Communist Party and the other democratic forces. . . . The work which we Communists are carrying on today is the very same work which was carried on earlier in America by Washington, Jefferson, and Lincoln; it will certainly obtain, and indeed has already obtained, the sympathy of democratic America.[48]

These expressions of pro-American sentiment, and the effort by Mao and his comrades to present themselves as the heirs of Washington and Lincoln, had a very practical and immediate goal. If the Chinese Communists were 'democratic', whereas the Kuomintang was reactionary and xenophobic, it would be appropriate for democratic America to give part of its aid to Yenan instead of channelling it all through the Nationalist Government in Chungking. (As we shall see in a moment, the Communists did in fact make concrete requests for such assistance.) But there may also have been some sincerity mixed with this obvious Machiavellism. The American journalists and other travellers who visited Yenan or the Communist guerrilla areas during the anti-Japanese war frequently felt themselves far more at home there than in Kuomintang China. They were attracted by the integrity, the concern for progress, and above all by the moral fervour they found there. They sometimes concluded that, since Mao and his comrades were equally serious-minded and idealistic, they must have the same ideals as themselves.*

It is not impossible that Mao and his comrades were the victims

* An interesting and provocative discussion of this problem is to be found in Tang Tsou, *America's Failure in China 1941–1950* (University of Chicago Press, 1963), pp. 176–236. Much of what the author says in this chapter, entitled 'The American Image of Chinese Communism and the American Political Tradition', is perceptive and well-founded. His presentation is marred, however, by extremely sharp and partly gratuitous personal attacks on some of the journalists and diplomats who shaped the American image of China in the 1930s and 1940s. If 'their misunderstanding of the nature and intentions of the Chinese Communist Party was simply a reflection of the climate of opinion at the time', and further grew out of 'the American political tradition itself' (p. 219), it seems hardly fair to treat them with such a mixture of hostility and contempt. But it would be ungenerous to complain too much of this minor blemish on what is unquestionably the most intelli-

of a similar misunderstanding. To be sure, they could not conceivably have imagined that the United States was committed to 'socialist democracy', as some Americans had concluded that the Chinese Communists were 'Jeffersonian Democrats'. But, as *Liberation Daily* pointed out in its Fourth of July editorial quoted above, Marxist-Leninists had always prized the historical role of the American revolution very highly. Mao had not yet, as he was to do in 1946, singled out the United States as the most viciously reactionary of all the imperialist powers and the greatest enemy of the world's peoples. This distinction he had attributed for many years to Germany and Japan – except for the period 1939–41, when it had gone to Britain. Within China Mao foresaw a long 'bourgeois-democratic' stage during which it would be necessary to collaborate (and struggle for hegemony) with a part at least of the Kuomintang. Under these circumstances there is no reason why he should, *a priori*, have ruled out cooperation with the least reactionary of the 'bourgeois' forces on a world scale. Moreover, as pointed out earlier, Mao's imagination was caught by Washington's struggle for independence long before he had even heard of Marx. There is more than one echo, in the 1944 salute to the Fourth of July, of Mao's admiring remarks more than thirty years earlier on a victory won only by 'eight long, bitter years' of struggle.

Whatever the proportion of sincerity and political calculation on both sides, the autumn of 1944 saw the first important contacts between Mao Tse-tung and representatives of the American Government. In August a U.S. military mission, accompanied by the diplomat John S. Service, established itself in Yenan. A *Liberation Daily* editorial, entitled 'Welcome to the Friends of the United States Army Observers Section', expressed the hope that this unit's reports would give the United States Army a good opinion of the Chinese Communists and of their resistance.[49] To Service himself Mao explained that, though the mission was useful in the fight against Japan, its chief importance lay in its 'political effect upon the Kuomintang'.[50]

gent, well-documented, and complete study of a very complex period. Moreover, Professor Tsou's forthright judgements are not directed exclusively at any one person or category; Chiang Kai-shek fares even worse than any of his American *bêtes noires*.

This mission was, in fact, closely bound up with the development of America's China policy as a whole. It followed and was in part a consequence of Vice-President Henry Wallace's mission to Chungking in June 1944, during which he had pressed on Chiang Kai-shek the importance of improving relations with the Chinese Communists in order to further the prosecution of the war against Japan. It accompanied the crisis opened by President Roosevelt's proposal to Chiang, early in July 1944, to place General Stillwell in direct command of all Chinese troops, both Nationalist and Communist.*

As a natural corollary of the plan for a unified command of all Chinese armies, the U.S. Government was shortly led to propose a coalition government for China, with Communist participation. This idea was picked up almost immediately by the Communist representative in Chungking, and then formally espoused by the leaders in Yenan.[51]

Only in the context of a close parallel between the policies of Washington and Yenan can there be explained an extraordinary incident which took place in October 1944, when Mao in person defended the leaders of the Anglo-Saxon democracies against Chiang Kai-shek's rudeness and xenophobia. On 10 October Chiang chose the occasion of the Chinese national holiday (the 'Double Tenth', the anniversary of the 1911 revolution) to proclaim categorically in a speech before the Kuomintang Central Committee that he would not yield on the Stillwell issue. Mao Tse-tung immediately attacked him in an editorial, declaring that, by behaving in this way, Chiang had 'forfeited his position as leader of the war of resistance against Japan':

When the 'leader of one of the four great powers', on a national holiday such as this, characterizes the serious criticisms of foreigners (including Prime Minister Churchill and President Roosevelt) as 'giving credence to the malicious inventions of the enemy and of the traitors to the nation', he really harms the dignity of the Republic of China. Particularly interesting is the demand, put forward by Mr Chiang in his speech, according to which we should 'be able to stand by ourselves, strengthen ourselves, and take upon ourselves the responsibility of fighting the war independently.' [Chiang had threatened to do this if the United States insisted on imposing Stillwell as a condition of further

* For this background, see Tang Tsou, op. cit., pp. 162–75.

aid.] . . . This is a manifestation of Mr Chiang's xenophobia. The situation is like this: the war having reached its present point, and the Kuomintang government and high command being corrupt and incompetent as they are, some people think that it is necessary to establish a unified allied high command in the Chinese theatre. In order to reject this demand, people who habitually rely exclusively on foreign aid as their magic wand have also begun talking about 'independence and self-strengthening', and are moreover demanding that we 'take upon ourselves the responsibility of fighting the war independently'.*

The conflict over the appointment of Stillwell to command the Chinese forces ended, as everyone knows, in a complete victory for Chiang Kai-shek. Stillwell was recalled and replaced by General Wedemeyer, for whom Roosevelt no longer demanded authority over Chinese troops. Ambassador Gauss, who had been the first to propose a coalition government that would include the Communists, was replaced by Patrick Hurley, who had recommended Stillwell's recall to the President.†

Though Hurley took as the keystone of his policy in China support for Chiang Kai-shek, he was by no means hostile at first to the Communists. His relatively favourable attitude, as Tang Tsou clearly shows, was rooted in a double illusion: he did not believe that Mao and his comrades were 'real' Communists, and he thought them so weak that ultimately they would have to co-operate with Chiang, especially as Soviet policy would also push them in the same direction.[52] Even before his formal appointment as ambassador‡ he therefore flew up to Yenan to negotiate directly with Mao Tse-tung.

Hurley's political style must at first have been disconcerting to Mao. At the airport in Yenan he greeted the Chinese Communist

* This article, originally attributed to an anonymous 'Yenan Observer', was published in *Chieh-fang Jih-pao* on 12 October 1944. It has now been identified as written by Mao (*Selected Works*, Peking, Vol. III, pp. 229–33), but understandably the whole of the passage cited above has been eliminated.

† Tang Tsou, op. cit., pp. 100–124, gives an excellent summary of the Stillwell crisis.

‡ Hurley arrived in Chungking on 6 September 1944, as Personal Representative of President Roosevelt to Chiang Kai-shek. Ambassador Gauss resigned on 1 November and Hurley was nominated for the post on 30 November. See *United States Relations with China* (Washington, Department of State, 1949), p. 59.

leaders with an Indian warwhoop.[53] But in his pocket he carried a five-point draft agreement between the Communists and the Kuomintang which was more in harmony with the mentality of his hosts. Two days and nights of negotiations with Mao, which Hurley described as 'most strenuous and most friendly', resulted in a revised draft which was formally signed by Mao on 10 November. Hurley also signed, to indicate his personal approval of the text.*

For Hurley the decisive provision which justified the whole agreement was the one which stipulated: 'All anti-Japanese forces will observe and carry out the orders of the coalition National Government and its United National Military Council.' Mao and his comrades were quite willing to give this pledge in exchange for representation in the proposed coalition government and in the reorganized military council, the obligation on the new government to establish democratic freedoms, and the provisions for equal distribution of supplies from foreign powers. But these provisions were by no means acceptable to Chiang Kai-shek. The best that Hurley could extract from Chiang was a three-point plan offering a vague and conditional promise of civil rights and no change whatever in the government, which was to remain the possession of the Kuomintang alone, in exchange for the incorporation of the Chinese Communist forces into the National Army. Under the circumstances it is not surprising that on 24 December 1944 Mao Tse-tung should have telegraphed Hurley that the National Government had not shown sufficient sincerity to warrant continuing the negotiations.†

Though they were entirely disillusioned about Chiang's intentions, the Chinese Communist leaders still hoped to obtain useful support from the United States. Thus, in February 1945, Chu Te requested a loan of twenty million dollars for the purpose of encouraging the defection of troops from the puppet Nanking Government. But the plan was rejected by Hurley, who insisted

* According to the State Department White Book (ibid., p. 74) Hurley signed only as a witness, but Lord Lindsay's statement to the effect that his signature indicated approval and support for the agreement seems more convincing. See Tang Tsou, op. cit., p. 290.

† On Hurley's negotiations with Mao, see Tang Tsou, op. cit., pp. 289–95; United States Relations with China, pp. 74–5, contains the texts of the five-point agreement signed by Mao, and Chiang's counter-proposals.

that all American resources should be channelled through the National Government.* The brief interlude of limited political cooperation between the United States and the Chinese Communists thus led to no concrete results, but for the moment Mao continued to adopt a relatively friendly attitude towards America. On 13 April 1945 he signed a telegram to President Truman, expressing his 'profound sympathy' on the occasion of Roosevelt's death,[54] and his report to the Seventh Congress of the Chinese Communist Party ten days later contained numerous passages (now eliminated) praising the American contribution to the war effort and China's own struggle.†

The change in this climate can be dated very precisely; it begins with the *Amerasia* case in June 1945, when John S. Service was arrested for having given copies of his reports to Philip Jaffe, the editor of this publication.‡ It will be recalled that in late 1944 and early 1945 Service had spent six months in Yenan with the U.S. military mission. Jaffe, who had prepared the unexpurgated translation of Chiang Kai-shek's *China's Destiny*, was no doubt also favourably known in Yenan. The Communist press therefore reacted with extreme violence. A *Liberation Daily* editorial interpreted the arrests as a turning point in American policy towards China. The Communists, said the article, were not hostile to the American people, nor to the American Government, but they were hostile to American imperialists like Hurley and his ilk. The editorial, which was written throughout in a sharp and almost aggressive tone, concluded with the warning that, if the American

* Tang Tsou, op. cit., p. 178. The idea of appealing for American support had perhaps been suggested to Chu Te by the projects on foot in U.S. military and diplomatic circles, in late 1944 and early 1945, for sending American paratroopers and technicians to the Communist bases, and for supplying military aid to the Communists provided they cooperated with these American forces. See Tang Tsou, op. cit., pp. 152–3, 178, 298; Feis, *The China Tangle*, pp. 205–6.

† For one friendly reference see Schram, op. cit., p. 74. Mao also declared that the 'three great democratic countries, England, America and the Soviet Union', remained united and would continue to unite in the future, despite occasional differences of opinion. See the translation from the original text in Stuart Gelder, *The Chinese Communists* (London, Gollancz, 1946), p. 2; *Lun lien-ho cheng-fu* (Chung-kuo Ch'u-pan-she, 1946), p. 2.

‡ For a brief account see Tang Tsou, op. cit., p. 541.

authorities chose to support the Chinese reactionaries, they would receive from the Chinese people the lesson they deserved.[55]

There is no evidence on whether Mao wrote this editorial himself, but in the middle of July he contributed three anonymous articles in ten days to the Yenan press denouncing the 'bankruptcy' of the 'Hurley–Chiang Kai-shek duo'.* To be sure, he still emphasized the community of interests between the Chinese and American peoples, and recognized that 'quite a few politicians and military men', as well as a large part of American public opinion, understood the aspiration of the Chinese people to freedom and independence. But the tone was as peremptory as that of the editorial on the *Amerasia* case, and the lesson the same: if the United States Government adopted the policy of exclusive support for Chiang against the Communists, a line being advocated by Hurley since the failure of his efforts at conciliation, it would 'with its own hands place a crushing burden on its back'.[56] In a conversation on 4 July 1945 with two representatives of the Democratic League then visiting Yenan, Mao was even more vehement.'Since I have been able to fight Japan with these few rusty rifles,' he said, 'I can fight the Americans too. The first step is to get rid of Hurley, then we'll see.'† The climate in relations between Yenan and Washington was thus stormy as the inevitable crisis in Chinese internal politics, provoked by Japanese surrender, approached.

As he made ready for this new phase of his political career Mao Tse-tung strengthened and consolidated his hold on the organization and thinking of the Chinese Communist Party. At the Seventh Congress in April 1945 a new party constitution was adopted, containing a preamble in which 'The Thought of Mao Tse-tung' was officially enshrined as necessary to 'guide the entire work' of the party.[57] Liu Shao-ch'i praised Mao as 'not only the

---

* Only two of these, dated 10 and 12 July, are included in the *Selected Works* (Peking, Vol. III). For the identification of a third editorial dated 19 July as by Mao, see Schram, op. cit., pp. 264–5.

† Tso Shun-sheng, *Chin san-shih nien chien-wen tsa-chi* (*Interesting Events in the Past Thirty Years*) (Hongkong, Freedom Press, 1954), p. 90. The fact that the interview took place on the fourth of July was the result of sheer chance, but it serves to underscore the contrast with the climate just a year earlier, marked by the editorial already mentioned on the U.S. national day.

greatest revolutionary and statesman in Chinese history, but also its greatest theoretician and scientist'.[58]

The veritable cult of Mao Tse-tung which burst forth into blossom at this congress, and which has continued to flower ever since,* can be traced back to the period of the rectification campaign in 1942–4. On 23 January 1942, just before the official opening of the campaign, Mao issued an order to political workers in the armies of the Border Region, directing them to print several thousand copies of the 'Resolution of the Ninth Congress of the Fourth Army' for distribution down to the level of company commander. The recipients were to take it as a textbook and 'read it thoroughly'. 'Cadres at all levels should all read it thoroughly', Mao added.[59] This resolution, which dated from 1929, had in fact been written by Mao himself;† he was therefore ordering the attentive study of his own works. Further manifestations of the cult are to be found in the messages from labour heroes which began to arrive in Yenan towards the end of 1943, saluting Mao as 'the star of salvation (*chiu-hsing*) of the Chinese people' and such.‡

It was entirely natural that the blossoming of the Mao cult in 1945, like the rectification campaign which provided the soil for its early growth, should have been marked by a strong insistence on the 'sinification' of Marxism. Liu Shao-ch'i praised Mao for this in his report at the Seventh Congress; he also described Mao's thought as an 'admirable example of the nationalization of Marxism'.§

On this occasion Liu moved beyond the sphere of pure ideology to assert a claim for the role of Mao's thought, and by implica-

---

* Not without Mao's participation. In July 1945 he contributed an anonymous article to the Yenan press stating that the Chinese people wanted to follow 'Mao Tse-tung's way'. See Schram, op. cit., p. 265.

† Only a brief extract from this long text is included in the *Selected Works*. See Schram, op. cit., pp. 199–201.

‡ The above phrase is from a message published in the *Chieh-fang Jih-pao* of 21 November 1943. See also the letter of 29 November 1943 from Yenan labour heroes, published in ibid., 2 December 1943, and the request to Mao to plant some grain in person (ibid., 11 February 1944).

§ These expressions are somewhat attenuated in the official English translation cited above. For a French translation based on the Chinese text, see Carrère d'Encausse and Schram, *Le Marxisme et l'Asie*, pp. 361–5.

tion of the Chinese Communist leadership, in the world revolutionary movement as a whole. 'The thought of Mao Tse-tung', he wrote, '. . . will make great and useful contributions to the struggle for the emancipation of the peoples of all countries in general, and of the peoples of the East in particular.'[60] In 1936 Mao had declared that when the Chinese revolution was victorious, 'the masses of many colonial countries' would 'follow the example of China and win a similar victory of their own', and in 1940 he had written that all revolutions in colonial and semi-colonial countries would have a 'new-democratic' character like that of China.[61] But these statements could be interpreted simply as an affirmation of the solidarity uniting all dependent countries, and of the family resemblance characterizing their political forms. Now, for the first time, Liu Shao-ch'i put forward the idea, which he was to develop in ever sharper and more explicit terms in subsequent years, that China was not merely the pioneer but the leader and ideological mentor of anti-imperialist revolutions throughout Asia and Africa.

This claim to leadership was even less welcome in Moscow than were Mao's pretensions to having 'sinified' Marxism. Joined to the differences in political interests between the Chinese Communists and the Soviets, it repeatedly troubled the relations between Mao and his Russian comrades.

For the moment, in the spring of 1945, the explicitly proclaimed policy goal of Mao Tse-tung and Stalin was the same: a coalition government in China with Communist participation. 'On Coalition Government' was precisely the title which Mao had given his long report to the Seventh Congress in April, and the Soviets also supported the slogan. But the order of priorities in Moscow and Yenan was no doubt somewhat different. In the long run both Mao and Stalin naturally expected Communism to triumph in China. Neither of them thought that the Chinese Communists were strong enough to take over leadership of the government immediately. (All suggestions of this in Mao's report as published in the current edition of his *Selected Works* were added after the fact.*) But Mao was vitally interested in laying the foundations

* For one example of such an addition regarding the 'leadership of the Communist Party' in the political system proposed by Mao, see Schram, op. cit., p. 218. For other affirmations, added in the current edition,

for his ultimate triumph, whereas Stalin was more sensitive to the effect which open civil war might have on Chiang's willingness to satisfy Russian demands accepted by Roosevelt at Yalta, such as the creation of a naval base at Port Arthur.*

The sudden collapse of Japanese resistance, on the bombing of Hiroshima and Nagasaki, obliged both Mao and Stalin to make concrete decisions of enormous import for the future much sooner than either had anticipated. The vital question involved the acceptance of surrender by Japanese and puppet troops in China. On the very day of the Japanese surrender (14 August) Stalin had concluded, as agreed at Yalta, a treaty of friendship and alliance with the Chungking Government which, if faithfully carried out, would have given a large measure of satisfaction to Chiang Kai-shek on this point. A supplementary agreement provided that, as soon as any part of the territory in Manchuria occupied by the Soviet troops in their advance ceased to be 'a zone of immediate military operations', it would be turned over to representatives of the Chinese National Government for administration. In the course of the negotiations in Moscow Stalin had cynically underscored the point at issue, warning Chiang's Foreign Minister T. V. Soong on 10 August that if the treaty was not signed immediately, the Chinese Communists would get into Manchuria first.†

Mao's intention was effectively that the military forces under his command should accept the surrender of the Japanese and puppet troops everywhere possible, so as to obtain important stocks of arms while expanding and consolidating the Communist territorial basis. On 10 August Chu Te gave concrete expression to this policy by ordering all armed forces under his command to occupy cities, towns, and communication lines held by the enemy, and to demand the surrender of Japanese and puppet

---

regarding the leadership of the Communist Party or of the proletariat, see *Lun Lien-ho cheng-fu*, pp. 24–6, or Stuart Gelder, op. cit., pp. 25–7, compared with *Selected Works* (Peking), Vol. III, pp. 281–3.

* For an analysis of Soviet motives at the time, in which this line of reasoning is developed, see MacLane, op. cit., pp. 180–82.

† For an excellent summary of these negotiations see Tang Tsou, op. cit., pp. 270–87.

troops. Chiang immediately issued a counter-order, instructing all units of the Communist forces to stay where they were, pending further instructions, and forbidding them to disarm the enemy. This drew from Mao another anonymous article in the Yenan press, violently denouncing Chiang as a 'fascist ringleader, autocrat and traitor' who preferred to cooperate with the enemy rather than with his own countrymen. (On 11 August Chiang had issued a proclamation inciting the puppet commanders to shift their allegiance to Chungking.) If one considered the effective contributions of the various armies to the struggle against Japan, wrote Mao, only the Communist forces had earned the right to accept the surrender of the enemy troops. He ended with an appeal to 'the three Allied Powers' for Chu Te's right to send representatives to the inter-allied negotiations on Japanese surrender.*

Barely two weeks later Mao arrived in Chungking, in response to an invitation from the man whom he had just been calling (anonymously, it is true) a fascist and a traitor. On the eve of his departure he drafted an inner-party circular to explain his policy to the cadres. His starting-point was the balance of power in the country. 'The speedy surrender of the Japanese invaders', he wrote, 'has changed the whole situation. Chiang Kai-shek has monopolized the right to accept the surrender, and for the time being (for a stage) the big cities and important lines of communication will not be in our hands.' The Communist forces should by no means accept the situation passively; on the contrary, they should still fight to take all they could, especially in North China. But there were possibilities other than the military ones. Since the Soviet Union, the United States, and Great Britain all disapproved of civil war in China, it was possible that the Kuomintang might 'conditionally recognize' the status of the Chinese Communist Party, and that the Communists might likewise 'conditionally recognize' the status of the Kuomintang. This would open the way to a period of peaceful cooperation during

---

* On these events see Tang Tsou, op. cit., pp. 303–11. The text of Mao's 12 August editorial published in the *Selected Works* (Peking), Vol. IV, pp. 27–31, is substantially identical with that which appeared at the time (*Chieh-fang Jih-pao*, 13 August 1945).

which the Communists should strive to extend their influence in the cities by legal means. In exchange, they might accept some of the demands that Chiang would undoubtedly make for reductions in the 'Liberated Areas', in order to gain the sympathy of public opinion in China and abroad. The basic principles of their strategy, however, remained the same: 'to wage struggles with good reason, with advantage, and with restraint; to make use of contradictions, win over the many, oppose the few and crush our enemies one by one.'[62]

The face-to-face encounter between the two old enemies in Chungking was their first since 1926. Both Mao and Chiang were aware that the nation would have to judge not only two opposing conceptions of China's future, but the men that incarnated them. A few weeks earlier in Yenan Mao had remarked to a visitor: 'Mr Chiang considers that, in general, there cannot be two suns in the sky, nor can a people have two sovereigns.* But I don't care; I'm determined to give him two suns to look at.'[63] Shortly after his arrival he gave to an acquaintance a copy of his poem 'Snow', which found its way into the newspapers. The effect produced by this poem, one of Mao's best, may have helped to compensate for his relatively awkward appearance, beside the dignified and well-groomed Chiang.

The forty-three days of negotiations, punctuated by banquets, which took place during Mao's stay in Chungking showed that, while neither side wished to incur the opprobrium of having stood in the way of a peaceful solution, there was in fact no ground for agreement. Realizing that for the moment they were not strong enough to insist on participation in the National Government, the Communists dropped their coalition demand and limited themselves to requiring the democratization of the existing government under Chiang Kai-shek. They agreed, as Mao had predicted they would find it advisable to do, to abandon certain of the 'Liberated Areas', and to reduce the size of their army to twenty divisions, if the government would reduce its forces in the same proportions. But they insisted on maintaining a solid nucleus of military power, and a territorial basis for that power – on remaining a state within a state. Mao's reasoning was clearly that, if he could maintain the *status quo* in one form or another, the corrup-

* This is a quotation from *Mencius*, Part I, Ch. 5.

tion and incompetence of the Kuomintang, with the consequent disintegration of its political and military apparatus, would eventually enable the Communists to take over leadership of the country by default, through a suitable combination of political and military means. For his part, Chiang was determined not to entrust himself to time, but to crush the military power of the Communists immediately. He therefore intensified his attacks on the base areas almost before the ink was dry on the provisional agreement of 10 October 1945, which marked the end of Mao's stay in Chungking. It was not so much that he shared Mao's view that events were moving in a direction favourable to the Communists as that he overestimated his own strength at the moment.*

He was encouraged in his assurance not only by growing, though cautious, United States support, especially in transporting his armies northwards so that they could occupy parts of Manchuria ahead of the Communists. He was also persuaded that he enjoyed some degree of Soviet backing or, at the least, benevolent neutrality, in his efforts to unite the country politically. Stalin had said as much during the July negotiations with T. V. Soong in Moscow.[64] That he was not altogether insincere is confirmed by the well-known report from Yugoslav sources that Stalin in 1948 told Eduard Kardelj how after the war he had advised the Chinese comrades to 'join the Chiang Kai-shek government and dissolve their army' because 'the development of the uprising in China had no prospect'.†

In practice, Soviet policy in China appeared hesitant and contradictory. During the first three months following the Japanese surrender the Red Army gave substantial aid to the Chinese Communists in consolidating their position in Manchuria, by allowing Chu Te's forces to move into important areas, and denying access to the Nationalists. In November Moscow reversed this attitude, and allowed Chiang's armies to occupy

* The above account of the negotiations and the analysis of Chiang's policy in this and the following paragraph is largely based on Tang Tsou, op. cit., pp. 316–24.

† Vladimir Dedijer, *Tito Speaks* (Weidenfeld and Nicolson, 1953), p. 331. Numerous scarcely veiled references from the Chinese side in recent years, some of which will be referred to below, tend to confirm that this declaration is authentic. It is repeated, with slightly different words, in Milovan Djilas, *Conversations with Stalin* (Rupert Hart-Davis, 1962), p. 182.

certain places previously held by the Chinese Communists. Then, in March and April 1946, on the eve of the final Soviet withdrawal, support was once more given to the Chinese Communists. No doubt, as Tang Tsou suggests, 'the apparently contradictory and at the same time perplexing behaviour of the Soviet Union towards China represented an adjustment of her basic policy to the prevailing circumstances as viewed in Moscow.' But it is perhaps doubtful whether the 'basic policy' was simply and clearly, as he affirms it to have been, 'hostility towards the Nationalists and support for the Communists'.* The pattern continued to be obscured both by Stalin's preoccupation with the security of the Soviet state, and by his lack of enthusiasm for a dynamic revolutionary movement which he might not be able to control.

It must be admitted that the prevailing circumstances to which Stalin had to adjust his policy were themselves extraordinarily complex and obscure. Since neither Communists nor Nationalists wanted to take public responsibility for a rupture which both of them believed inevitable, frankness and clarity were hardly likely to characterize the policy of either Mao or Chiang. The resulting confusion was compounded by the unrealistic, contradictory, and vacillating character of American policy at this time.† It is not necessary here to try to analyse in detail the complicated negotiations which took place during the year of General George C. Marshall's mission in China, especially as these events have recently been the subject of a precise, well-documented, and objective account.‡ During the first half of 1946 Mao remained more confident than Chiang about the consequences for his own position of a period of political rather than military struggle. This confidence was buttressed by the recommendations, largely favourable to the Communist position, of the Political Consultative Conference, which had been agreed upon during Mao's visit to Chungking in October and which finally met in January 1946. But despite this nuance, both Mao and Chiang continued to talk negotiations and resort in fact to military force. It is there-

* Tang Tsou, op. cit., pp. 338–9. On Soviet policy in Manchuria in 1945–6, see ibid., pp. 327–40, and MacLane, op. cit., pp. 206–17.
† See Tang Tsou, op. cit., pp. 349–400, for an examination of the reasons for this confusion and lack of realism.
‡ ibid., pp. 401–40.

fore not surprising that, by the middle of 1946, the country found itself involved in full-scale civil war.

An important factor in Mao's decision not to shrink from war, even though he did not seek it and would have preferred to advance his cause by political action, was his conviction that American military power in general and American nuclear armaments in particular were not of decisive importance compared with the resolute struggle of the masses. On 9 August 1945 the story in *Liberation Daily*, dealing with the atomic bomb that had been dropped on Hiroshima the day before, was entitled 'A Revolution in the Art of War'. But only four days later a further article stated that, though the atomic bomb was powerful, one should not exaggerate its effects. The decisive factor in the Japanese surrender had been the Soviet entry into the war, and the ground forces of the other allies.[65]

This second viewpoint was directly inspired by Mao's position, as he expounded it on 13 August 1945 at a meeting of cadres in Yenan:

Can atom bombs decide wars? No, they can't. Atom bombs could not make Japan surrender. Without the struggles waged by the people, atom bombs by themselves would be of no avail. If atom bombs could decide the war, then why was it necessary to ask the Soviet Union to send its troops? Why didn't Japan surrender when the two bombs were dropped on her, and why did she surrender as soon as the Soviet Union sent troops? Some of our comrades, too, believe that the atom bomb is all-powerful; that is a big mistake. . . . What influence has made these comrades look upon the atom bomb as something miraculous? Bourgeois influence. . . . The theory that 'weapons decide everything', the purely military viewpoint, a bureaucratic style of work divorced from the masses, individualist thinking and the like – all these are bourgeois influences in our ranks. We must constantly sweep these bourgeois things out of our ranks just as we sweep dust.[66]

This viewpoint was reaffirmed categorically and colourfully in Mao's famous interview of August 1946 with Anna Louise Strong, when he declared that the atom bomb was a 'paper tiger' which looked terrible, but in fact was not.[67] This corresponded, of course, to the official Soviet view of the time, though Stalin's language was not quite so picturesque. But by early 1948, when the Soviet Union had made progress in its own nuclear weapons programme, Stalin proclaimed his admiration of the atomic bomb, declaring that it was 'a powerful thing, pow-er-ful!'* Mao has not significantly changed his views even today, as we shall have occasion to see. A man's personality, as shaped by tradition and experience, is at least as important as objective reality in determining his view of the world.

The course of the civil war, from 1946 to 1949, is largely a matter for military historians, and need not concern us in detail here. On the strictly military level Mao's overall strategy was largely like that which he had recommended for the anti-Japanese war in his writings of 1938; it was to move from a stage in which mobile warfare against a stronger enemy would play the essential role to a stage of positional warfare involving direct assaults on the principal Kuomintang strong points. On Christmas Day 1947 he summarized the operational principles of the Red Army as follows:

1. Attack dispersed, isolated enemy forces first; attack concentrated, strong enemy forces later.

2. Take small and medium cities and extensive rural areas first; take big cities later.

3. Make wiping out the enemy's effective strength our main objective; do not make holding or seizing a city or place our main objective. . . .

4. In every battle, concentrate an absolutely superior force (two, three, four and sometimes even five or six times the enemy's strength), encircle the enemy forces completely, strive to wipe them out thoroughly, and do not let any escape from the net. . . . Strive to avoid battles of attrition in which we lose more than we gain or only break even. . . . In this way, although we are inferior as a whole (in terms of numbers), we are absolutely superior in every . . . specific campaign. . . . As time goes on, we shall become superior as a whole and eventually wipe out all the enemy. . . . [68]

Now, as always in the past, Mao regarded political mobilization and winning the sympathy of the masses as indispensable to the success of the military struggle. To be sure, he still believed, as he had proclaimed in 1938, that 'political power grows out of the barrel of a gun'.[69] He called on the Red Army, in 1947, to

---

* At a banquet with the Yugoslavs in January 1948. Djilas, op. cit., p. 153.

'give full play to our style of fighting – courage in battle, no fear of sacrifice, no fear of fatigue. . . .'[70] But he also issued numerous directives on the policies which should be followed in the 'liberated areas' in order to consolidate the territorial basis of the revolution. The agrarian policy followed was more radical than that of the 1937–45 period, which had involved interest and rent reduction rather than immediate land reform, but the tactics were to be gradual and adapted to local conditions. Mao still intended to include the 'patriotic gentry' in the 'very broad united front' he was determined to maintain.[71] Only after several years of Communist control in a given area would all land be re-distributed; for the moment the reform should not affect more than ten per cent of the population.[72] Mao also caused to be reissued the 'three rules of discipline' and the 'eight points for attention'; these, in one form or another, had expressed for nearly twenty years a respect for the civil population and abstinence from plunder which distinguished the Red Army from all the armies which the Chinese peasantry had seen in the past, and contributed greatly to winning the support of the population.[73]

The Chinese civil war of 1946–9 is unquestionably one of the most striking examples in history of the victory of a smaller but dedicated and well-organized force enjoying popular support over a larger but unpopular force with poor morale and incompetent leadership. During 1945 the military forces under the command of the Eighth Route Army and the New Fourth Army had expanded from a total of about half a million to a total of about one million men. The Kuomintang forces were approximately four times that size.* By the middle of 1947, after a year of large-scale civil war, the proportion had shifted from one in four to one in two, and Mao was able to affirm in a directive that if destruction of the enemy forces continued at the same rate, within one more year the advantage would pass to the Communists.[74] By the end of 1947 Mao proclaimed that an historic turning point had been reached – 'the turning point from growth to extinction for Chiang Kai-shek's twenty-year counter-revolutionary rule.'[75] Six months later the numerical balance had shifted further, to a point where Mao's forces were only slightly inferior to those of Chiang.[76] Given the superior morale and generalship

---

* See the table in J. Ch'ên, op. cit., p. 374.

of their armies, this meant that the Communists were in fact now the stronger, and indeed the last half of 1948 was marked by a series of clear Communist victories which decided the issue of the war.*

The rampant inflation, waste, and corruption which prevailed in Kuomintang China contributed to this result, as did the war-weariness and hostility to Chiang Kai-shek virtually universal among the population. Mao's hand was strengthened by blunders such as the outlawing of the Democratic League in September 1947, which drove a large part of the 'third force' in China (to the extent that it was ever a force at all) into the arms of the Communists.[77] But the war was also lost by the Nationalists on the purely military plane, not least because Chiang, in the appointment of key commanders, continued obstinately to put personal loyalty to himself before competence.

In November 1948 Mao wrote that the war would be much shorter than he had originally estimated, and would probably be over in another year or so.[78] By early December virtually all Manchuria and Northern China had fallen to the Communists, with the exception of an area around Peiping and Tientsin containing something over half a million Nationalist troops. Faithful to his principle that the destruction of the enemy forces was the decisive aim, Mao directed Lin Piao not to attack immediately, but to devote the next two weeks to encircling and cutting off the various Kuomintang formations so that they could not escape southwards by sea.[79] The plan was carried out with dispatch and the five principal strong points of the Kuomintang position were then taken one by one. Tientsin was captured in the middle of January after twenty-nine hours of fighting; Peiping was peacefully surrendered by General Fu Tso-i to avoid useless destruction.† The whole of China was clearly at the mercy of the Communists if they pursued their offensive.

At this point peace moves began to emanate from Nanking. On 1 January 1949 Chiang himself issued a statement, expressing

---

* For a succinct analysis of these battles and the reasons for the Nationalist defeats, see Tang Tsou, op. cit., pp. 482–4.

† General Fu was rewarded by an appointment as Minister of Water Conservancy in the government of the Chinese People's Republic formed in the autumn.

his desire for peace and affirming that he did not care whether he remained in office or retired, and on 21 January he handed over the presidency to the vice-president, Li Tsung-jen. But on the one hand he retained for himself most of the effective levers of control, and used them to further his own policy, which now consisted in abandoning the mainland and establishing himself on Taiwan.* And on the other hand, Mao's terms, understandably enough, were very little short of total surrender, and ultimately were not accepted even by the peace faction within the Kuomintang.

On 20 April the final version of the Communist terms was rejected by the Nanking Government.† The next day Mao issued an order to the People's Liberation Army for a 'country-wide advance'.[80] The same day the Communist forces crossed the Yangtze on a three-hundred-mile front, and on 23 April they captured Nanking. Mao expressed his feelings in a poem written a few days after the event:

> Around Mount Chung‡ a sudden storm has arisen,
> A million courageous warriors cross the great river.
> The crouching tiger and the coiled dragon§ are more majestic
>   than ever in the past,
> The universe is in turmoil, we are all exalted and resolute.
> Let us gather up our courage and pursue the broken foe,
> It is not fit to seek praise by imitating the Tyrant of Ch'u.
> If heaven had feelings, heaven too would grow old,
> The true way that governs the world of men is that of radical
>   change.[81]

Although, as we have seen, Mao was a great admirer of Sun Tzu, he here took issue with the classic master of military strategy, who had written: 'When you surround an army, leave an outlet free. Do not press a desperate foe too hard.'[82] Mao justified a contrary course by suggesting that the Communists should not repeat the error of 'the Tyrant of Ch'u' – a reference to Hsiang

* See Tang Tsou, op. cit., pp. 497–8.

† For the text of the proposed agreement, which had been worked out in two weeks of negotiations in Peiping with representatives of the Kuomintang, see *Selected Works* (Peking), Vol. IV, pp. 390–96.

‡ Mount Chung is situated to the east of Nanking.

§ In speaking of Nanking, Chinese military strategists have traditionally likened it to a crouching tiger overshadowed by a coiled dragon (Mount Chung).

Yü, who spared the life of Liu Pang, the future founder of the Han dynasty, when he had him in his power, only to be ultimately destroyed by the rival he had allowed to live. Mao had no intention of risking his own victory over the man with whom he had contended for nearly a quarter of a century for the mastery of China.

In taking this position he was not differing only with Sun Tzu. An authoritative commentary on this poem, published in 1964, states:

Before and after the campaign for the liberation of Nanking, there were some well-meaning friends, both within and without the country, who said that we should be content with separate régimes in North and South China and should not provoke the intervention of imperialism, especially of American imperialism.*

It is virtually certain that among these 'well-meaning friends' was none other than Stalin himself. Within the country they apparently included some members of the Chinese Communist Party. In a New Year's message entitled 'Carry the Revolution Through to the End', written on 30 December 1948, Mao denounced the efforts of the United States Government to organize 'an opposition faction within the revolutionary camp to strive with might and main to halt the revolution where it is, or if it must advance, to moderate it and prevent it from encroaching too far on the interests of the imperialists and their running dogs.'[83]

In any case, whoever the well-meaning and peace-loving people may have been who, in early 1949, advised Mao to stop the revolution half way, their counsels did not prevail. As his armies continued their triumphant advance Mao Tse-tung began to turn his thoughts once more to political questions – to the preparations for establishing a new government, and to the basic outlines of the policy to be followed by the new régime.

* Kuo Mo-jo, commentary on the above poem published in *Hung Ch'i*, No. 1, 1964. The poem itself was first published at the same time, together with nine more recent ones.

Notes

1. 'Appeal of the Central Soviet Government to the Muslims' (signed by Mao alone), in *Tou-cheng*, No. 105, 12 July 1936, pp. 1–3.
2. For extracts see Schram, *Political Thought of Mao*, pp. 189–90. A complete translation is appended to my article 'Mao Tse-tung and Secret Societies', *China Quarterly*, No. 27, 1966, pp. 11–13.
3. *Selected Works* (Peking), Vol. I, p. 264. Text of the telegram of 5 May in ibid., pp. 279–80.
4. Snow, *Red Star Over China*, pp. 33–6.
5. *Chung-kuo ti hsin hsi-pei*, n.p., May 1937, pp. 74–5.
6. *Mao Tse-tung, kuan-yü t'ing-chan k'ang-Jih chih t'an-hua* (*A Talk by Mao Tse-tung on Ceasing Hostilities and Resisting Japan*) (Yenan, 1936 – mimeographed pamphlet).
7. 'A Statement on Chiang Kai-shek's Statement', 28 December 1936, *Selected Works* (Peking), Vol. I, p. 257.
8. ibid., pp. 281–2.
9. ibid., p. 261.
10. *Shih-lun hsüan chi* (Shanghai, May 1937), pp. 359–72.
11. *Selected Works* (Peking), Vol. I, pp. 270–1.
12. Schram, op. cit., pp. 159–60; see also *Lun hsin chieh-tuan*, Ch. IV, par. 2.
13. *Lun hsin chieh-tuan*, Ch. III, par. 18, and Ch. V, par. 5.
14. Chiang Kai-shek, *Soviet Russia in China: A Summing-Up at Seventy* (London, Harrap, 1957), pp. 87–8.
15. Schram, op. cit., pp. 138, 160.
16. Johnson, *Peasant Nationalism and Communist Power*, p. 99.
17. ibid., pp. 73–4.
18. ibid., passim, especially p. 93.
19. ibid., pp. 74–6, 123–32.
20. *Mu-ch'ien hsing-shih ti fen-hsi* (n.p., Li-lun yü Shih-chien She, 1936), pp. 62–4.
21. Snow, *Red Star Over China*, pp. 107–13.
22. *Basic Tactics*, pp. 56–7.
23. ibid., p. 136.
24. *Selected Works* (Peking), Vol. II, p. 148; *Chieh-fang*, No. 43–4, p. 20.

25. See, in particular, 'On Protracted War', pars. 35–54, in *Selected Works* (Peking), Vol. II, pp. 136–47.
26. Schram, op. cit., pp. 269–70.
27. ibid., pp. 271–4.
28. Cited in MacLane, op. cit., p. 113.
29. Hu Chiao-Mu, *Thirty Years of the Communist Party of China* (Lawrence & Wishart, 1951), p. 53.
30. For an extract showing the variants between the two editions, see Schram, op. cit., pp. 161–4.
31. See the passages translated in Carrère d'Encausse and Schram, *Le Marxisme et l'Asie*, pp. 350–60.
32. Schram, op. cit., p. 161.
33. MacLane, op. cit., pp. 118–23.
34. *Selected Works* (Peking), Vol. II, pp. 431–2.
35. Johnson, op. cit., p. 139.
36. ibid., pp. 132–6.
37. ibid., pp. 138–9.
38. ibid., pp. 144–5.
39. ibid., p. 139.
40. ibid., p. 58.
41. Schram, op. cit., pp. 113–14.
42. Boyd Compton, *Mao's China*, pp. 33–53.
43. Ho Kan-chih, *A History of the Modern Chinese Revolution*, p. 377.
44. See Boyd Compton, op. cit.
45. Schram, op. cit., pp. 289–90.
46. Broadcast talk of 7 November 1941, *Chieh-fang Jih-pao*, 7 November 1941.
47. *Chieh-fang Jih-pao*, 13 December 1941. ('Directive on the Anti-Japanese United Front in the Pacific', dated 9 December.)
48. *Chieh-fang Jih-pao*, 4 July 1944.
49. ibid., 15 August 1944.
50. Herbert Feis, *The China Tangle* (Princeton, Princeton University Press, 1953), p. 162.
51. Tang Tsou, *America's Failure in China*, pp. 172–5.
52. ibid., pp. 176–95.
53. Feis, op. cit., p. 214.
54. *Chieh-fang Jih-pao*, 14 April 1945.
55. *Chieh-fang Jih-pao*, 25 June 1945.
56. Schram, op. cit., pp. 276–9.
57. Liu Shao-ch'i, *On the Party* (Peking, Foreign Languages Press, 1950), p. 157.
58. ibid., p. 35.
59. *Chieh-fang Jih-pao*, 15 April 1942.

60. Liu Shao-ch'i, op. cit., p. 33.
61. Schram, op. cit., pp. 256, 259.
62. *Selected Works* (Peking), Vol. IV, pp. 47–51.
63. Tso Shun-sheng, *Interesting Events in the Past Thirty Years*, p. 90.
64. Tang Tsou, op. cit., p. 275.
65. *Chieh-fang Jih-pao*, 9 and 13 August 1945.
66. *Selected Works* (Peking), Vol. IV, pp. 21–2.
67. ibid., pp. 97–101.
68. *Selected Works* (Peking), Vol. IV, pp. 161–2; see also Schram, op. cit., pp. 210–11.
69. Schram, op. cit., pp. 209–10.
70. *Selected Works* (Peking), Vol. IV, p. 161.
71. ibid., p. 136.
72. ibid., pp. 193–6, 201–2.
73. ibid., pp. 155–6.
74. ibid., p. 144.
75. ibid., p. 157.
76. J. Ch'ên, *Mao and the Chinese Revolution*, p. 374.
77. Tang Tsou, op. cit., p. 461.
78. *Selected Works* (Peking), Vol. IV, p. 288.
79. ibid., p. 292.
80. ibid., pp. 387–9.
81. My translation, as published in *Problems of Communism*, September–October 1964, p. 39.
82. Sun Tzu, *The Art of War*, Chap. VII.
83. *Selected Works* (Peking), Vol. IV, p. 301.

Chapter 9
The Foundation of the Chinese People's Republic

In 1948, at Mao's suggestion, one of the May Day slogans proclaimed by the Chinese Communist Party had been: 'All democratic parties, people's organizations and public personages should quickly call a Political Consultative Conference to discuss and carry out the convening of a people's congress and the formation of a democratic coalition government.'[1] A preparatory committee for organizing just such a conference was established in June 1949 by the Communist leaders. (At this stage it was named the 'New Political Consultative Conference', to suggest continuity with the Political Consultative Conference of January 1946, which had finally failed because of Chiang Kai-shek's hostility to its recommendations for a coalition government.) In his address to the preparatory committee Mao announced that it was necessary to 'convene a Political Consultative Conference, proclaim the founding of the People's Republic of China, and elect a democratic coalition government to represent it.' And he added, repeating the definition of the future China which he had been using constantly since 1945: 'Only thus can our great motherland free herself from a semi-colonial and semi-feudal fate and take the road of independence, freedom, peace, unity, wealth and power (*fu-ch'iang*).'*

In the same speech Mao declared:

We are willing to discuss with any foreign government the establishment of diplomatic relations on the basis of the principles of equality, mutual benefit, and mutual respect for territorial integrity and sovereignty, provided it is willing to sever relations with the Chinese reactionaries, stops conspiring with them or helping them, and adopts an attitude of genuine, and not hypocritical, friendship towards People's China. The Chinese people wish to have friendly cooperation with the

---

* *Selected Works* (Peking), Vol. IV, pp. 405–6. Mao had proclaimed as his goal 'independence, freedom, peace, unity, wealth and power' in his report 'On Coalition Government' to the Seventh Party Congress in April 1945. He frequently employed the phrase in writings at that time. For an example see Schram, *Political Thought of Mao*, pp. 276–9.

people of all countries and to resume and expand international trade in order to develop production and promote economic prosperity.[2]

This did not mean, however, that he proposed to adopt a neutral attitude between the 'imperialists' (obviously the American imperialists), who were supporting 'their running dogs, the Chinese reactionaries', on the one hand, and the Soviet Union on the other. Even in the unlikely event that the United States were to 'stop conspiring' with the Chinese reactionaries, Mao clearly did not believe in the possibility of any real cooperation with an 'imperialist' country. This he stated sharply in his celebrated essay 'On the People's Democratic Dictatorship', published on 30 June 1949, in commemoration of the twenty-eighth anniversary of the Chinese Communist Party:

'You are leaning to one side.' Exactly. . . . All Chinese without exception must lean either to the side of imperialism or to the side of socialism. Sitting on the fence will not do, nor is there a third road.[3]

The essay 'On the People's Democratic Dictatorship' is one of the most important and seminal of all Mao's writings. In it Mao extracted the quintessence of such earlier and much more voluminous productions as 'On New Democracy' and 'On Coalition Government' and presented it in concise and pungent form. At the same time the total power which he already possessed, and with which he would soon be formally invested, made it possible for him to speak with far greater frankness than he had done during the years 1937–45. Then Mao had felt himself obliged, in courting (or appearing to court) the collaboration of the Kuomintang, to recognize the latter's leading role for an indefinite period – even if it was perfectly clear to anyone choosing to read between the lines that he was bent on gathering into his own hands the real sources of power. Now he was in a position to abandon all such compromises with expediency, and clearly to affirm both the reality of his own power, and the manner in which he proposed to use it in order to re-shape Chinese society.

The pattern of the future projected in this article was both flexible and rigid. It allowed for the participation of a very broad sector of the Chinese population in the political and economic life of the country. The petty bourgeoisie and 'national' bourgeoisie were not merely to be allied with the workers and peasants in a united front; they were to be included indefinitely among the 'people'. And as part of the people, they would enjoy rights denied to the 'reactionaries' – the landlords and the 'bureaucrat bourgeoisie' (the segment of the bourgeoisie linked to the Kuomintang). They would even participate in the 'people's democratic dictatorship', directed against the reactionaries. 'Democracy', wrote Mao, 'is practised within the ranks of the people, who enjoy the rights of freedom of speech, assembly, association, and so on. The right to vote belongs only to the people, not to the reactionaries. The combination of these two aspects, democracy for the people and dictatorship over the reactionaries, is the people's democratic dictatorship.' But at the same time Mao made it quite clear that these privileges of full citizenship were extended to the bourgeoisie only on condition of good behaviour. He called for the immediate strengthening of the 'people's state apparatus', which he defined frankly as 'an instrument for the oppression of antagonistic classes'. In the immediate future this instrument was to be used for the transformation and elimination of the landlord class and the 'bureaucrat-bourgeoisie'. As for the national bourgeoisie, Mao wrote:

When the time comes to realize socialism, that is, to nationalize private enterprise, we shall carry the work of educating and remoulding them a step further. The people have a powerful state apparatus in their hands – there is no need to fear rebellion by the national bourgeoisie.[4]

On 21 September 1949 the Chinese People's Political Consultative Conference (as it was finally named) began its first session in Peking.* The session lasted until 30 September, and gave its approval to the organization and composition of the new régime, to be presided over by Mao Tse-tung. Mao himself opened the first meeting with a speech of vibrant national pride, affirming that the Chinese had 'stood up', and would 'never again be an insulted nation'.† On 30 September Chu Te closed the proceedings with a speech ending 'Long Live Chairman Mao!'[5] And

---

* The name of the city was officially changed from Peiping ('Northern Plain') to Peking ('Northern Capital') on 27 September, when it was designated as the capital in place of Nanking ('Southern Capital').

† For extracts see Schram, op. cit., pp. 109–10.

on 1 October Mao stood on the T'ien An Men and formally proclaimed the establishment of the Chinese People's Republic.

Thus the 'coalition government' about which Mao had been talking and writing for the past five years was established. But it was a very different kind of coalition government from that which he had first envisaged in 1944. To be sure, the Political Consultative Conference included representatives of fourteen 'parties' and a variety of other groups and categories, as well as independent 'democratic personalities', and there were non-Communists both in the government and among the deputy heads of state. (The six vice-chairmen serving under Mao included Sung Ching-ling, the widow of Sun Yat-sen, and one representative each from the two most important minor parties, the 'Revolutionary Committee of the Chinese Kuomintang' and the 'Chinese Democratic League'.) But no one had any illusions over where the real power lay. The 'Common Programme' adopted by the Political Consultative Conference as a basis for the united front and the coalition government incorporated all the essential principles laid down by Mao in his article 'On the People's Democratic Dictatorship', including the definition of the 'people' as the four-class bloc and the foreign policy of 'leaning to one side'.[6] But the presence of non-Communists in the government could not reasonably be regarded as altogether meaningless. It was a concrete symbol of the broad support that the new régime enjoyed at the moment of its establishment, thanks to the widespread disgust with the Kuomintang and the widespread longing for a stable and efficient government at any price. It also corresponded, in Mao's own eyes, to the relatively moderate nature of the policies that he intended to pursue in the immediate future.

Two great tasks faced Mao and his colleagues: to consolidate their political control over the country; and, beginning with the reconstruction of the war-damaged economy, to lay the basis for future economic development. The complex human being who undertook to guide China towards political stability and economic progress was in many respects extremely well qualified for his task. He had behind him three decades of political experience. The years of guerrilla warfare against Japan, and of civil war against the Kuomintang, had enabled him further to develop both his gifts of leadership and the streak of ruthlessness noted by Edgar Snow in 1936. At the same time, these years of struggle had taught him the importance of mass support, and the means for mobilizing it. Having himself issued from the Chinese peasantry, he understood the needs and the intellectual world of the masses. Tso Shun-sheng has written: 'Mao Tse-tung is fundamentally a character from a Chinese novel or a [Peking] opera.'* This is not the whole truth, but it is an important aspect of the truth, and it undoubtedly added to his appeal for a large segment of the population. But as Mao set about his task in 1949, there was not one China, but several, and those aspects of his personality which were attractive to one part of the Chinese people often repelled another. Mao's leadership style, strongly marked by the conventions of popular literature, brought him closer to the peasantry, but might easily appear unsophisticated to the intellectuals and the bourgeoisie. Conversely, his aim of westernizing and modernizing China (though in Marxist-Leninist terms) met with the approval of a large part of the urban population, but was vaguely disturbing to the peasants.

Only one trait in Mao's character met with a universal response among his compatriots: his inflexible resolve to defend China's dignity and national interests. He would display this quality in full measure when he went to Moscow in December 1949 to seek aid and support for launching the new régime. Mao arrived in the Soviet capital on 16 December, just in time for the celebration of Stalin's seventieth birthday on the twenty-second.

The circumstances of this first encounter between Mao and Stalin were not wholly favourable. Stalin had contributed nothing to Mao's rise to control of the Chinese Communist Party, and though, once Mao had triumphed, Stalin had accepted the fact – outwardly at least – he had shown no great zeal in aiding his Chinese comrades to seize power from the Kuomintang. One of the reasons for his reticence had no doubt been fear of possible large-scale American intervention in the event of a more aggressive policy. And this fear had now been revealed as groundless.

* Tso Shun-sheng, *Interesting Events in the Past Thirty Years*, p. 86. Tso's observation was inspired in part by the enthusiasm with which Mao had followed a Peking opera based on an episode from *Water Margin* during the visit of the delegation of the Democratic League to Yenan in July 1945.

But the second, and possibly even more important, reason – distaste for the emergence of another Communist great power, which might not prove as tractable as his European satellites – remained. Indeed, the attitude of the Chinese in recent years had been well calculated to increase it.

We have already noted the significance of Liu Shao-ch'i's report at the Seventh Congress of the Chinese Communist Party in 1945, hailing Mao's thought as a guide for the peoples of other oriental countries. The following year, in a resounding interview with Anna Louise Strong, Liu had propounded similar ideas even more sharply:

> Mao Tse-tung's great accomplishment has been to change Marxism from a European to an Asiatic form. . . . China is a semi-feudal, semi-colonial country in which vast numbers of people live at the edge of starvation, tilling small bits of soil. . . . In attempting the transition to a more industrialized economy, China faces . . . the pressures . . . of advanced industrial lands. . . . There are similar conditions in other lands of southeast Asia. The courses chosen by China will influence them all.*

At the World Federation of Trade Unions meeting in Peking in November 1949 Liu Shao-ch'i had declared: 'The way which has been followed by the Chinese people . . . is the way which should be followed by the peoples of many colonial and semi-colonial countries in their struggle for national independence and people's democracy.'[7] But, far from accepting the originality of the Chinese 'way', Soviet spokesmen continued to emphasize that Mao and his comrades owed everything to the example and the support of the Russians.†

For his part Stalin had continued to show by his actions how little he was inclined to grant wholehearted support to Mao and his comrades. In January 1949, during the provisional pre-

* Anna Louise Strong, 'The Thought of Mao Tse-tung', *Amerasia*, XI, No. 6 (June 1947), p. 161. The effect of this kind of statement on Stalin can be measured from the fact that he personally intervened to prevent the publication in the Soviet Union of Miss Strong's book on Mao, already in print in some of the East European countries, because it emphasized Mao's original contribution.

† For examples of the contrasting views in Moscow and Peking and a discussion of the ideological issues involved, see Carrère d'Encausse and Schram, *Le Marxisme et l'Asie*, pp. 93–100, 361–97.

sidency of Li Tsung-jen, the Soviet Embassy in China had elaborated, in concert with Li, a draft agreement between the two countries, providing for Chinese neutrality in any future conflict and the elimination of American influence in exchange for 'real cooperation between China [i.e. Kuomintang China] and Russia'.* In April, furthermore, the Soviet Ambassador had been the only foreign diplomatic representative to accompany the Nationalist Government to Canton, instead of remaining in Nanking when the Communist forces entered it. The Soviets had even continued negotiating throughout May with the Nationalist régime (now once more headed by Chiang Kai-shek) over commercial rights in Sinkiang.[8]

Given this background of rivalry, both ideological and political, the negotiations between Mao and Stalin inevitably proved to be long and laborious. In an interview with Tass on 2 January 1950 Mao expressed the belief that his stay in Moscow would last 'a few weeks'. The exact length, he added, would be partly determined by 'the time necessary to resolve all the questions relating to the interests of the Chinese People's Republic'.[9] In fact, he stayed in the Soviet Union for more than nine weeks, and in the end was able to obtain only partial satisfaction for the interests of his country.

The agreements signed on St Valentine's Day, 14 February 1950, included a Treaty of Friendship, Alliance, and Mutual Assistance which assured China of Soviet support against attack by Japan or any other state cooperating directly or indirectly with Japan. A separate agreement provided for Soviet credits to China of sixty million U.S. dollars annually for five years – a very small sum in comparison either with U.S. aid to the Nationalists, or with Soviet credits to the Eastern European satellites. But in exchange for these limited gains, Mao was compelled to accept continued Soviet presence in Port Arthur and Dairen until 1952. (In fact Stalin subsequently failed to honour this deadline, and the two ports were returned to China only after his death.) He also recognized the independence of the Mongolian People's

* *United States Relations with China*, p. 293. This agreement came to nothing because the United States Government found it 'incredible' that Li Tsung-jen should ask for American support and at the same time negotiate with Moscow on the elimination of American influence.

Republic, and by implication its inclusion in the Soviet sphere of influence. Although it was perfectly clear that the Mongolians wanted no part of either Chinese or Soviet suzerainty, this was a bitter pill to swallow for a man who had been obsessed since earliest boyhood with the disintegration of the Chinese empire, and who had always defined that empire in the broadest possible terms. In 1936 he had affirmed his belief that whenever the revolution was victorious in China, Outer Mongolia would of its own accord join the Chinese federation,[10] and in 1939 he had defined the frontiers of China in such a way as to include both Outer and Inner Mongolia.* There is no reason to believe that he had subsequently modified his views – but in this, as in many other respects, he was obliged to compromise with reality.

On 17 February, as he was about to leave Moscow, Mao issued a statement declaring that Sino-Soviet friendship, as fixed in the new treaties, was 'eternal and indestructible'. He also asserted his conviction, following visits to factories and collective farms, that Soviet economic and cultural achievements would 'in the future serve as models for construction in New China'.[11]

In this he was no doubt at least partly sincere. Mao's inflexible resolve to maintain control of his own revolution did not encompass a desire to be original at all costs, or a reluctance to learn from Soviet experience. On the contrary, just as he had long since assimilated many basic principles of Leninism on the

* In 'The Chinese Revolution and the Chinese Communist Party' he wrote: 'The present boundaries of China are contiguous in the northeast, the northwest, and in part in the west to the Union of Soviet Socialist Republics.' There follows an enumeration of the countries contiguous on the west, south, and east. In the current edition an additional sentence has been inserted immediately after the one just cited: 'The northern frontier is contiguous to the People's Republic of Mongolia.' *Selected Works* (Peking), Vol. II, p. 305. There is no mention at all either of Mongolia or of a northern frontier in the original version as published in 1939 by the official Chiehfang She in Yenan. If this was an 'omission', it had still not been rectified either in an edition published in January 1949 at Peiping by the Hsin-hua Agency, or in one published in June 1949 at Hongkong. At the very least, the 1939 version leaves the issue conspicuously open. (There is no other gap in Mao's meticulous country-by-country enumeration of all the bordering lands.) But it seems much more likely that the reference to the frontier in 'the northeast and the northwest' was meant to designate the whole semicircular sweep of the boundary with the Soviet Union, Mongolia being considered as part of the Chinese side.

organization of the Communist Party and ways of conducting the struggle for power, Mao now intended to take full advantage of the methods developed by Lenin's successors for building a state machine and developing a socialist economy. But as he had mixed Leninist tactics with his own guerrilla methods in order to achieve power, so now he intended to adapt Soviet techniques for building socialism to the peculiar conditions obtaining in China.

Central among these conditions were the economic and cultural backwardness of the country, and the overwhelming role of the peasantry. After affirming that the People's Democratic Dictatorship had no fears about its capacity to 'remould' the national bourgeoisie, Mao had declared, in his famous essay: 'The serious problem is the education of the peasantry.' This was indeed the problem which would dominate the whole history of the new régime. In Mao's own words, the country-wide victory already basically won was 'only the first step in a long march of ten thousand *li*',[12] the march toward economic development and modernization. In this task the peasantry, which had carried almost the whole burden of armed struggle for a quarter of a century, would still be the 'principal ally of the working class'. But it could play its full role politically only if it also played its full role economically, and this, for Mao, meant 'the socialization of agriculture . . . coordinated with the development of a powerful industry having state enterprise as its backbone'. To say that this implied the 'education' of the peasantry was an understatement. What was needed – as Mao well knew – was nothing less than a cultural and intellectual revolution in the countryside, involving both the assimilation of modern scientific techniques and knowledge, and a change in the individualist outlook which has always and everywhere characterized land-hungry peasants. Such an upheaval in traditional modes of thought and behaviour would be welcomed only by some of the Chinese peasants, and actively opposed by others. Moreover, having fought and suffered for so long, even the politically-conscious peasants in the People's Liberation Army wanted to rest and enjoy the fruits of their victory. As early as February 1948 Mao had denounced the view, propagated by news agencies and radio stations under Communist control, that 'the poor peasants, having conquered the country, should now rule the country and take their ease in it'.*

To persuade the peasants that, instead of taking their ease, they should submit to an arduous process of 'remoulding', while at the same time conserving enough of their support to maintain the stability of the régime, was to be an enormous task. That Mao and his colleagues have made some headway cannot properly be doubted; recent events show how far they still are from having achieved their objectives. In early 1950, however, as Mao settled down after his Moscow visit to fix the basic policies of the new government, there seemed no immediate difficulty in sight from this direction. The first step in the re-shaping of social and economic relations across the countryside was to be the completion of land reform, in progress throughout the 'liberated areas' since the start of the civil war. This was, of course, entirely in keeping with Leninist doctrine, which had always provided for individual land-ownership as a stage on the road to collectivization. But Mao applied this policy in a characteristic fashion, stressing the 'educative' function of expropriating the landlords.

In his 'Hunan Report' of 1927 Mao had written:

When the local bullies and evil gentry were at the height of their power, they killed peasants without batting an eyelid. . . . In view of these atrocities by the local bullies and evil gentry, as well as the white terror let loose by them in the rural areas, how can one say that the peasants should not now rise and shoot one or two of them, and bring about a small-scale reign of terror in suppressing the counter-revolutionaries?†

As land reform swept along in the wake of the Red Army's advances, both before and after the establishment of the new

* Mao's very succinct Chinese text merely says 'poor peasants conquer country, sit on country' (ta chiang-shan, tso chiang-shan). The 'sit' suggests both the idea of 'rule', which is the translation given in the Selected Works (Peking), Vol. IV, p. 197, and that of resting. A contemporary text of this directive in Hsin-wen cheng-ts'e shou-ts'e, a mimeographed manual for inner-party use distributed by the South China Propaganda Department of the Chinese Communist Party, is substantially identical with the current version. It contains, however, a complaint that the Central Committee's directives were frequently not obeyed.

† Schram, op. cit., p. 186. Finding this statement a bit too moderate, Mao in the 1951 edition changed 'one or two' to 'a few'.

régime, Mao urged the peasants to rise and kill not merely one or two, but a goodly number of landlords. At the same time he continued to warn cadres, as he had done since 1948, against 'indiscriminate killing'.[13] Though he did not shun violence, he regarded it primarily as a measure for altering the balance of forces and psychological relationships in the countryside. Apart from the fact that many landlords had arbitrarily beaten and killed their tenants in the past, and deserved the supreme punishment in Mao's eyes, it was only when they had denounced the crimes of their former exploiters in public meetings and then put them to death that the peasants would come to grasp that something had changed, and that they were now the masters.

The number of victims in this first revolutionary upheaval of the countryside was relatively limited; one of the most critical sources refers to 'tens of thousands',[14] which would correspond to no more than half of one per cent of the total number of landlords. The real wave of terror was still in the future. Politically and psychologically, this was nonetheless a very violent and traumatic experience for all those concerned in it. Economically, on the other hand, the policy pursued was exceedingly moderate. Contrary to what had been done during the closing stages of the civil war, only the land of landlords, and not that of rich peasants, was to be confiscated.* In a report to the Third Plenum of the Chinese Communist Party on 6 June 1950 Mao explained this change in policy on both political and economic grounds. Politically, the rich peasants no longer represented a danger, as they had done 'when the PLA was locked in a life-and-death struggle with the Kuomintang reactionaries'. Economically, it was desirable to follow a policy of 'preserving a rich peasant economy, in order to further the early restoration of production in the rural areas'.† On the surface there was a certain resem-

---

* Certain rich peasants, called 'semi-landlords', were to be deprived of that part of their land which they rented out instead of cultivating themselves with the aid of hired labour. See Article 6 of the Agrarian Reform Law. The dispossessed landlords were to be given allotments of the same size as the poor peasants.

† New China's Economic Achievements (Peking, 1952), p. 6. Liu Shaoch'i amplified the point in a speech of 14 June. The Agrarian Reform Law of the People's Republic of China (Peking, 1952), pp. 84–8.

blance between this policy and the encouragement of the *kulaks* to 'enrich themselves' during the NEP period in the Soviet Union. But the Chinese moved directly to this moderate line without passing through a preliminary and radical phase of 'war Communism', and they avoided, too, the extremes in the encouragement of private enterprise that characterized the NEP. While the rich peasants were allowed to retain their land, they were to be rigorously prevented from acquiring any more. In addition they were to be politically isolated by being forbidden to join the new peasant associations, which were limited to poor and middle peasants.

Almost as important as the Agrarian Reform Law in the plan for the transformation of existing social relationships was the Marriage Law, which Mao promulgated on 30 April 1950.[15] The principles and aims of the law were basically similar to those which Mao had defined nearly twenty years earlier in Kiangsi, when he had decreed a set of provisional marriage regulations. One of its aims was the establishment of equal rights for women, a cause which had been close to Mao's heart ever since his student days. 'Under feudal domination', he had written in 1931, 'the oppression and suffering borne by women is far greater than that of man.' At the same time, the purpose of the marriage reform (both in 1931 and in 1950) was to make 'free choice the basic principle of every marriage', and so strike a blow at the 'whole feudal system of marriage' and the family system built upon it, especially the power of parents over their children.*

Although Mao's constant attitude since 1919 amply demonstrates that he regards free choice in marriage as desirable in it-

---

* The above quotations are all from Mao's decree of 28 January 1931, putting into effect the Provisional Marriage Regulations of the Kiangsi Soviet Republic (Schram, op. cit., pp. 228–9). But the principles and aims of the two reforms were identical. For the text of the 1950 law, see *Hun-yin-fa wen-t'i chieh-ta hui-pien* (*A Compilation of Replies to Questions on the Marriage Law*) (Wen-hua Kung-ying She, 1951), pp. 1–8. The 'basic principles' defined in the very first sentences of the law are very similar to those stated by Mao in 1931: 'Abolish the feudal system of marriage characterized by the use of constraint, respect for men and contempt for women, and indifference to the interests of sons and daughters. Put into practice a system of marriage characterized by freedom to marry for men and women, one wife to one husband, equal rights for men and women, and the protection of the legal rights of wives, sons, and daughters.'

self, this reform also had the obvious political function of weakening the hold of the family on the individual and thus facilitating the incorporation of the citizens into new forms of social organization, directed towards the progressive transformation of traditional China. In this domain, as in the sphere of agrarian policy, Mao and his comrades were setting out to liberate the individualist forces latent in a pre-capitalist society, and use them to destroy the basis of that society, before proceeding to the further stage of collectivization.

Progress in both domains was slow, with the resistance particularly sharp in the sphere of family relationships. In the summer and autumn of 1951, a year and a half after Mao had promulgated the Marriage Law, the highest party authorities repeatedly denounced the persistence, among members of the Communist Party and even the cadres, of the 'feudal attitude that attaches importance to men and slights women'.[16] As for land reform, Liu Shao-ch'i had announced in his report of June 1950 that it would be completed – except for areas inhabited by national minorities – only towards the end of 1951. Moreover, he added that 'a rich peasant economy' would be preserved 'throughout the whole stage of New Democracy', being superseded 'only when the conditions are mature for the extensive application of mechanized farming, for the organization of collective farms and for the socialist reform of the rural areas'. This, he declared, 'will take a somewhat long time to achieve'.[17]

In fact, the whole process of transforming Chinese society was very soon to be markedly accelerated. This change was no doubt due in part to Mao's own impatient temperament, and to the natural tendency of any revolutionary movement, once launched, to gather momentum. But a very important influence on the internal policies of the Chinese People's Republic was unquestionably exercised by the Korean War, which broke out just as the Agrarian Reform Law was about to be put into effect.

The respective roles of Moscow, Peking, and Pyongyang in the decision to attempt the conquest of South Korea remain obscure and controversial. Every possible theory has been defended at one time or another in the course of the past fifteen years, from those attributing the whole responsibility to Stalin, to those blaming it all on Chinese bellicosity. Although the question cannot be adequately dealt with here, the weight of the evidence seems to support the conclusion that the war was basically a Soviet initiative. Mao may or may not have acquiesced in it (some evidence indicates that he was not even informed until a day or two before the attack); but if he did so, it can only have been on the supposition (undoubtedly shared by Stalin as well) that the United States would not intervene, and the North Koreans would be able to finish the job themselves. (Both these hypotheses appeared altogether reasonable; Secretary of State Acheson had clearly affirmed, on 12 January 1950, that Korea was outside the U.S. defence perimeter, and the first days of fighting left little doubt over what would have been the outcome if there had been no foreign intervention on either side.) It seems, in any event, virtually inconceivable that Mao should deliberately have courted a conflict with the United States. The civil war was still not over; isolated pockets of resistance remained to be wiped out in various remote provinces; control had not yet been established in Tibet; and, above all, Mao was resolved to defeat the remnants of Chiang's forces on Taiwan. Troops for this purpose were poised in the adjoining provinces, and there is clear evidence that the invasion was planned for the summer.*

Not only did the Chinese Communist régime have other military tasks on its hands, but for six months Mao and his colleagues had been urgently preoccupied by the necessity to demobilize part of the vast armies accumulated during the civil war, in order to reduce the drain on the public finances, release more workers for productive activity, and thus combat inflation. Only about one third of the rural population had as yet passed through the experience of land reform. Another hundred million peasants were to undergo it in the latter part of 1950, thus absorbing the energies of most available cadres. The whole emphasis at the June sessions of the party and in the state organs had been on economic problems. Mao's opening and closing speeches to the National Committee of the Political Consultative Conference had contained altogether only a single sentence on foreign affairs, affirm-

---

* Allen S. Whiting, *China Crosses the Yalu: The Decision to Enter the Korean War* (New York, Macmillan, 1960), pp. 21-2. Whiting's book remains the best study of the subject as a whole.

ing the importance of solidarity with the Soviet Union. In his closing speech on 23 June 1950, two days before the outbreak of the Korean War, Mao explained that in the course of the new democratic revolution the Chinese people would be tested by two great 'trials', war and land reform. 'The trial of war', he declared, 'already belongs basically to the past.' The second trial of land reform had now to be undergone, after which the way would be open for development to socialism.[18] Thus Mao deliberately assured his compatriots, two days before the attack on South Korea, that the trial of war was over. Moreover, even after the outbreak of the Korean War, there was no serious effort to mobilize opinion until several months had elapsed. The press devoted relatively little space to the conflict, and anti-American articles were hardly more numerous or more virulent than usual. A change became discernible only in late August, and did not reach its height until after Chinese troops were already fighting in Korea.*

In view of all of this, it is most unlikely that Mao Tse-tung deliberately engaged his country before 25 June in a policy that he believed might involve it in war. There is, indeed, a hint that he did not anticipate a direct confrontation with the United States in his indignant reaction to President Truman's order of 27 June 'neutralizing' the Taiwan Strait.

This decision, which represented (like the decision to intervene in Korea) a reversal of previous United States policy,† had very far-reaching implications. Henceforth Washington would be prepared to intervene directly if necessary to prevent action by the Chinese Communists to conquer Taiwan and liquidate Chiang Kai-shek's rival government. As a result, a question which had hitherto been considered by both Mao and Chiang as purely Chinese could no longer be resolved without a direct clash between China and the United States.

In the face of this new development, Mao declared angrily that

* This affirmation is based on my own examination of the *Jen-min Jih-pao* for the period in question. See also Whiting, *China Crosses the Yalu*, especially pp. 82–4, 139.

† See Tang Tsou, *America's Failure in China*, pp. 558–64, who correctly underscores the consequences of Truman's decision, and particularly the fact that it 'doomed to failure Acheson's attempt to turn Chinese nationalism against the Soviet Union'.

the American President had proved his own previous statements about not intervening in Taiwan to be 'fraudulent', and the United States had thus 'openly exposed its own imperialist face'. Repeating his familiar theme that imperialism was 'outwardly strong but inwardly rotten', he nonetheless called on the people of China and of the world to be on the alert to 'defeat any provocation by U.S. imperialism'.[19]

Mao's statement likewise reflected the fact that, for the moment, he was much more interested in Taiwan than in Korea. While he certainly regarded the unification of Korea under Communist control as good in itself, it is doubtful that he would have been inclined to promote the aims of the 'fraternal party' in Pyongyang at the risk of drawing American attention to the Far East, precisely at a moment when Washington appeared to be in the process of disengaging itself from its Chinese commitment. (The distinctly unfraternal attitude of the Korean Communists towards their Chinese comrades in the recent past would hardly have encouraged such altruism.*)

More broadly, the probable aims of the Korean adventure (to disrupt American influence in Japan by proof of American weakness, and to alter the global balance of power) corresponded to Soviet rather than Chinese interests at the time. To be sure, Mao was as interested as anyone in fighting 'American imperialism', but he could well have waited to do so until the internal situation was consolidated and he had settled accounts with Chiang Kaishek. There is more than a hint that the Chinese felt they were sacrificing themselves for Soviet aims in the bitter reflections on the Korean War published in recent years:

The leaders of the C.P.S.U. . . . . accuse us of hoping for a 'head-on clash' between the Soviet Union and the United States and trying to push them into a nuclear war. Our answer is: No, friends. You had better cut out your sensation-mongering calumny. The Chinese Communist Party is firmly opposed to a 'head-on clash' between the Soviet Union and the United States, and not in words only. In deeds too it has worked hard to avert direct armed conflict between them. Examples of this are the Korean war against U.S. aggression, in which we fought side by side with Korean comrades. . . . We ourselves preferred to

* On relations between the Korean and Chinese Communists prior to 1950, see Whiting, *China Crosses the Yalu*, pp. 42–5.

shoulder the heavy sacrifices necessary* and stood in the first line of defence of the socialist camp so that the Soviet Union might stay in the second line.[20]

Whether, as I have argued here, Mao played little or no part in planning the attack on South Korea, or whether he associated himself wholeheartedly with Stalin's decision on the false assumption that the war would remain a purely Korean affair, the vigour and effectiveness of the American military reaction soon obliged both Moscow and Peking to face difficult policy choices. During August and September the two Communist great powers endeavoured to secure a diplomatic settlement on a compromise basis. The Soviet delegate to the Security Council abandoned the boycott he had been conducting for seven months against the presence of a Nationalist Chinese delegate,† and began to press for negotiations with the participation of the Chinese People's Republic, as well as of North and South Korea. But by this time Washington was confident of total victory, and refused to settle for anything less than the unification of all Korea under United Nations auspices. Even the increasingly sharp and explicit warnings that China would intervene if her borders were threatened did not suffice to temper this optimism, or halt MacArthur's march to the Yalu.‡

So Mao found himself, only a year after the establishment of the Chinese People's Republic, involved in war with the United States. This new commitment inevitably created both economic

---

* Among these was a personal sacrifice on Mao's part. His son Mao An-ying was killed in combat. An-ying, Mao's son by his first wife Yang K'ai-hui, had studied in Moscow during the Second World War. (For a portrait of the young Mao at the time of his return to China, see the article in *Kung-jen Jih-pao*, 18 July 1961, translated in *Survey of the China Mainland Press*, No. 2575, 11 September 1961.) By an arrangement which seemed logical enough at the time, but appears curiously ironic, in view of Mao's violent attacks on 'revisionism' in recent years, he was the room-mate of Luigi Longo's son.

† It was his absence, it will be recalled, which had made it possible to obtain Security Council approval for President Truman's decision to inter-vene in Korea, and to conduct the intervention under the auspices of the United Nations, with the participation of other countries.

‡ On the evolution of Chinese policy from August to November, the failure of attempts at compromise, and the steps leading to Peking's inter-vention, see Whiting, *China Crosses the Yalu*, pp. 68–172.

and political stresses. We have seen that Mao's own attitude towards America, though basically hostile, was not without ambiguity. The sentiment in China as a whole was still less unequivocally anti-American. Even after the two countries were actually at war, periodicals addressed to cadres of the Chinese Communist Party still found it necessary to publish articles explaining patiently why America should be considered an enemy and not a friend.[21]

Mao himself, in a telegram of 2 December 1950 to representatives of industrial and commercial circles in Tientsin, praising them for their contribution to the 'Resist America and Aid Korea' movement launched at the end of October, implicitly recognized the persistence of pro-American feeling. 'The American imperialists', he declared, 'have addressed a great deal of deceptive propaganda to the Chinese people. All patriots should refuse to believe in these deceptive discourses.'[22]

The recognition by Mao and his colleagues that a part of the population regarded with mixed feelings the conflict on which they had embarked undoubtedly further strengthened the natural tendency towards the imposition of a harsher discipline and the repression of dissent that manifests itself in all nations at war. The result was a sharp increase in the amount of terror and coercion employed by the régime in its effort to establish its own power and root out or disarm all opposition, actual or virtual.

This process was carried out in stages, marked by a succession of mass campaigns: the campaign against counter-revolutionaries; the thought reform movement; the 'Three-Antis' campaign; and the 'Five-Antis' campaign. The first was launched on the basis of the very severe 'Regulations regarding the punishment of Counter-Revolutionaries', promulgated by Mao on 21 February 1951, which provided death or long prison sentences for very broadly defined offences.[23] By March a succession of mass rallies had begun in the major cities, at which important 'counter-revolutionaries' were sentenced to death after a public denunciation of their crimes. According to an article of the Minister of Public Security published at the time, nearly thirty thousand meetings were held in Peking alone within a few months, attended by a total of more than three million people. In large areas of

Central and Southern China a majority of the population had been subjected to such 'education'.[24] Long lists of the names of executed 'counter-revolutionaries' appeared day after day in the newspapers.

As for the number of victims, it was officially stated in October 1951 that 800,000 cases of counter-revolutionaries had been dealt with during the first six months of 1951 by the people's courts alone. Chou En-lai later declared that 16·8 per cent of the counter-revolutionaries tried had been sentenced to death, mostly prior to 1952. The combination of these two pieces of information would give a figure of 135,000 executions during the first half of 1951. The actual number was undoubtedly much higher. Hostile estimates have ranged as high as ten or fifteen million victims. A reasonable estimate would appear to be from one to three million executions all told.*

If we take the middle figure of two million victims, this amounts to about 0·3 per cent of the total population of China. (It thus corresponds to about 150,000 executions in a country the size of France or Britain, or 600,000 in the United States.) This is not an enormously large toll for a social revolution of this magnitude, carried out in the wake of a long and cruel civil war which had taken even more victims on both sides. Moreover, there is no doubt that among the 'counter-revolutionaries' thus repressed were included many individuals who had in fact engaged in organized or clandestine activities against the régime – though others perished merely because of their class origins. But at the same time, the picture is very far from that of a revolution carried out exclusively by education and gentle persuasion, which is sometimes propagated nowadays by some of the more

* The figure of 800,000 cases is given by Peter Tang, op. cit., p. 240; Chou En-lai's speech of 26 June 1957 can be found in *Communist China 1955–1959: Policy Documents with Analysis.* With a Foreword by Robert R. Bowie and John K. Fairbank (Cambridge, Mass., Harvard University Press, 1962), p. 303. The estimate of from one to three million victims is given by Jacques Guillermaz, *La Chine populaire*, third edition (Paris, Presses Universitaires de France, 1964), p. 47. General Guillermaz, currently French Military Attaché in Peking, spent nearly ten years in China in the same capacity between 1937 and 1949. His evaluation carries considerable weight as that of an honest and impartial observer with access to a great deal of information.

naïve among the defenders of Peking (though hardly by the Chinese themselves). Throughout most of 1951 a real climate of terror reigned throughout the country.

In October 1951 Mao Tse-tung announced in a speech before the third meeting of the National Committee of the People's Consultative Political Conference that the three great campaigns which had taken place during the past year – agrarian reform, 'Resist America and Aid Korea', and the suppression of counter-revolutionaries – had all achieved victory. Land reform would be completed by the end of 1952, except in areas inhabited by national minorities. The American invasion of the People's Democratic Republic of Korea had been thown back. As for the 'counter-revolutionary remnants', they would 'very soon be basically eliminated on the mainland'.[25]

The campaign against counter-revolutionaries was substantially different from those which followed it, because it was aimed essentially at eliminating a segment of the population. Only some of the 'reactionary elements' were to be physically destroyed, but the others were to be re-shaped by the harsh methods of labour camps, and they were to play no important role in the new China Mao was intent on building. The party cadres, the objects of the 'Three-Antis' campaign, and the 'national bourgeois', the objects of the 'Five-Antis' campaign, were, in contrast, regarded as full participants in the new society, though a minority of undesirable elements was to be eliminated from both groups. They were therefore to be re-shaped more by persuasion and less by constraint.

The contrast in methods appears even more clearly in examining another campaign, which had begun immediately after liberation but which only reached its peak in 1951 – the 'thought reform' movement among the intellectuals. A large-scale effort in this sphere had been advocated by Mao in his closing speech to the Political Consultative Conference on the eve of the Korean War; now, in October 1951, he expressed his satisfaction that it was already in progress. In advocating this campaign Mao had emphasized that thought reform by criticism and self-criticism was one of the 'democratic methods' for 'educating and convincing' those who belonged to the people, fundamentally different from the treatment meted out to the enemy classes, who would be 'forced to transform themselves through labour in order to become new people'.[26] It was indeed a less violent method than the open terror applied against counter-revolutionaries; but it was 'democratic' only in the sense of broad participation, and not in the sense that those involved in it possessed any significant degree of control over their destinies.

The thought reform campaign was of considerable practical importance for the functioning of the cadres on which the revolutionary régime was to rely in its effort to change the face of China. But it was also important for revealing Mao's style of leadership in general; it displayed in relatively pure form a goal and a technique for attaining it which have belonged, in greater or lesser degree, to every act of the Peking régime since its establishment. It was a natural outgrowth of the 'rectification campaigns' in 1942–4, and manifested Mao's unquenchable desire to harmonize the two conflicting imperatives of 'conscious action' by individuals and impeccable social discipline. To one standing outside the system the two imperatives appear contradictory; it is possible to compromise between them, but not to satisfy both completely. And yet Mao will accept nothing less. Some observers have concluded that he cannot be serious about this; he must want simply robots or 'ants' and nothing more. To hold this view, however, is not only to oversimplify Mao's own ideas, but to lose sight of a whole dimension of the drama of the Chinese intellectuals in the twentieth century.

They had rejected the past, with its respect for tradition and constituted authority; but they had found no real solace or satisfaction in a purely individualist protest. The old monarchical system had been overthrown; but parliamentary democracy had not emerged as an effective alternative. Psychologically, they aspired to certainty, and politically, to stability and efficiency. Their identity had already been challenged historically, by the collapse of the old Chinese universe and the humiliating incursions of the West. These facts determined in large measure their reaction to the shattering experience of identity-destruction and identity-reconstruction which is the essence of thought reform as practised by the Chinese Communists.

Robert Jay Lifton, in his admirably restrained and thoughtful study of this problem, has suggested that an important aspect of thought reform was the replacement of the filial piety which constituted the central moral value of traditional China by a new form of filial piety directed towards the régime.* In his view such a reorientation of the personality had strong attractions even for those intellectuals politically unsympathetic to the régime, since it filled the void created by the violent rejection of paternal authority almost universal in the modernizing China of the preceding half century.

Dr Lifton does not speculate on the possible relationship between the transfer of filial piety from the father to the Communist régime and the growing cult of Mao Tse-tung as an omnipotent, omniscient, and solicitous father figure; and it will be as well not to do so here. There can be no doubt, on the other hand, that Mao, as one who had personally known the void created by the rejection both of paternal authority and of a father-centred tradition, understood intuitively the nostalgia for certainty and authority which characterized the Chinese intellectuals. But there were severe limitations on this understanding. Never having been really exposed to modern scientific and technical knowledge, he could not grasp that certain questioning attitudes were more than a manifestation of the 'liberalism' he condemned;† they were an integral part of modern culture. Being utterly convinced that his own thought was correct, he supposed that everyone else would be better off thinking exactly as he did. In his effort to conciliate spontaneity and discipline he was therefore inclined to place the emphasis heavily on conformity.

Logically enough, the thought reform campaign was preceded and accompanied, like the rectification campaign of 1942-4, by a movement for the study of 'Mao Tse-tung's Thought'. For some time, a special 'Committee for the Publication of the Selected

* Robert Jay Lifton, *Thought Reform and the Psychology of Totalism: A Study of 'Brainwashing' in China* (New York, Norton, 1961), Ch. 19, especially pp. 371, 379. This volume, based on extensive interviews in Hongkong with both Chinese and Westerners who had undergone 'thought reform', is the most interesting and suggestive study of the subject. See also E. H. Schein, *Coercive Persuasion* (New York, Norton, 1961).

† See 'Combat Liberalism', *Selected Works* (Peking), Vol. II, pp. 31-3. As Lifton points out very justly, Mao's definition of 'liberalism' in fact embraces a great many principles for safeguarding the feelings of others and facilitating social relationships which are drawn from the ethics of traditional China, and have nothing to do with liberalism in the usual sense. Lifton, op. cit., p. 383.

Works of Mao Tse-tung', attached to the Central Committee of the Chinese Communist Party, had been hard at work preparing a new canon of the leader's works, which still remains the basis of all editions or translations published in China. As we have seen repeatedly in the course of this biography the texts were all thoroughly rewritten by Mao himself, so as to remove both youthful errors and any points of fact or doctrine which did not suit the current orthodoxy. The first volume of this bowdlerized edition was published in October 1951, but, beginning on the thirtieth anniversary of the Chinese Communist Party, 1 July 1951, the principal texts were published one after another in the press. At the same time a series of articles by leading ideologists vaunted the significance of Mao Tse-tung's thought.*

The consequence of thought reform promoted in the spirit described was tragedy for some; happiness, emotional and intellectual security for others; and for still others (probably the majority) a mixture of acceptance and inner revolt.† In human terms the undertaking thus appears as a highly ambiguous and generally repressive one, even though the students, professors, and other intellectuals who experienced it were not normally subjected to the violence inflicted on the Westerners who were 'reformed' in prison.‡ On the other hand, for the functioning of the régime itself, the result was undeniably positive.

In his speech of 23 October 1951 Mao declared:

Thought reform, and especially thought reform of all categories of intellectuals, is one of the important conditions for the thoroughgoing

* See in particular Lu Ting-i, 'The World Significance of the Chinese Revolution', *Jen-min Jih-pao*, 23 June 1951; Ch'en Po-ta, 'Mao Tse-tung's Thought is the Synthesis of Marxism-Leninism and the Chinese Revolution', ibid., 13 July 1951.

† As Lifton points out, the Chinese whom he had the opportunity to interview were all, by the nature of things, the 'failures' of the system, since they had chosen to leave the country. (The Westerners had been forcibly expelled.) But from the degree of acceptance shown for some of the values inculcated, even by the most hostile among these, it is possible to imagine the spectrum of attitudes that must characterize those who remained behind.

‡ All of the latter without exception were shackled for varying periods with heavy and very painful irons. Though this phase of mental and physical suffering had a definite function in the technique of thought reform, it is easy to see it also as a symbolic revenge for the humiliation inflicted on China by the West.

democratic transformation and the progressive industrialization of our country.[27]

Indeed, given the immensity of the tasks confronting them and the limited number of experienced Communist cadres available, Mao and his comrades could not afford to waste trained man-power, except when they were confronted with irreconcilable political opposition. The thought reform movement, which enabled them to remould sufficiently students of diversified class origin and political background, former Kuomintang officials, and the like, so that they could be employed by the new régime at least for a time, was therefore in fact, as Mao declared, a decisive contribution to the economic and social transformation of the country.

The 'Three-Antis' campaign, into which the thought reform movement blended at the end of 1951, and the 'Five-Antis' campaign, during the spring of 1952, put a greater stress on social utility and a lesser stress on inner transformation; but they none-theless called upon the techniques used in thought reform, as well as on the mass denunciations utilized against the counter-revolutionaries, though these were conducted in a less hostile spirit. The first campaign, which affected party and state cadres, was directed against corruption, waste, and bureaucracy. It re-sulted in the purging of some people for one or other of these three offences, and in inciting the remainder of those affected to rid themselves energetically of 'bourgeois thinking'. The 'Five-Antis' campaign, directed against the 'five poisons' of bribery, tax-evasion, fraud, theft of government property, and theft of state economic secrets, affected primarily the merchants and in-dustrialists of the 'national bourgeoisie' who were still operating their firms in a semi-autonomous manner. It was linked to the preceding campaign because bribery by businessmen meant cor-ruption of civil servants. The national bourgeoisie being an inter-mediate class (part of the people, but attached to its own peculiar interests nonetheless), the methods and aims of this campaign were intermediate also. Its purpose was not, as in the case of agrarian reform, to eliminate a class altogether, reducing its members to the status of manual workers. Peasants did not re-quire landlords to work the land; but the skills of the factory owners and businessmen were still required to direct their enterprises. Consequently the aim of the campaign was in considerable part to remould their thinking, and to prepare them for their future role as salaried managers. But at the same time it was directed at destroying what still remained to them of an independent power position rooted in the resources of their enterprises. They were therefore fined large sums and then saved from bankruptcy only by loans which made them financially dependent on the government. There was also a greater use of mass denunciation and constraint than in the case of the cadres affected by the 'Three-Antis' movement. The pressure was largely psychological – the threat of professional ruin drove a large number of businessmen to suicide – though the menace of prison hovered in the background for the worst offenders. Thought reform techniques were employed in requiring businessmen to 'confess their sins' before their employees.*

Once more, Mao was in the forefront of these campaigns. In his New Year's message for 1952, after calling for victory by the Chinese people on the 'Resist America and Aid Korea Front', on the front of national defence, land reform, repression of counter-revolutionaries, economic construction, thought reform, etc., he drew special attention to the 'newly-opened front' of the 'Three-Antis' movement:

> I want to call upon all the people and all the political workers of our country to unite, and, with great waving of banners and beating of drums, to develop a mighty and irresistible large-scale struggle against corruption, waste, and the bureaucratic spirit.[28]

At the same time that Mao made this highly characteristic utterance, with its accent on struggle and dramatic effects, the Chinese economy appeared to be moving towards planned growth along Soviet lines. The 'Mutual Aid Teams' organized in the course of 1952 represented a loose form of cooperation, with no exact parallel in Russian experience; but these were presented as merely a transitional stage to collective farms of the Soviet type.† At the end of 1952 it was announced that the first five-year

* For a vivid description of the 'Five-Antis' movement, see A. Doak Barnett, *Communist China: The Early Years, 1949–1955* (Pall Mall Press, 1964), pp. 135–71.

† On the early phase of collectivization, see Barnett, op. cit., pp. 172–88.

plan would start in 1953, and in February 1953 Mao called on his compatriots to study Soviet experience and learn from it. Stalin's death a month later drew from Mao one of his most eloquent and effusive tributes to the man who had stood at the head of the international Communist movement for nearly as long as Mao had been active in it.* For the moment Mao Tse-tung appeared to the world as Stalin's faithful disciple, and Communist China as a copy of the Soviet Union. But events were soon to show that the reality was much more complex, and that Mao's warlike temperament was to be at least as important as Soviet planning techniques in shaping the methods by which he undertook to lead his country towards the goal of 'wealth and power'.

* Regarding the degree of sincerity which can be attributed to this article, see Chapter 10.

Notes

1. *Selected Works* (Peking), Vol. IV, p. 276.
2. ibid., p. 408.
3. ibid., p. 415.
4. ibid., pp. 417–19.
5. *Chung-hua Jen-min Kung-ho-kuo k'ai-kuo wen-hsien* (*Documents on the Establishment of the Chinese People's Republic*) (Hongkong, Hsin-min-chu Ch'u-pan-she, 1949), p. 58.
6. See the text of the 'Common Programme' itself, and of Chou En-lai's report on it, in *Chung-hua Jen-min Kung-ho-kuo k'ai-kuo wen-hsien*, pp. 253–74.
7. *Hsin-hua Yüeh-pao*, Vol. I, No. 2, pp. 440–41.
8. Tang Tsou, *America's Failure in China*, p. 502.
9. Text in *Hsin-hua Yüeh-pao*, Vol. I, No. 3, p. 579.
10. Schram, *Political Thought of Mao*, p. 287.
11. *Chung-Su Yu-hao ho-tso ti hsin shih-tai* (Tientsin, Sino-Soviet Friendship Association, February 1950), pp. 4–5.
12. 'On the People's Democratic Dictatorship', *Selected Works* (Peking), Vol. IV, p. 422.
13. See, for example, Mao's order of 15 February 1948, *Selected Works* (Peking), Vol. IV, pp. 201–2.
14. Peter Tang, *Communist China Today* (New York, Praeger, 1957), p. 267.
15. For the text of Mao's decree putting the Marriage Law into effect, see *Jen-min Jih-pao*, 1 May 1950. As Chairman of the Chinese People's Republic, he also promulgated the land reform law on 30 June. *Hsin-hua Yüeh-pao*, 2 (4), 1950, p. 967.
16. See *Kan-pu hsüeh-hsi tzu-liao*, No. 40, passim, and especially pp. 11, 25.
17. *The Agrarian Reform Law of the People's Republic of China*, p. 88.
18. *Kan-pu hsüeh-hsi tzu-liao*, No. 11, July 1950, pp. 7–12.
19. *Imperialism and all Reactionaries are Paper Tigers* (Peking, Foreign Languages Press, 1958), p. 31. (Translation slightly modified.)
20. 'Two Different Lines on the Question of War and Peace', *Peking Review*, No. 47, 1963, pp. 12–13.
21. See, for example, *Kan-pu hsüeh-hsi tzu-liao*, No. 21, pp. 11–26.
22. *Shih-shih shou-ts'e*, No. 5, 20 December 1950, p. 3.
23. Text in *Jen-min Jih-pao*, 22 February 1951.

24. Lo Jui-ch'ing, article in *Jen-min Jih-pao*, 11 October 1951.
25. *Jen-min Jih-pao*, 24 October 1951.
26. *Kan-pu hsüeh-hsi tzu-liao*, No. 11, p. 12.
27. *Jen-min Jih-pao*, 24 October 1951.
28. *Shih-shih shou-ts'e*, No. 1, 1952, p. 4.

Chapter 10
In Search of a Chinese Way

If one were to summarize the evolution of Chinese policy since 1949, and more particularly since 1958, one might describe it as the progressive exaltation of the human will over the rational analysis of the facts. But this definition, though suggestive, oversimplifies an extremely complex problem. One can get somewhat nearer to a balanced view by taking into account three factors: the means employed to re-shape the mentality of the population; the ideas of the leadership regarding their capacity to abolish the natural limits on human actions; and the concrete methods and objectives of economic and social policy. In these terms, one might say that the methods of 'thought reform' – described in the previous chapter – have been constantly utilized in the sixteen years since the establishment of the Chinese People's Republic, on a scale and with an intensity which make the process an essential attribute of Mao's China. The spirit and the effectiveness of these methods tend to give the leaders an exaggerated idea of their power over men and materials, thus encouraging the further development of that voluntarism which has been so distinguishing a trait of Mao's thought from the beginning. But this extreme voluntarism collides with the methods of economic development employed by the régime, which have, on the whole, remained rational (except for the interlude of the 'Great Leap Forward' in 1958–60); and it is also mitigated by the realism and common sense which have continued to coexist in Mao's own mind and personality with his warlike enthusiasm – at least until very recently. The overall result is a zigzag evolution in which the emphasis is periodically shifted from the difficulties hindering the industrialization of a backward country to the extraordinary force inherent in all newly liberated peoples, and especially in the Chinese people, which makes them capable of transforming the world at will.

Down to 1955, the economic policy of the Peking Government was characterized on the whole by considerable moderation. In the cities, beginning with the 'Five-Antis' campaign in 1952, the businessmen and industrialists of the 'national bourgeoisie' were progressively deprived of effective control over their own enterprises. But, at the same time, they retained all or part of their property rights, and continued to draw a share of profits.

After three years of preparation, the first five-year plan was launched at the start of 1953, though the details of the plan were finally spelled out only in 1955. This first step in the transition to a socialist economy received formal consecration in the Constitution of the Chinese People's Republic, adopted in September 1954, which announced in its preamble that the necessary conditions had now been created for 'planned economic construction and gradual transition to socialism'. But the pace of transformation envisaged remained moderate. The new constitution stipulated that the state remained a 'people's democratic dictatorship'. And in his opening address to the First Session of the National People's Congress, summoned to consider the constitution, Mao stated that it would require 'several five-year plans' to turn China into a highly industrialized country.[1]

Needless to say, the establishment of the new governmental structure did not change the realities of power. Mao Tse-tung was Chairman of the Chinese People's Republic under the new constitution, as he had been under the provisional order adopted in 1949, and though the 'united front' and its expression, the Chinese People's Political Consultative Conference, were maintained in existence side by side with the new National People's Congress, effective control of the state apparatus remained vested in the Chinese Communist Party.

Though Mao discharged his responsibilities as Chief of State conscientiously, receiving, for example, the credentials of nearly all ambassadors himself,* he often did so in a somewhat whimsical and off-hand manner. On one occasion, receiving for the first time a particularly lanky Western diplomat, Mao greeted him abruptly with the exclamation, '*Ne-mo kao a!*' which might be freely rendered: 'My God. As tall as that!'

---

* From the establishment of the People's Republic in October 1949, until his retirement at the end of 1958, Mao personally granted nearly seventy formal audiences to diplomats of the appropriate rank for the handing over of their credentials.

On the whole, the planning techniques and priorities employed during the first five-year plan were based on the Soviet model. Heavy industry was emphasized even more than in the first Soviet plan of 1929–34, but this choice was explained by the extreme imbalance of the pre-1949 Chinese economy, in which virtually the whole modern sector was composed of light industry in or near the great coastal cities.[2]

In the countryside, the beginning of the five-year plan in 1953 was marked by the passage from 'mutual-aid teams' to 'semi-socialist' agricultural producers' cooperatives. The new form was called semi-socialist because the income of the cooperative was distributed among its members partly on the basis of the amount of land and capital which they had contributed at the time of its formation, and not merely according to their labour. This programme was carried out at first with considerable restraint. As late as July 1955 only some 15 per cent of the peasant families in China belonged to such cooperatives. Then suddenly, in his speech of 31 July 1955, Mao called for a sharp increase in the rate of collectivization. As a result of this appeal (which was published only in October), the process of collectivization accelerated so rapidly that even Mao's optimistic prophecies were outstripped.

Simultaneously with this speed-up in agricultural collectivization, virtually all the industrial and commercial enterprises remaining in private hands were nationalized. To be sure, many members of the 'national bourgeoisie' were retained as salaried managers. But although theoretically a place was still reserved for them in the 'four-class dictatorship', this formula was henceforth void of any real content.

The step thus taken was further underscored by a campaign against the writer Hu Feng, who was presented as the ideological representative of the bourgeoisie. During the winter of 1955–6, there was also a call for a 'leap forward' in industry, which clearly prefigured the 'Great Leap' of two years later.[3]

But it is above all the wave of collectivization in the countryside which marked the great turning-point during the winter of 1955–6. The concrete reality of this evolution and the state of mind which accompanied it were strikingly conveyed in a volume entitled *Socialist Upsurge in China's Countryside*, published early

in 1956.* This was composed of reports from the national and
regional press on various aspects of collectivization; its great
interest lies in the fact that Mao himself contributed not only
a general introduction, but also comments on much of the
material included.

There was, to be sure, a rational justification for the campaign
in favour of collectivization, which had been set forth by Mao in
his speech of 31 July 1955. In order to meet the needs of industrialization, it was essential to increase agricultural production, and Mao was convinced that the cooperatives provided
the best framework for introducing the mechanization and other
techniques required to achieve this end.[4] But in his eyes, the
changes in mentality and atmosphere which accompanied the
collectivization process were far more important:

> In China, 1955 was the year of decision in the struggle between
> socialism and capitalism. . . . The first half of 1955 was murky and ob-
> scured by dark clouds. But in the second half, the atmosphere changed
> completely. Tens of millions of peasant households swung into action.
> In response to the call of the Central Committee, they adopted co-
> operative methods. As this is being written more than 60 million
> peasant households in various parts of the country have already joined
> cooperatives. It is as if a raging tidal wave has swept away all the
> demons and ghosts. . . . By the end of this year the victory of socialism
> will be practically assured. Of course, many more battles lie ahead.
> We must continue to fight hard.[5]

This combative temperament, this imagination peopled with
heroes and demons, was accompanied as usual by an unshakeable
faith in the 'conscious action' of human beings. As he had been
doing ever since the 'land verification movement' of 1933, Mao
stressed that the transformation of the Chinese countryside was
above all a political and ideological process:

> Political work is the lifeline of economic work. This is par-
> ticularly true at a time when the economic and social system is under-
> going a fundamental change. The agricultural cooperative movement,
> from the very beginning, has been a severe ideological and political

* The original publication was in three volumes, but a one-volume ab-
ridgement appeared shortly afterwards. An English translation, with the
above title, was published by the Foreign Languages Press in Peking in 1957.

struggle. No cooperative can be established without going through such a struggle. Before a brand-new social system can be built on the site of the old, the site must first be swept clean. Old ideas reflecting the old system unavoidably remain in people's heads for a long time. They do not easily give way.[6]

As in 1933, another aspect of Mao's policy was to curtail the influence of the representatives of the formerly more affluent strata among the peasantry who had contrived to regain something of their former social prestige or economic advantages in a disguised form. The articles included in *Socialist Upsurge in China's Countryside*, and Mao's comments on them, denounce both the infiltration of 'counter-revolutionaries' and 'rascals' into the management of the cooperatives, and the passive or negative attitude toward collectivization of the former 'upper middle peasants'. Mao also insisted on the importance of going forward rapidly from semi-socialist to fully socialist cooperatives, in which the share of each member would be determined exclusively by his labour.* He stated that three years was a reasonable period for accomplishing this transition. In fact, it was largely completed during the first half of 1956, and within three years from the time Mao edited *Socialist Upsurge in China's Countryside* the Chinese peasants were to find themselves plunged into the adventure of the communes.

During the latter half of 1956 Mao's impatience was once more muted, and a more prudent line adopted, for a whole series of reasons: the resistance of the peasants to hasty collectivization, a poor harvest, industrial disorganization caused by the 'leap forward', and the anxiety inspired first by Khrushchev's secret speech at the Soviet Twentieth Party Congress, and then by the Hungarian Uprising. But Mao's optimistic temperament soon found a new outlet. Convinced that on the whole the people of China accepted the basic principles of the régime, Mao caused the slogan 'Let a Hundred Flowers Bloom' to be proclaimed in the

---

* For Mao's views on class differentiation and the role of the various strata in the cooperatives, see especially *Socialist Upsurge in China's Countryside*, pp. 235–9. (His formula was 'to give control to the poor peasants, and go on from there to strengthen unity with the middle peasants'.) On the issues referred to above, see also ibid., pp. 336, 357, 367–9, 460–61, 477–8, 498.

middle of 1956, and then intervened personally in February 1957 to carry this tendency further forward with his famous speech, 'On the Correct Handling of Contradictions Among the People'. While this policy grew in part out of Chinese conditions, it was also an attempt to apply the lessons learned from Soviet and Hungarian experience. Before analysing it, we must therefore re-trace the course of Chinese foreign relations, and more particularly of relations with Moscow, since the death of Stalin.

In foreign policy, as in domestic policy, the years 1953–5 corresponded to a period of moderation on the part of the Peking Government. Like the policy of the Soviet Union, but independently and in an original fashion, Chinese policy evolved towards a line of 'peaceful coexistence'. Nehru's initiatives towards putting an end to the Korean War had convinced both Stalin and Mao that a positive role in world affairs could be played by countries which were not socialist, but desirous of remaining neutral in the Soviet-American conflict. As early as 1951 India and China exchanged cultural delegations, and in April 1954 the two countries concluded a treaty incorporating the famous 'five principles' of peaceful coexistence, or *Panch Sheel*. Two of these reproduced textually the principles put forward by Mao Tse-tung in his speech of June 1949 as a basis for the foreign relations of the new régime.[7]

This tendency to cultivate good relations with the 'bourgeois' leaders of the countries bordering on China, instead of calling for their overthrow (as Mao had done in 1949 precisely in the case of Nehru),[8] found its culmination in the Bandung Conference of 1955. Here Chou En-lai, who had already made a reputation as an urbane and skilful diplomat at the Geneva Conference of 1954 on Indochina, appeared side by side with Nehru as one of the two principal representatives of the non-European world, divided by ideology, but united by the fact that they were Asians.

Like the Soviet leaders, Mao Tse-tung and his comrades are perpetually obliged to choose between a policy primarily diplomatic, involving support for the 'bourgeois' governments of Asia and Africa provided that they are non-aligned, and a policy primarily revolutionary, aiming at the overthrow of these same bourgeois allies. For its own part, the Soviet Union embarked definitively, in 1955–6, on a policy which made the 'national bourgeoisie' the bearer of progress in the underdeveloped countries for a long time to come. And during the Bandung era, China displayed a tendency to make similar concessions. Thus, in March 1957, Chou En-lai recognized that the peoples of many Asian and African countries had 'embarked on the road of independence and development' under 'the leadership of the nationalists'.[9] But in fact such a policy contradicted very profound tendencies growing out of the experience of Mao and his comrades. It was therefore normal that, when Mao's internal policy entered a more radical phase during the second half of 1957, this militant attitude should find its expression in foreign affairs as well.

But if the turn to the left in Mao's foreign and domestic policy which took place during the winter of 1957–8 is different from all the other zigzags before or since, it is primarily because these years were also characterized by a decisive change in Peking's relations with the Soviet Union. This period constitutes a major watershed in the history of the contemporary world. Before it, China's economic and other policies appeared basically similar to those of the Soviet Union, and the monolithic unity of the Communist bloc was taken for granted by most observers, despite the Yugoslav precedent. After it, China was embarked on a series of policies radically different from those of the Soviets both in style and content, and an evolution was in progress that would soon lead to an open clash between Europe-centred and Asia-centred forms of Communism.

On the eve of Stalin's death, despite the causes of tension accumulated over the years, relations between Moscow and Peking were superficially close and harmonious. Early in February 1953, at the Fourth Meeting of the National Committee of the People's Political Consultative Conference, Mao attached special importance in his closing remarks to the importance of studying Soviet experience.[10] Stalin's death a month later drew from him an article entitled 'The Greatest Friendship', which hailed his late comrade as 'the greatest genius of the present age'. In the light both of what we know about Stalin's success in dealing with the Chinese revolution, and of what Chinese spokesmen have since said about his blunders, this passage in Mao's parting eulogy must have been inspired either by deliberate irony, or by deference to the new Soviet leadership. No doubt shock caused by the death of a man who, whatever his failings, had been the central figure in Mao's political universe throughout nearly the whole of his adult life also played a role, but this is hardly sufficient to explain statements such as these:

Everyone knows that Comrade Stalin had an ardent love for the Chinese people and believed the might of the Chinese revolution to be immeasurable. [As we have seen he had such confidence in it that he was still negotiating with the Nationalists in early 1949.] To the prob- lems of the Chinese revolution he contributed his sublime wisdom [!!]. And it was by following the theories of Lenin and Stalin and with the support of the great Soviet Union and all the revolutionary forces of all other countries that the Chinese Communist Party and the Chinese people a few years ago won their historic victory.[11]

On 1 April Molotov announced Soviet approval of the initiative taken by Chou En-lai on the previous day over the exchange of sick and wounded prisoners, an initiative which ultimately opened the way to the ending of the Korean War. In September 1953 the Soviet leaders concluded a new agreement on economic aid to China providing for the construction of ninety-one plants in addition to the fifty stipulated in the 1950 agreements. This drew from Mao Tse-tung a telegram thanking the Soviet Government and people for their 'great, many-sided, long-term and unselfish aid'.[12] And in September and October 1954 Khrushchev visited Peking and made further gestures, including increased economic aid, the return of Port Arthur and Dairen to China by the follow- ing May, and the abandonment of the Soviet share in the joint- stock companies formed in 1950.

Thus in concrete fashion the new Soviet leaders showed their desire to improve relations with China and to treat her on a basis of equality. The obscure episode of Kao Kang's fall was probably also a manifestation of greater independence from the Soviet Union on Mao's part. As we saw earlier, Kao was one of the founders of the Soviet movement in North-Western China. At the time of the decisive civil war with the Kuomintang, Mao sent him to Manchuria with Lin Piao's army. In 1949 Kao became the chief leader of both the party and state apparatus in the North- Eastern Region (one of the six regions into which the country was divided prior to the adoption of the 1954 constitution). Late in 1952, he was made the head of the newly created State Planning Committee. Towards the beginning of 1954, however, he dropped from view, and it was subsequently claimed that he had committed suicide in February 1954, after being criticized for his 'anti-party activities'. There is reason to believe that he was intimately linked with the Soviets,* and it is doubtful that Mao could have removed him during Stalin's lifetime without arousing stronger objections than were apparently voiced by the new Soviet leadership.

The starting-point in the disintegration of these new close ties between Moscow and Peking was unquestionably Khrushchev's secret speech at the Twentieth Party Congress denouncing Stalin's crimes. This does not mean that the Chinese objected as violently to Khrushchev's action as they have recently claimed. On the contrary, the Chinese press at the time indicated strong approval of certain aspects of 'destalinization'. Nevertheless, the Twentieth Party Congress marked the beginning of an evolution in Soviet policy, the ultimate consequences of which were to prove unacceptable to Mao, even if he did not clearly foresee them at the time. Above all, the manner in which 'destalinization' was initiated could not fail to irritate and alarm the Chinese leaders. The reasons for their disapproval were essentially those which they now claim to have expounded to their Soviet comrades at the time: 'total lack of an overall analysis' of Stalin; 'lack of self-criticism'; and 'failure to consult with the fraternal parties in advance'.[13]

The last of these three complaints is the most obvious, though not necessarily the least important. Mao unquestionably regarded himself after Stalin's death as one of the very few dominant figures in the world Communist movement; it is not improbable that he regarded himself as the leading figure, indeed. That Khrushchev should have ventured on a policy which might undermine the very basis on which both of them stood without even consulting him was certainly felt as an unpardonable slight.

The reproach of 'lack of self-criticism' is also easily understandable. While at the time observations on this score, if they

* This is confirmed even by a Kuomintang account of the affair, although such an admission tends to weaken Taipei's case that Mao himself is a Soviet puppet. See Cheng Hsieuh-chai, *An Interpretation of the Purge of Kao and Jao in the Chinese Communist Party* (Asian People's Anti-Communist League, Republic of China, 1955), pp. 12–13.

were actually made, were undoubtedly couched in courteous terms, Mao must already have felt something of the contempt for Stalin's henchman, now turning so bitterly against his dead master, which was to be expressed later in language of biting scorn:

In what position does Khrushchev, who participated in the leadership of the Party and the state during Stalin's period, place himself when he beats his breast, pounds the table and shouts abuse of Stalin at the top of his voice? In the position of an accomplice to a 'murderer' or a 'bandit'? Or in the same position as a 'fool' or an 'idiot'?[14]

The objection to the 'total lack of an overall analysis of Stalin' combined two ideas. On the one hand, it meant that the evaluation of Stalin should not be unilateral. Mao is now reported to have told Mikoyan as early as April 1956, two months after Khrushchev's speech, that 'Stalin's merits outweighed his faults'.[15] Whether or not he was quite so categorical at the time, this is entirely in harmony with the viewpoint expressed in the editorial of 5 April 1956 – 'On the Historical Experience of the Dictatorship of the Proletariat' – which was probably written and certainly inspired by Mao:

Some people consider that Stalin was wrong in everything. This is a grave misconception. Stalin was a great Marxist-Leninist, yet at the same time a Marxist-Leninist who committed several gross errors without realizing that they were errors. We should view Stalin from an historical standpoint, make a proper and all-round analysis to see where he was right and where he was wrong, and draw useful lessons therefrom.[16]

But the 'overall analysis' demanded by Mao was not merely a balanced analysis; it was an analysis which made some attempt to relate the negative aspects of Stalin's rule to their historical context, and did not merely dismiss them as the crimes of one man. Thus the passage just cited from the April 1956 editorial of *People's Daily* continues:

Both the things he did right and the things he did wrong were phenomena of the international Communist movement and bore the imprint of the times. Taken as a whole, the international Communist movement is only a little over a hundred years old, and it is only thirty-nine years since the victory of the October Revolution. Experience in many fields of revolutionary work is still inadequate. Great achievements have been made, but there are still shortcomings and mistakes.

In the course of recent Sino-Soviet polemics, Moscow has suggested that, if Mao objected to the campaign against the 'personality cult', it was because he saw in it an implied attack on the cult of his own personality. That there exists in China a cult of Mao and of his thought is not subject to question. We have seen how this cult has developed since the 'rectification campaign' of 1942, and we shall shortly have occasion to consider some of its manifestations in recent years. Though it has its own peculiarities, this cult and its consequences do bear a certain resemblance to the phenomena observable in the Soviet Union during Stalin's later years, and in this context the criticisms of the Stalin cult may indeed have caused Mao some uneasiness. But the main source of his hostility to 'destalinization' as promoted by Khrushchev undoubtedly lay in his profound historical sense. As he put it at the time, Stalin's actions 'were phenomena of the international Communist movement and bore the imprint of the times'. To affirm that Stalin's whole policy during the last two decades of his life was basically a series of crimes and errors was therefore to cast discredit on the world Communist movement as a whole, and indirectly on the Chinese Communists who had come to power with the support – albeit the grudging support – of this movement. Khrushchev's attempt to establish a complete separation between Stalin the man and the system he had dominated for a generation appeared to Mao logically absurd and politically explosive – as indeed it was.

Mao was thus unquestionably hostile from the beginning to the manner in which 'destalinization' was pursued, though not to criticism of Stalin as such. On the other hand, there is no contemporary evidence to support the claim, made since 1963, that the Chinese also immediately indicated their opposition to Khrushchev's two other major doctrinal innovations at the Twentieth Party Congress, 'peaceful transition to socialism' through the conquest of a parliamentary majority, and the possibility not merely of coexistence but of cooperation with 'imperialist' countries. These questions were to be discussed at the 1957 Moscow meeting of Communist and workers' parties, which marked a further important stage in Sino-Soviet relations; but meanwhile the context in which the debate took place was profoundly affected by the events of October and November 1956 in Poland and Hungary.

The Chinese now maintain that they adopted diametrically opposite attitudes towards the problems of the two countries, opposing the 'great-power chauvinism' of the Soviet leaders in attempting to cow their Polish comrades by armed force, and at the same time counselling immediate intervention to smash the 'counter-revolutionary rebellion' in Hungary.[17] The first of these affirmations is very likely true; the second is clearly over-simplified. In contemporary texts – such as the editorial, 'More on the Historical Experience of the Dictatorship of the Proletariat', published on 29 December 1956 – the importance of maintaining an attitude of equality in relations between the socialist countries and avoiding 'great-nation chauvinism' is repeatedly stressed.[18] As regards the Hungarian affair, on the other hand, Mao at the time did not simply attribute it to the 'counter-revolutionaries'; he declared that the latter had 'taken advantage of the discontent of the masses' and thus induced 'a section of the people' to revolt against the people's government.[19]

Why was it possible for the counter-revolutionaries to succeed in this enterprise? The two reasons given in the editorial of December 1956 clearly foreshadow not only Mao's policy of the controlled release of tension announced in his speech of the following February, but the ultimate issue of that policy. On the one hand, 'the democratic rights and revolutionary enthusiasm of the Hungarian working people were impaired' as a result of errors by the leadership; on the other hand, 'the counter-revolutionaries were not dealt the blow they deserved', and 'Hungary had not yet made a serious enough effort to build up its dictatorship of the proletariat'.[20] In his speech of 27 February 1957 Mao applied the Hungarian lesson to China:

Within the ranks of the people, we cannot do without freedom, nor can we do without discipline; we cannot do without democracy, nor can we do without centralism. . . . Under democratic centralism, the people enjoy a wide measure of democracy and freedom, but at the same time they have to keep themselves within the bounds of socialist discipline. All this is well understood by the masses of the people.[21]

The conviction that, in China, the masses of the people understood the limits of the freedom offered to them was the foundation of Mao's whole policy in the spring of 1957, when the 'Hundred Flowers' campaign reached its high point. Contrary to the illusions widely entertained abroad at the time, this policy was never a policy of 'liberalization', in the sense that it was intended to encourage diversity as desirable in itself. Mao's idea was rather that, Marxism being the only true form of thought, it would eventually triumph over all others if, within certain limits, a debate was encouraged among the partisans of various contending viewpoints. In the long run, this process would serve to educate non-Communist intellectuals, writers, and cadres, and transform them into socialist intellectuals. In the immediate future, the criticisms of the non-Marxists, even if they were partly false, would have the advantage of obliging the Marxists to re-think their own position. 'Correct ideas, if pampered in hothouses without exposure to the elements or immunization against disease, will not win out against wrong ones', said Mao.[22]

Mao's hopes were disappointed. The critics did not stop at the criticism of particular abuses, but called into question the basic principle of the whole system – the monopoly of power in the hands of the Communist Party. Mao was therefore led to put the emphasis on the other half of his formula for avoiding a Hungarian-type uprising in China – on discipline rather than freedom, and on strengthening the 'proletarian dictatorship' rather than on 'getting rid of the root cause of disturbances' by stamping out abuses by the authorities. The new 'rectification campaign' in progress since the beginning of 1957, which had originally been directed against bureaucratic tendencies among the cadres of the party and the state, was transformed into a violent campaign against 'rightist elements', who were sent to the countryside for reform through participation in suitable work. To be sure, in a country such as China, where the intellectuals had always been distinguished by a profound contempt for manual labour, some concrete experience of productive work was undoubtedly a useful educational measure. But in many cases the 'hsia-fang' (literally 'send down') policy involved imposing either excessively hard physical labour, which ruined the health of men who were not accustomed to it, or humiliating tasks (such as

cleaning the lavatories in the university where one had previously been a professor) which were more calculated to break the
spirit of those involved than to re-educate them.*

These harsher measures of social control were accompanied by
a growing radicalism in both economic policy at home, and
attitudes towards the outside world. It is in this context that the
meeting of Communist and workers' parties took place at Moscow in November 1957.

Mao Tse-tung himself led the Chinese delegation. This was his
second journey outside China, and the first occasion on which
he appeared at length before a forum composed of a large number
of leading figures in the world Communist movement. (During
his visit in the winter of 1949–50 he had spoken at the meeting
in honour of Stalin's seventieth birthday, but otherwise his stay
had been taken up with private negotiations.) The Chinese now
claim that at the 1957 meeting they sharply disagreed with the
Soviet position on the issue of peaceful transition to socialism,
declaring that, while it might be possible in some instances to
seize power by winning a parliamentary majority, such a possibility should not be over-stressed, lest the result be to 'weaken
the revolutionary will of the proletariat'.[23] But it appears that in
fact they were basically satisfied with the final decisions of the
conference on this and most other points.

The most colourful debate was that dealing with international
relations and the question of war and peace. It was here also that
Mao Tse-tung personally intervened in the most striking fashion.
Regarding the international balance of power in general, he
coined the famous slogan 'The east wind prevails over the west
wind'. At the time, the 'East' in this slogan was the *political* East,
the socialist camp headed by the Soviet Union, whose power had
just been dramatized by the launching of the first sputnik. It was

---

* The turn to the left in the second half of 1957 also put an end to the
evolution in China, towards the protection of 'socialist legality', which had
paralleled that of the Soviet Union since 1953, although it did not lead to
completely arbitrary treatment of suspects. For an admirably balanced
pioneering study tracing the roots of current legal procedures both in the
Soviet example (Stalinist, and post-Stalinist) and in the Chinese tradition,
see Jerome A. Cohen, 'The Criminal Process in the People's Republic of
China: An Introduction', *Harvard Law Review*, 79 (3), January 1966, pp.
469–533.

only later that the expression came to have geographical and racial connotations.

But it was Mao's intervention on the subject of thermonuclear war that undoubtedly caused the greatest sensation. Emphasizing that it was the 'war maniacs' who might 'drop atomic and hydrogen bombs everywhere', and that the socialist countries would never be the aggressors in a world war, Mao asserted that the issue should nonetheless be considered at its worst – and proceeded to do so:

I debated this question with a foreign statesman.* He believed that if an atomic war was fought, the whole of mankind would be annihilated. I said that if the worst came to the worst and half of mankind died, the other half would remain while imperialism would be razed to the ground and the whole world would become socialist; in a number of years there would be 2,700 million people again and definitely more. We Chinese have not yet completed our construction and we desire peace. However, if imperialism insists on fighting a war, we will have no alternative but to make up our minds and fight to the finish before going ahead with our construction. If every day you are afraid of war and war eventually comes, what will you do then?[24]

According to recent Chinese accounts, an agreement had been signed a month earlier, in October 1957, under which the Soviet Government was to provide China with a sample atomic bomb and technical data on its manufacture.† Listening to Mao's intervention at the November meeting, Khrushchev and his comrades, sobered by the knowledge of the destructive capacities of their own bombs, may well have felt some anxiety regarding the consequences of putting such weapons into Chinese hands. But if the problems of a prudent and responsible attitude in foreign affairs undoubtedly played an important part in the genesis of the Sino-

* According to the Soviet version of this statement, and to later Chinese accounts, the 'foreign statesman' involved was Jawaharlal Nehru.

† Chinese statement of 15 August 1963, *Peking Review*, No. 33, 1963, p. 14. The Soviets have neither confirmed nor denied this affirmation, but have limited themselves to complaining that the Chinese were revealing secret information regarding the defence arrangements of the Socialist camp. It is possible that there was no promise or intention of handing over a sample atomic bomb to Peking, but there was unquestionably some kind of agreement on new defence technology.

Soviet conflict, the evolution of Chinese internal policy and the reactions aroused by it were no less decisive.

It was in the spring of 1958 that a new and radical policy, characterized by the slogans of the 'Great Leap Forward' and the 'permanent revolution',* took definite shape. In April, on the eve of the second session of the Eighth Congress of the Chinese Communist Party at which these slogans were proclaimed, Mao Tse-tung expressed his implacable determination to re-shape the Chinese people, and his conviction that it was possible to do so, in particularly striking terms:

Apart from their other characteristics, China's 600 million people have two remarkable peculiarities; they are, first of all, poor, and secondly blank. That may seem like a bad thing, but it is really a good thing. Poor people want change, want to do things, want revolution. A clean sheet of paper has no blotches, and so the newest and most beautiful words can be written on it, the newest and most beautiful pictures can be painted on it.[25]

Thanks to these advantages, Mao concluded that China 'might not need as much time as previously thought to catch up with the big capitalist countries in industrial and agricultural production'. It was Liu Shao-ch'i who, at the Party Congress in May 1958, made himself the advocate of the Great Leap Forward and the permanent revolution, but he did so in the name of Mao Tse-tung. When in the course of the summer of 1958 'people's communes' began to be formed on an experimental basis, Mao personally took a hand in encouraging this development. In September he returned from an inspection trip around the country, and declared that, in order to achieve a leap forward in agriculture even greater than that which had been accomplished in 1958, it was necessary to extend the system of people's communes throughout the whole country. At the same time, he defended the programme for producing home-made steel in small quantities throughout the countryside in terms which clearly revealed his whole mentality and his approach to economic problems:

During this trip, I have witnessed the tremendous energy of the masses.

* On the substance of this policy and the reasons for translating the term *pu-tuan ko-ming* as 'permanent revolution' rather than 'uninterrupted revolution', see my monograph *La 'révolution permanente' en Chine* (Paris, Mouton, 1963).

On this foundation it is possible to accomplish any task whatsoever. We must first complete the tasks on the iron and steel front. In these sectors, the masses have already been mobilized. Nevertheless, in the country as a whole, there are a few places, a few enterprises, where the work of mobilizing the masses has still not been properly carried out. . . . There are still a few comrades who are unwilling to undertake a large-scale mass movement in the industrial sphere. They call the mass movement on the industrial front 'irregular' and disparage it as 'a rural style of work' and 'a guerrilla habit'. This is obviously incorrect.[26]

It would be hard to find a more striking manifestation of what I have called Mao's 'military romanticism'.* Not only does he regard war as the supreme adventure and the supreme test of human courage and human will, but the warlike quality of his temperament and imagination are such that he tends to pose economic and even scientific and philosophical problems in these terms.

The Chinese themselves, though they do not call Mao a 'military romanticist', have characterized him, since 1958, as a 'revolutionary romanticist'. Until the time of the Great Leap Forward, romanticism was regarded in China, as it now is in the Soviet Union, as an inherently reactionary current of thought. But 'revolutionary romanticism' is something else; it is defined as above all an attitude, consisting in '. . . trying to see the new things in life, trying to be good at reflecting the new things and helping them to develop.'[27] The two greatest examples of this attitude are considered to be Mao Tse-tung and Lenin, both of whom held that, while one must take account of objective conditions, one must not overestimate them or find in them an excuse for failing to act.

There is certainly no doubt that Mao is 'romantic' in this sense. The system of the people's communes, which had been extended virtually to the whole of the Chinese population by the end of 1958, appears to have been directly inspired by this tendency in his personality. Economically, the aim of the communes was to speed up the development of the country by providing an administrative framework for the organization of human labour to replace scarce machines, and by providing for

* See my article 'The "Military Deviation" of Mao Tse-tung', *Problems of Communism*, No. 1, 1964, pp. 49–56.

the cultivation of the land in units large enough to facilitate the introduction of tractors and other machinery. Politically, their purpose was to strengthen the control of the state over the life of the individual, by combining the governmental, administrative, economic, and military apparatus into a single entity on the basic level, and by weakening rival organizations such as the family. Ideologically, they would serve to demonstrate, in the context of the growing Chinese resentment at the attitude of superiority exhibited by the Soviets, that China had discovered and put into practice social forms even more advanced than those of Russia, though her revolution was infinitely younger.

All of these objectives were 'romantic' in the sense they they tended to exalt the power of the human will. A cold analysis of the facts should have told Mao that the communes as originally conceived could not achieve either their economic objective of increasing production or their political objective of social control without a far larger number of skilled administrators and much more effective statistical services than the country then possessed. He nonetheless pushed ahead with his policy, on the assumption that revolutionary enthusiasm and ideological purity could make up for the lack of technical competence and material means. As for the ideological aspect of the communes, it was characterized by a touching but naïve utopianism, which finds expression in the following jingle composed by the peasants of Hunan in the summer of 1958:

> Setting up a people's commune is like going to heaven,
> The achievements of a single night surpass those of several millennia
> The sharp knife severs the roots of private property
> Opening a new historical era.*

It is true that Mao's extreme voluntarism stems naturally from Lenin's adaptation of Marxism to conditions in a relatively backward country. But Lenin's faith in the power of the human will to transform reality manifested itself primarily in the domain of politics. It was left to the Chinese, during the period of the Great

* Quoted by Kuan Feng, article in *Che-hsüeh yen-chiu*, No. 5, 1958, pp. 1–8. For a full translation of this article, which reflects the measureless optimism which prevailed when the communes were first being set up, see Schram, *La 'révolution permanente' en Chine*, pp. 6–18.

Leap Forward, to extend it to the realm of nature. Thus a volume on the study of Mao Tse-tung's thought declared in 1958:

Many living examples show that there is only unproductive thought, there are no unproductive regions. There are only poor methods for cultivating the land, there is no such thing as poor land. Provided only that people manifest in full measure their subjective capacities for action, it is possible to modify natural conditions.[28]

Stalin, too, had entertained extravagantly ambitious ideas on the possibility of transforming deserts into gardens, but he had never suggested that deserts did not exist. And, in any event, those years in the Soviet Union were past. Will had yielded to rationality as the principle on which the building of the Soviet economy was based. Mao's idea that it was possible to leap over objective difficulties and catch up with the most advanced industrial countries therefore aroused no little annoyance in Moscow, as did the Chinese claim that, thanks to the people's communes, they could begin laying the foundations of Communism immediately.

Simultaneously with the emergence of these subjects of conflict in the realm of Chinese domestic policy, Mao and Khrushchev for the first time came into direct collision in the summer of 1958 on a concrete issue of foreign policy. On 28 July Khrushchev proposed a summit meeting of the United States, Britain, France, the Soviet Union, and India to settle the Middle Eastern crisis. This was his most precise gesture so far towards the method of direct dealings with the leaders of the Western powers that was increasingly to characterize his policy. But three days later he suddenly made an unplanned and secret visit to Peking, and while there he withdrew his offer of a summit. It seemed obvious that this brusque turnabout was due to pressure from Mao Tse-tung, hostile to such dealings with the 'American imperialists' who were still maintaining an alternative Chinese Government on Taiwan. To drive home his point, Mao began shelling the off-shore islands at the end of August 1958.

All the particular subjects of discord which thus emerged in the period from the autumn of 1957 to the autumn of 1958, while important in themselves, were at the same time symptoms of a fundamental discrepancy; the mentality of the Chinese and Soviet leaders had been shaped by such very different experience that, even with the best of will, it would have been extremely difficult for them to meet in the same intellectual universe.

The extreme voluntarism and intransigence which have characterized every aspect of Mao's policy, both foreign and domestic, since the end of 1957 are clearly inspired by a return to the sources of his own experience. To proclaim that wars are decided not by thermonuclear weapons, but by the courage and perseverance of those who fight them, even if the life of half of humanity is at stake, is to reaffirm the value of the lessons learned during the war against Chiang Kai-shek, when Mao's 'millet plus rifles' effectively triumphed over his adversary's airplanes and tanks.* To deny categorically that the problem of the revolution in the advanced countries presents itself differently today from what it did in the time of Marx or Lenin, as the Chinese denied in their statement at the 1957 conference on the issue of peaceful transition, and to denounce, as they subsequently did, not merely the substance of attempts by European Communists such as Togliatti to adapt themselves to new conditions, but the very fact of attempting such an *aggiornamento*, is to shut oneself up within the limits of Chinese experience and to view the rest of the world through the prism of this experience.

These tendencies of Mao's were manifested precisely at the moment when the Soviet Union – as a result both of its economic and social evolution and of new thinking by Khrushchev himself – was moving towards a greater awareness of the problems facing the modern world, and beginning to view the capitalist countries in a less doctrinaire spirit. Inevitably, Khrushchev's Russia appeared to Mao to be '*embourgeoisé*' and 'revisionist', while Mao's China appeared to Khrushchev as fanatical and primitive.

But if Khrushchev had, or was shortly to acquire, a better understanding of Europe and America than any of his predecessors, he displayed an ignorance of Chinese psychology even crasser than that of Stalin. It is true that the Chinese were giving him some cause for complaint. The over-hasty creation of the people's communes, with their attempt to organize every aspect of eco-

---

* It is true that similar views are held by certain people in the West who do not have the excuse either of ignorance regarding nuclear weapons, or of having been marked by a quarter of a century of guerrilla warfare.

nomic and social life on a very large scale, led to confusion which eventually engendered disastrous economic failure. Nevertheless, the Chinese persevered in their policy, and in their claims to have found a short-cut to Communism (with only limited concessions on both points), despite strong Soviet opposition. It appeared to Khrushchev, and undoubtedly to the vast majority of Soviet citizens, that the Chinese were wasting precious resources which the Soviet Union had given them and which could have instead been used productively at home.

All this made Khrushchev's exasperation understandable. His reaction was characterized, however, by two crude blunders. On the one hand, he indulged in deliberate mockery of the communes, first in private, in his famous declarations of December 1958 to Senator (now Vice-President) Humphrey, and then openly and publicly. On the other hand, he ultimately moved towards the use of economic pressure as a means for bringing the Chinese to heel. The result was to rekindle Mao's resentment against the past arrogance of the Europeans, and to confirm by a visceral reaction what he was already coming to believe on political grounds, that Khrushchev's Russia was more and more difficult to distinguish from the capitalist countries.

Mao was unquestionably rendered even more sensitive to Khrushchev's mockery of the communes and to his efforts to make political capital out of China's economic difficulties by the fact that he had staked his whole prestige on this effort to overcome China's backwardness at one bold stroke. It is surely no accident that Mao's decision not to stand as a candidate for another four-year term as Chairman of the Chinese People's Republic was announced at the Sixth Plenum of the Central Committee of the Chinese Communist Party in December 1958. For this same plenum also marked the first reluctant step towards recognition that it was not possible to leap headlong into Communism overnight. Whether discontent within the party at the reckless way in which the communes had been organized led to any actual pressure on Mao to resign, or whether he took this decision entirely on his own initiative, it clearly reflected his lack of enthusiasm for administering policies which were no longer entirely after his own heart.

To be sure, Mao Tse-tung retained the chairmanship of the

party, as well as his unique position as supreme charismatic leader of the Chinese revolution. But henceforth, as Peking's economic policy gradually moved, in 1959 and 1960, towards greater prudence and rationality, Mao devoted himself primarily to foreign affairs. Nature and technology had proved more difficult to bend to his will than he had imagined. Now he would try bending China's enemies instead – both the open enemy, 'American imperialism', and the more insidious enemy, 'modern revisionism'.

In the summer of 1959 he visited his native village of Shaoshan for the first time since he had carried out his famous investigation of the peasant movement in 1927. Musing on the past, he found in it a source of confidence for the present:

> My memories of the past are unchanging, and I curse the
> inexorable flux of time.
> I am in my native place thirty-two years ago.
> Red flags fly from the spears of the enslaved peasants,
> Black hands raise high the lash of the tyrannical landowner.
> Only because so many sacrificed themselves did our wills become
> strong,
> So that we dared command the sun and moon to bring a new
> day.
> I love to look at the multiple waves of rice and beans,
> While on every side the heroes return through the evening haze.*

During this same month of June, while Mao was in Hunan, the Soviets, according to later Chinese statements, unilaterally tore up the 1957 agreement on new military technology.[29] This was one of the first explicit gestures in Khrushchev's attempt at forcing the Chinese to modify their policies through economic and political pressure. It might have been expected that one whose will was strong enough to 'command the sun and moon' would not yield to such tactics. Mao expressed his feelings in another poem written at the beginning of July 1959 while driving up Lu Shan, the mountain in Kiangsi province where the Left Kuomintang leaders had conferred in July 1927, on the eve of the Nanchang Uprising, over ways of stamping out Communism, and where the

---

* Poem dated June 1959. My translation, from 'Mao as a Poet', *Problems of Communism*, September–October 1964, p. 39. The heroes in the last line are the labour heroes of the present, i.e., the peasants returning from work.

Eighth Plenum of the Chinese Communist Party was shortly to
be held:

> The lonely peak soars abruptly beside the great river,
> I leap toward the summit, winding four hundred times through the
>   verdant nature.
> Calmly and coldly I turn towards the sea and gaze at the world,
> A hot wind scatters rain over the river to the horizon,
> Clouds cross the Nine Branches* and float over the Yellow Crane
>   Pavillion†
> White mist rises from the waves rolling toward the coast.
> Where has Magistrate T'ao gone now?
> Can one cultivate the land in the Peach-blossom Spring?[30]

The key to this poem is to be found in the next-to-last line.
'Magistrate T'ao' refers to the great poet T'ao Yüan-ming (died
AD 427), who in fact was a magistrate only once in his life, and
then only for a period of some eighty days. His official career
ended abruptly when he was called upon to receive a superior
official with all the appropriate pomp and obsequiousness, and
refused to do so, preferring to abandon his office. 'How can I bow
my back before a petty rustic fellow for five pecks of rice?'
he is reported to have said. It seems highly probable that in this
poem Mao saw himself in the role of T'ao, and Khrushchev in
that of the 'petty rustic fellow'.

The last line is an allusion to the title of a famous work by T'ao
describing a kind of Utopia. Mao was clearly thinking of the
China of the communes and the 'Great Leap Forward' as an
earthly paradise, superior to T'ao's imaginary refuge from the
world. The Lu Shan plenum of the Central Committee, only a
month later, dealt a rude blow to this vision. The official com-
muniqué of 26 August recognized that the figures published previously for economic achievement during the first year of the 'Great
Leap' were exaggerated by 40 to 50 per cent. In particular, the grain
harvest had been only 250 million tons instead of 375. (The

---

* Near Lu Shan, to the east, a number of tributaries flow into the Yangtze. These are commonly referred to in classical poetry as the 'Nine
Branches'.

† This pavillion, situated on a height overlooking the Yangtze, to the
west of Wuchang, is linked to a legend about a Taoist who is said to have
passed the spot mounted on a yellow crane.

real figure was undoubtedly still lower.*) Another bad harvest in 1959, due in part to natural calamities, but also to the disorganization of the economy and the resistance of the peasantry to the extremes of collectivization and discipline practised in the communes, was to lead to a lean and bitter winter. This crisis was met and surmounted by maintaining the existence of the communes as administrative units, but introducing a greater and greater degree of decentralization, by which the effective control of economic activity was handed over first to the 'production brigade', corresponding to the former cooperatives, and then to the 'production team', corresponding on the whole to the mutual aid teams introduced in 1950–51. The overly heavy accent on industry in economic planning was corrected, and it was recognized that for some time to come agriculture would constitute the basis of the national economy.

From these problems, Mao turned increasingly to foreign affairs, not only because he was depressed by internal developments, but because he found growing cause for annoyance in the behaviour of foreign powers, and especially of the Soviet Union, towards China. The first of such incidents had its beginning at the Lu Shan plenum of August 1959, which marked the beginning of a full-scale retreat from the Great Leap Forward. On this occasion, Mao's old comrade from Chingkangshan days, P'eng Te-huai, then Minister of Defence, launched a violent assault on the policies that had prevailed during the past year and a half, an act which led to his dismissal and disgrace. According to several accounts, P'eng had communicated his views to the Communist Party of the Soviet Union in a letter delivered during a 'military good-will' tour of the Warsaw Pact countries in the spring of 1959.† This undoubtedly seemed to Mao a clear case of Soviet interference in Chinese internal affairs.

Two other clashes followed in quick succession. On 9 September 1959 Tass issued a statement on the Sino-Indian border con-

* See Conclusion.

† The most detailed of these is the article of David A. Charles, 'The Dismissal of Marshal P'eng Te-huai', *China Quarterly*, No. 88, 1961, pp. 63–76. According to this and other accounts, Khrushchev, at the Bucharest meeting in 1960, defended not only P'eng but Kao Kang, saying that both had committed no offence save opposing Mao's incorrect policies towards the Soviet Union (ibid., p. 75).

flict which had broken out in August, expressing regret at the incidents and advising 'both governments' involved to make efforts to resolve this 'misunderstanding'. The Chinese not unnaturally regarded it as a breach of proletarian solidarity to put a fellow socialist state and a 'bourgeois' power on the same plane. Their resentment was increased by the fact that the statement had been shown to them at the last minute, on the morning of the ninth, and the Soviets had gone ahead and published it the same evening despite a formal request from Peking not to do so.[31] Both the Chinese and the Soviets now date the beginning of open discord between them from the appearance of this statement.

On 30 September Khrushchev arrived in Peking, fresh from his visit to the United States and from his conversations with Eisenhower at Camp David. This in itself was not calculated to ingratiate him with Mao, but in the course of his visit he gave his hosts still greater offence. At the banquet in his honour offered by Mao Tse-tung on the evening of his arrival, he sharply attacked those who wanted to 'test by force the stability of the capitalist system'. According to a later Chinese statement, 'the whole world recognized this as an insinuation that China was being "bellicose" regarding Taiwan and the Sino-Indian boundary.'[32]

In the course of conversations with the Chinese leaders, Khrushchev went so far as to remind Mao and his comrades that after the October revolution Lenin had made a 'temporary concession' in recognizing the establishment of the Far Eastern Republic in Siberia. The implication was that, similarly, Mao should recognize the temporary separation of Taiwan from China.[33] The analogy was not a very apt one; the Far Eastern Republic was a pro-Soviet buffer state, and not the refuge of a bitterly hostile government committed to reconquering the whole of the national territory. But even if there had been a genuine parallel, Mao would not have admitted it. To his emotions, if not to his intellect, to sacrifice Russian territory was one thing; to sacrifice Chinese territory was another. Under no circumstances could he have tolerated Khrushchev's implicit support of the idea of 'Two Chinas', even as a transitional phase.

From this point onward, Sino-Soviet relations moved, as everyone knows, towards ever sharper and more open conflict. In the context of the present volume it would be superfluous to analyse in detail the stages in this process and the exceedingly voluminous documents produced by both sides to justify their positions. In the concluding pages of this biography I shall deal with the conflict only in relation to Mao, considering both how he has placed his stamp on the Chinese attitude, and how the lessons of the conflict have shaped his political credo, as it can be discerned today.

On 16 April 1960, *Hung Ch'i* (*Red Flag*), the theoretical organ of the Chinese Communist Party, published the editorial 'Long Live Leninism!'. It is generally assumed that Mao Tse-tung wrote portions of this and the other important anonymous texts published subsequently in Peking, or at least inspired them and added a few characteristic touches to the final version. The ostensible reason for the appearance of this article was the ninetieth anniversary of Lenin's birth, but the Chinese subsequently indicated that its real purpose was to combat the tendency towards coming to an agreement with the imperialists which found expression in the forthcoming (and ultimately abortive) summit conference in Paris. In this editorial Mao's writings during the war of resistance against Japan were cited in support of the thesis that 'an awakened people will always find new ways to counteract reactionaries' superiority in arms'. Ideas similar to those expounded by Mao at the 1957 Moscow meeting were put forward regarding the eventual consequences of an atomic war:

Should the imperialists impose such sacrifices on the peoples of various countries, we believe that, just as the experience of the Russian revolution and the Chinese revolution shows, those sacrifices would be rewarded. On the debris of imperialism, the victorious people would create very swiftly a civilization thousands of times higher than the capitalist system and a truly beautiful future for themselves.[34]

Once more, Mao was reaching back to his own experience as the supreme touchstone for judging every kind of problem. Most of the article was devoted to a denunciation of the idea that economic and technological change had in any way invalidated the political analyses of Marx and Lenin.

Although in this article, as in all of their other ideological productions down to this point, the Chinese had carefully maintained the fiction of attacking only 'revisionism' in general, or the Yugoslavs, such an open and systematic refutation of the whole Soviet position apparently convinced Khrushchev that the time had come to put an end once and for all to Chinese dissidence in the world Communist movement. At the Rumanian Communist Party Congress in June 1960 he made a speech on the question of war and peace in which he denounced the Chinese as 'madmen' who wanted to unleash nuclear war, as 'pure nationalists' in their relations with India, etc. In what Mao must have understood as a deliberate insult, he declared that nowadays militia were 'not troops but just human flesh' in the face of atomic weapons.[35] And he repeated the affirmation, galling to Chinese national pride, that the future development of the international situation would depend primarily on relations between the Soviet Union and the United States.

All the Communist parties represented in Bucharest except that of Albania joined in Khrushchev's abuse of the Chinese. This verbal and organizational attack was followed immediately by direct economic pressure. In July 1960, the Soviets unilaterally and brusquely decided to withdraw all the Soviet experts in China within one month, thus tearing up, according to later Chinese accounts, 'hundreds of agreements and contracts'.[36]

In this context, Chinese tactics at the second Moscow meeting of November 1960 were different from what they had been in 1957. Then it was Mao who had insisted on the leading role of the Soviet Union in the socialist camp. Now his representatives emphasized instead the fact that all Communist parties were equal and entitled to determine their own policy in their own way, being bound only by the general principles of Marxism-Leninism and statements agreed upon by all the parties. They also held out for the recognition of the Soviet Union as the head of the socialist camp, but the primacy thus accorded to Moscow was to be purely formal and symbolic. Most probably, as many observers have suggested, it was intended to safeguard the unity of the camp in preparation for the day when the Chinese would be strong enough to assert their own leadership.

Side by side with these organizational issues, the Chinese insisted throughout the discussions on points related to national dignity, and to the susceptibilities of the former colonial and semi-colonial peoples in the face of the slightest hint of superiority on the part of European Communists. Thus they succeeded in getting removed from the final text of the 'Statement' issued at the end of the conference the declaration that the 'democratic forces' in the advanced capitalist countries should place 'their experience and knowledge' at the service of the formerly dependent countries.* In the face of the abrupt withdrawal of Soviet aid, Mao and his comrades were moving towards the position that every country must develop its economy primarily through the heroic efforts of its own people, and they were in no mood to recognize the need for any 'comradely aid' from those whom they regarded as traitors and revisionists.

The winter of 1960–61 was a bitter one in China. An extremely efficient system of rationing spread the hunger equally over the entire population, but in order to attenuate the famine it was necessary to make large grain purchases from Canada and Australia in February 1961. In these grim circumstances, Mao reached back into China's past for reasons to have confidence in her future:

> Above Mount Chiu-i white clouds are flying,
> Borne by the wind, the daughters of Emperor Yao drift down
>   from the blue-green hills,
> Each branch of dappled bamboo has been sprinkled with a
>   thousand tears,
> Innumerable rose-coloured clouds cover the slopes with manifold
>   garments.
> The waves on Lake Tungt'ing boil up like snow reaching to the
>   heavens,
> The people of the Long Island sing earth-shaking songs.
> Inspired by all this, I would dream a dream equally vast,
> And see the Land of the Hibiscus wholly illuminated by the light
>   of dawn.[37]

The two daughters of the legendary Emperor Yao are supposed to have been the wives of his successor, Emperor Shun. During a journey to the south, Shun died near the Hsiang River, in Hunan, in the vicinity of Mount Chiu-i, and his two wives came and wept on the banks of the river. The leaves of the bamboo were permanently discoloured by their tears, thus giving rise to

* On this point, and on the 1960 conference in general, see Carrère d'Encausse and Schram, *Le Marxisme et l'Asie*, pp. 114–20 and 422–33.

the species known as dappled bamboo. The Long Island referred to in the poem is located in the Hsiang River near Changsha, and in evoking the 'earth-shaking songs' of its inhabitants Mao is obviously thinking of the revolutionary experiences of his youth. The 'Land of the Hibiscus is a literary name for Hunan, drawn from a T'ang poem. According to commentaries published in China, Mao's dream of seeing his native province 'wholly illuminated by the light of dawn' refers to his hope of seeing full Communism in China one day. Just as in the past the beauty of the dappled bamboo grew out of sorrow, so today the most difficult time is the time when things are about to change for the better. We are told that this is a 'law of materialist dialectics',[38] but it is obvious that Mao was thinking and feeling here not in terms of Marxism-Leninism but in terms of the historic destiny of China, from the great Yao and Shun to his own day.

Following the Twenty-Second Congress of the Communist Party of the Soviet Union in October 1961, at which Chou En-lai had reacted to Khrushchev's attacks on Stalin by ostentatiously laying a wreath at the latter's tomb, Sino-Soviet relations continued to degenerate. Two episodes in the autumn of 1962 marked a new climax in this process: the Sino-Indian border war and the Soviet-American crisis over Cuba. The announcement in August 1962 that the Soviets were to furnish Migs to India aroused anger in Peking; the Chinese offensive in October, which may have been calculated to embarrass the Soviets as well as to humiliate Nehru, brought tempers in Moscow to the point where officials did not hesitate to denounce 'Chinese nationalism' bitterly even in conversations with Western Europeans and Americans. The Cuban crisis, in which Khrushchev was guilty, in Chinese eyes, first of 'adventurism' and then of 'capitulationism', brought further proof, in Mao's view, that the Soviet Union under its existing leadership was unfit for its position at the head of the socialist camp. From this point forward, Mao, who had hitherto regarded Khrushchev as an erring comrade who might conceivably return to the fold, appears to have considered him as a renegade past hope of redemption. Mao would still, almost down to the eve of the Soviet leader's fall, address occasional appeals to him to repent and recognize the error of his ways, but it was clear that he did not have the slightest hope that these appeals would be heeded.

Henceforth the Soviet leaders would be the subject of an antagonism virtually as great as that visited on the American 'imperialists', and of even greater contempt. The fact that he now faced as enemies both of the world's two strongest powers, far from alarming Mao, appeared to exhilarate him, judging by the poem he wrote on his sixty-ninth birthday:

> Winter clouds are heavy with snow, which flies like willow catkins,
> The myriad blossoms have all faded, and flowers are rare for the moment.
> High in the heavens frozen currents whirl impatiently,
> On the earth warm breezes are blowing imperceptibly.
> Only the hero dares pursue the tiger,
> Still less does any brave fellow fear the bear.
> The plum blossoms rejoice in the snow that fills the sky,
> Small wonder that the cold should kill the flies.[39]

In these lines written on his birthday, Mao's thoughts certainly turned towards himself, and we may assume that he is the 'hero' (or at least one of the heroes) cited. The 'tiger' is plainly the (paper) tiger of imperialism, and the 'bear', the Russian bear. Thus for Mao, Khrushchev was even less frightening than a paper tiger.

On 7 January 1963 *Pravda* for the first time attacked the Chinese openly and explicitly, instead of referring to the Albanians, to 'dogmatists', 'leftists', etc. Two days later Mao penned the last of his poems published to date, and the one which best conveys his overwhelming pride in his country and his overwhelming confidence in himself. These verses may have been partly inspired by the *Pravda* attack two days before; in any case, they were a direct reaction to the international situation as a whole. According to one of the Chinese commentaries, '1963 was a year in which the great anti-Chinese chorus of all reactionary forces in the world reached a high point, and yet it was also the year in which the revolutionary wave of the peoples of Asia, Africa, and Latin America reached a climax. It was a time of a vast anti-Marxist-Leninist ideological current, and yet a time when courageous fighters for truth manifested themselves without ceasing.'[40] This situation Mao described in an extraordinary mixture of revolutionary and traditional imagery which epitomizes the whole of his thought and action:

On our tiny globe
A few flies smash into the walls,
They buzz,
Some loudly complaining,
Others weeping.
The ants climb the flowering locust, boasting of their great country*
But for ants to shake a tree is easier said than done.
Just now the west wind drops leaves on Ch'angan,
Whistling arrows fly through the air.

How many urgent tasks
Have arisen one after another!
Heaven and earth revolve,
Time presses,
Ten thousand years is too long,
We must seize the day.
The four seas rise high, the clouds and the waters rage,
The five continents tremble, wind and thunder are unleashed,
We must sweep away all the harmful insects
Until not a single enemy remains.[41]

The world is 'tiny', according to the official commentaries, because it is small in the eyes of the proletarian revolutionary, who is capable of transforming it by his action. The walls are those of true Marxism-Leninism, against which the flies of 'imperialism' and/or 'modern revisionism' hurl themselves in vain. The ants of the following two lines are the Soviets, as the Chinese themselves have confirmed. Denouncing Soviet attacks on Mao Tse-tung and his thought, an editorial in *Jen-min Jih-pao* declared in April 1964: 'In their vain effort to vilify Comrade Mao Tse-tung and his ideas, the leaders of the CPSU are like ants trying to shake a great tree, ridiculously overrating themselves.'[42]

Ch'angan, the ancient capital, presumably stands for China in general. According to the Chinese commentaries, the 'leaves' represent the reactionary forces withered by the revolutionary blast, but Mao was probably thinking also of the propaganda of

---

* Allusion to a story by a T'ang dynasty writer about a man who dreamt that he visited the country of the Great Flowering Locust Tree. The king of this country gave him his daughter's hand and showered him with honours. Upon awakening, he discovered that there was an ant-hill at the foot of the locust tree outside his window, which represented the reality behind his dream.

the Soviet and other 'revisionists'. The 'sharp arrows' represent the Marxist-Leninist truth with which China replies to her enemies. The last lines of the poem, we are told, signify that the revolutionary peoples of the world do not want to wait for their deliverance from without, but are determined to liberate themselves.[43]

This last idea prefigures the central theme in all of the pronouncements on the international situation which have come out of Peking in the course of the past three years – that the peoples of Asia, Africa, and Latin America are the principal revolutionary forces in the world today. This vision of the world revolution, which had been hinted at by Mao and other Chinese spokesmen, especially since 1960,* was first spelled out explicitly in the Chinese letter of 14 June 1963. It took on a new sharpness and precision following the signing of the nuclear test ban treaty, which led Mao to renounce even the appearance of comradely solidarity with the Soviet leaders.

In Mao's view, the European proletariat, both in Russia and Eastern Europe where it held power and in the advanced capitalist countries where it did not, had surrendered to false prophets such as Khrushchev and Togliatti and betrayed its revolutionary mission both at home and abroad. Only the peoples of Asia, Africa, and Latin America remained truly revolutionary, and their struggles would ultimately redeem both a decadent Europe and themselves. Mao saw himself as the incarnation of these forces, the only truly dynamic and progressive forces on the stage of world history, and therefore as the natural leader of the world revolution.

* See Schram, *Political Thought of Mao*, p. 256.

Notes

1. *Jen-min Jih-pao*, 16 September 1954. For the text of the constitution, see Liu Shao-ch'i, *Report on the Draft Constitution of the People's Republic of China*, followed by *Constitution of the People's Republic of China* (Peking, Foreign Languages Press, 1954).
2. Li Fu-ch'un, speech of 5–6 July 1955, in *Communist China 1955–1959* (Cambridge, Mass., Harvard University Press, 1962), pp. 59–60.
3. ibid., p. 2.
4. Schram, *Political Thought of Mao*, pp. 247–9.
5. *Socialist Upsurge in China's Countryside*, pp. 159–60.
6. ibid., p. 302.
7. See Chapter 9.
8. Schram, op. cit., p. 260.
9. Carrère d'Encausse and Schram, *Le Marxisme et l'Asie*, pp. 414–15.
10. *Jen-min Jih-pao*, 8 February 1953.
11. Schram, op. cit., pp. 294–5.
12. *Jen-min Jih-pao*, 16 September 1953.
13. Editorial of 6 September 1963, *Peking Review*, No. 37, 13 September 1963, p. 9.
14. Editorial 'On the Question of Stalin', *Peking Review*, No. 38, 20 September 1963, p. 11.
15. Editorial of 6 September 1963, *Peking Review*, No. 37, 13 September 1963, p. 8.
16. Schram, op. cit., p. 298.
17. *Peking Review*, No. 37, 13 September 1963, p. 10.
18. *The Historical Experience of the Dictatorship of the Proletariat* (Peking, Foreign Languages Press, 1959), pp. 24, 33, 47, 58, etc.
19. ibid., p. 50; Schram, op. cit., p. 239.
20. *The Historical Experience of the Dictatorship of the Proletariat*, p. 50.
21. Schram, op. cit., p. 239.
22. ibid., p. 242.
23. 'Outline of Views on the Question of Peaceful Transition', 10 November 1957, published as Appendix I to the editorial of 6 September 1963, *Peking Review*, No. 37, 13 September 1963, pp. 11, 21.
24. Cited in the Chinese Government statement of 1 September 1963, *Peking Review*, No. 36, 6 September 1963, p. 10.

25. Schram, op. cit., p. 253.

26. ibid., pp. 253–4.

27. Lin Mo-han, *Keng kao-ti chü-ch'i Mao Tse-tung wen-i szu-hsiang ti ch'i-chih* (*Let us raise even higher the banner of Mao Tse-tung's literary and artistic thought*) (Shanghai, Shanghai Wen-i Ch'u-pan-she, 1960), p. 24.

28. *Hsüeh-hsi Mao Tse-tung ti szu-hsiang fang-fa ho kung-tso fang-fa* (Peking, Chung-kuo Ch'ing-nien Ch'u-pan-she, 1958), p. 73.

29. *Peking Review*, No. 33, 1963, p. 14, and No. 37, 1963, p. 12.

30. My translation, from *Problems of Communism*, September–October 1964, p. 40.

31. 'The Truth About How the Leaders of the C.P.S.U. Have Allied Themselves With India Against China', *Peking Review*, No. 45, 1963, p. 19.

32. ibid., p. 19.

33. Statement of 1 September 1963, *Peking Review*, No. 36, 1963, p. 13.

34. *Long Live Leninism!* (Peking, Foreign Languages Press, 1960), p. 22.

35. *Peking Review*, No. 36, 1963, p. 13, and No. 37, 1963, p. 13.

36. *Peking Review*, No. 37, 1963, p. 14.

37. My translation, *Problems of Communism*, September–October 1964, pp. 40–41.

38. See the commentary by An Ch'i in *Wen-i-pao*, No. 3, 1964.

39. My translation, *Problems of Communism*, September–October 1964, p. 42.

40. An Ch'i, in *Wen-i-pao*, No. 3, 1964.

41. My translation, *Problems of Communism*, September–October 1964, pp. 42–3.

42. Translation in *Peking Review*, No. 18, 1964, p. 14.

43. An Ch'i, *Wen-i-pao*, No. 3, 1964; Ch'en Chih-hsien, commentary in *Szech'uan Wen-hsüeh*, March 1964, p. 50.

Conclusion

The orientation adopted by Mao Tse-tung in late 1962 and early 1963 has continued to determine Chinese policy down to the present time. As this book goes to press, in January 1967, there is some uncertainty as to the role still played by Mao as an individual. It is obvious that he fully approves of the radical policies adopted during the past year, but he may no longer have the physical and intellectual energy to supervise their day-by-day execution. In a sense, this is only of minor importance when we try to evaluate his impact on China and on the world. For whether the most recent tendencies in Peking represent primarily the last effort of a dying man to fix the shape of things to come, or the first phase in a struggle for power among his successors, the context within which these events develop has been set by Mao. Even if he should disappear tomorrow, through death or incapacity, the situation in China will continue to bear his mark in a double sense: the remaining leaders have been his followers for decades, and the problems with which they must deal are the outgrowth of his policies.

In this perspective, Mao's political and ideological heritage and its probable future influence, rather than his current activities, are the decisive factors in judging his place in history, though Mao himself obviously does not believe this, and is busy trying to erect a monument to himself in his own lifetime. At the same time, recent events, although in many respects they grow out of the traits which have characterized Mao's aims and methods in the past, are so singular and so fraught with consequences for China and the world that it is impossible to terminate this account of Mao's life without discussing them in some detail. But before doing so, let us try to sum up developments down to the beginning of 1966.

The tendencies which characterize the evolution of Chinese policy over the past few years are for the most part the expression of a single dominant concern: to establish the continuing validity of Mao's thought and Mao's example both within and without the country.

An early manifestation of the tendency to transpose Mao's experience in China to the world scene is to be found in the editorial of October 1963, entitled 'Apologists of Neo-Colonialism', which contains such deliberate parallels with Mao's 1927 report on the peasant movement in Hunan that we may assume he had a hand in writing it. In 1927 Mao had declared that, while reactionaries thought the situation in Hunan was 'an awful mess' (*tsao ti hen*), revolutionaries should judge it 'very good indeed' (*hao ti hen*).[1] In 1963 he made of the attitude toward the wars of national liberation being conducted by the peoples of Asia, Africa, and Latin America the touchstone for distinguishing a true revolutionary. This 'mighty revolutionary storm', he wrote – in 1927 the peasants had been likened to 'a tornado or tempest, a force so extraordinarily swift and violent that no power, however great, will be able to suppress it' – 'makes the imperialists and colonialists tremble and the revolutionary people of the world rejoice. The imperialists and colonialists say, "An awful mess!" The revolutionary people say, "Very good indeed". '[2]

In this editorial, which thus bears Mao's stamp, we read:

Today the national-liberation revolutions in Asia, Africa and Latin America are the most important forces dealing imperialism direct blows. The contradictions of the world are concentrated in Asia, Africa and Latin America.

The centre of world contradictions, of world political struggles, is not fixed but shifts with changes in the international struggles and the revolutionary situation. We believe that, with the development of the contradiction and struggle between the proletariat and bourgeoisie in Western Europe and North America, the momentous day of battle will arrive in these homes of capitalism and heartlands of imperialism. When that day comes, Western Europe and North America will undoubtedly become the centre of world political struggles, of world contradictions.

Despite the patronizing recognition that one day Europe may once more play a significant revolutionary role, this statement expresses only extremely limited confidence in the willingness of the people or even of the proletariat either in the Soviet Union or in the advanced countries of Europe and America to risk their own tranquillity in order to support the struggles of their less fortunate fellow men in Asia and Africa. Mao occasionally pays lip service to Marxist class analysis by declaring the contrary, but it is difficult to believe that he is sincere.*

In Mao's eyes, the relevance of his thought and experience to the Asian, African, and Latin American revolutions today is dual. The conquest of China by revolutionary war based on the countryside may serve as a model both for the national liberation struggles within other countries and for an over-all revolutionary strategy against imperialism. This point has been made most explicitly and graphically in Lin Piao's 1965 article, 'Long Live the Victory of People's War!' On the one hand, Lin Piao emphasizes – as did the editorial note accompanying the republication, a few days earlier, of the complete text of Mao's 1938 essay on strategy in the guerrilla war against Japan – that the theory of a people's war developed by Mao three decades ago is of vital practical importance for the oppressed people of Asia, Africa, and Latin America in their struggle for liberation.[3] But he also transposes Chinese experience to the world scene.† In China, he reminds us, the revolution – unlike that in Russia – started in the countryside and triumphed only at the end in the cities. Similarly, the cause of world revolution as a whole depends finally on the struggles of the peoples of Asia, Africa, and Latin America, which constitute the 'rural areas of the world'.[4]

Here we see, carried to the extreme, the tendency of the Chinese leaders to see the whole world through the prism of their own experience. We have passed through a century in which the Europeans, each in his own way, have understood and misunderstood Asia in terms of concepts shaped in a different cultural and historical environment. Some endeavoured to transplant there parliamentary democracy and others proletarian dictatorship – political forms about equally discordant with the reality to which they were to be applied. Today it seems that we are in the process of moving from the century of the incomprehension of Asia by Europe to the century of the incomprehension of Europe by Asia. Mao Tse-tung, in the name of an experience which he himself

* His statement of 8 August 1963, see *Peking Review*, No. 33, 1963, p. 7.
† In fact, this analogy was coined by the Indonesian Communist leader Aidit. See *Peking Review*, No. 24, 1965, p. 11.

has frequently declared to be unique, now proclaims resoundingly that armed struggle is an indispensable aspect of any true revolution and excommunicates with contempt comrades in Europe and America who presume to seek other ways. Just as the United States has often sought to force the peoples of other lands and other continents to be free in the American fashion, Mao now seems bent on forcing the peoples of the world to be revolutionary in the Chinese fashion. It is highly likely that these enterprises are equally doomed to failure.

If Mao interprets the world as a whole in terms of China, this is not merely because, like everyone else, he sees life through the glasses of his own experience. His attitude is also related to the role which his compatriots have always attributed to their own civilization. This book opened on the Chinese vision of China as the 'central country'; it closes on the same theme. We first caught sight of the youthful Mao Tse-tung as he deplored the humiliations inflicted on China by the Europeans, and conjured his fellow citizens to strengthen their bodies so that they might resist the foreigners. We take leave of him as he denounces the anti-Chinese insolence of the Soviets, and hails the inevitable triumph of the masses in arms, despite the atomic bombs of the 'imperialists'. To be sure, both China and Mao are very different from what they were half a century ago. The way they propose to the world is no longer the way of Confucius, but the way of Mao Tse-tung. And yet there remains a profound similarity between the political and intellectual universe of Mao's youth and that which exists in China today. Then, as now, the Chinese way was seen as distinctively Chinese and at the same time universal.

This double pretension to uniqueness and to universality emerges with particular clarity in the claims made for the thought of Mao Tse-tung. We have seen in the course of this biography in how many respects the theory and practice of the Chinese revolution as developed by Mao over the past forty years diverge from strict Leninist orthodoxy. And yet the ideologists writing in Peking are determined at all costs to present this history as orthodox. Thus the first text in the current canon of the *Selected Works*, dating from a period when Mao was more inclined than almost any other leading figure in the Chinese Communist Party to see in Chiang Kai-shek the true leader of the national revolution, has been rewritten and reinterpreted to make of it an expression of intransigent insistence on Communist leadership in the revolution. There is, of course, a particular reason for this audacious imposture: to show that since Mao is by definition infallible, his thought has been, from beginning to end, correct and consistent. But this reshaping of the past also serves the larger aim of demonstrating that the thought and action of the Chinese revolutionaries have always been orthodox and therefore universal.

This tendency finds striking expression in Lin Piao's article, mentioned above, which describes 'the contemporary world revolution' as the encirclement of 'the cities of the world' (North America and Western Europe) by the 'rural areas of the world' (Asia, Africa, and Latin America). For at the same time, Lin emphasizes repeatedly that Mao Tse-tung's way, like the way of the October Revolution, is based on the hegemony of the proletariat. To be sure, the idea that, while the peasantry constitutes the principal revolutionary *force* in agrarian countries, this force must be subordinated to the *leadership* of the workers is perfectly orthodox Leninism. But it is difficult to see what 'working-class leadership' means at all concretely when the decisive phases in the revolutionary process take place far from the urban centres where the real proletariat is concentrated. Lin Piao's use of this orthodox formula would appear to serve no other purpose than to affirm that the Chinese revolution is a true Communist revolution, and therefore a proper rallying point for genuine revolutionaries throughout the world.

The same desire to show that China is both unique and universal can be seen in the categories applied by Chinese historians to their country's past. Rejecting both the Marxist concept of the 'Asiatic mode of production', with its implications of the inferiority of the Asian civilizations,* and Stalin's viewpoint according to which traditional China was simply 'feudal' and therefore similar to other societies at a corresponding stage in

* On this concept and its use today by European Communists against the Chinese, see Carrère d'Encausse and Schram, *Le Marxisme et l'Asie*, pp. 13–21, 130–31. Since the volume just cited was published, the Soviets themselves have been moving toward resurrecting the 'Asiatic mode of production'. Mme Carrère d'Encausse and I propose to deal with this development in the forthcoming English edition of our book.

their development, historians in Peking now proclaim that imperial China was characterized by a 'peculiar Asiatic form of feudalism', in some ways more typical than European feudalism. Thus the China of the past belonged to universal history, and was at the same time distinctive and superior, just as today China marches in the forefront of world progress because of her exemplary fidelity to the universal truth of Marxism-Leninism, in its specific Chinese incarnation. Mao's thought and action are 'entirely Marxist', but they are also 'entirely Chinese', and 'the highest expression . . . of Chinese national wisdom'.[5] It is not merely that they are both Marxist *and* Chinese; they are all the more Marxist *because* they are Chinese.

This double claim to universality has taken on strange and paradoxical forms today, in the era of the 'Great Proletarian Cultural Revolution'. But before discussing these, let us try to take the measure of Mao's achievement in modernizing and strengthening his country. It is easy to see why the style of Chinese economic development since 1958, with the emphasis on mobilizing human beings in place of scarce machines and on combining traditional handicrafts with large-scale industry, would appear relevant to the peoples of other non-European countries. The impact of such real and spectacular achievements as the production of an atomic bomb virtually without foreign assistance is also entirely understandable. Nothing is more difficult, on the other hand, than to provide an accurate over-all evaluation of the state of the Chinese economy today, for the crucial facts are almost totally lacking. Since 1960, no absolute figures whatever have been published in Peking, and the specialists in other countries who endeavour to piece together the available evidence are even more uncertain in their conclusions than those who worked on the Soviet economy fifteen years ago, when official statistics were similarly vague. Certain broad generalizations can, however, be made.

It is clear, first of all, that the greatest single weakness of the Chinese economy remains the agricultural sector. Mao and his comrades began by dealing more tactfully and successfully with the peasantry than the Soviets had done three decades earlier. Despite some resistance from the peasants, collectivization was achieved at far less cost in lives and suffering than in the Soviet

Union. Then, after having been more prudent in this domain, the Chinese leaders became more adventurous and plunged the countryside into confusion by the frantic creation of the communes. The result was near-disaster for the economy and the population. In 1957, the grain harvest had reached at last a figure of 187 million metric tons, corresponding approximately to the average level in the pre-1937 period. The harvest of 1958, which did not yet show the full effects of the administrative upheaval begun during the summer, was probably the best in China's history. It did not amount, however, even to the revised figure of 250 million tons announced in August 1959 – still less to the 375 million tons originally claimed – but rather to some 200 or 210 million tons. There followed a precipitous decline which may have fallen as low as 150 million tons in 1961, followed by a return to a level of about 200 million tons in 1963 and 1964. Given the increase in population, this corresponds to a *per capita* supply slightly under that in pre-war China. In the worst years, the calorie intake probably fell below 1,500 per day, and only an extremely efficient rationing system prevented mass starvation.*

In other words, despite the divergence from Soviet practice in the agricultural sphere, first in one sense and then in the other, the net result in China is very similar to that in Russia – namely, stabilization of production at a level roughly equivalent to that prevailing before the revolution under normal conditions. It may be added that the recovery of agriculture in the years after 1962 was due in large part to measures which in fact brought Chinese practice closer to that of the Soviet Union: re-establishment of the private plot, which had been entirely suppressed during the first phase of the communes,† and decentralization of production to a level comparable with that of the kolkhoz.

The achievements in industry are, of course, of quite a different

* For a brief summary of the available evidence, see Colin Clark, 'Economic Growth in Communist China', *China Quarterly*, No. 21, 1965, pp. 148–9. See also John Lossing Buck, 'Food Grain Production in Mainland China', in J. L. Buck, O. L. Dawson, and Y. L. Wu, *Food and Agriculture in Red China* (Stanford: Stanford University Press, 1965).

† See Kenneth R. Walker, *Socialisation and the Private Sector, 1956–1962* (London: Frank Cass, 1965). The importance of the problem is indicated by the fact that in 1956 the private plot had furnished as much as one-seventh of the total calorie intake of the average peasant.

magnitude. Here, too, the attempt, during the period of the 'Great Leap Forward', to increase production by an all-out effort in all directions at once, without adequate centralized planning to ensure that these various activities were compatible with one another, led to serious reverses. The resulting disorganization, plus the shift in emphasis from industry to agriculture necessitated by the food crisis of 1960–62, led to a veritable collapse in industrial production, which has been compared by several authors to a depression in a capitalist country.* The level of industrial production, which had continued to grow in 1959 and the first half of 1960, suddenly dropped far below what it had been in 1957. By 1963 the measures adopted in 1961 and 1962 to create a more balanced economy, with agriculture as the basis, and heavy industry as the guiding factor, had brought about a resumption of progress in industrial production; but the rate of growth probably remained lower than it was under the first five-year plan, and lower than it would have been if the second five-year plan had not been jettisoned in favour of the 'Great Leap'.

There remain the solid accomplishments: a steel production two or three times that of India, though the two countries started from the same extremely low level,† and the capacity to produce not only atomic bombs but a whole series of items such as trucks, locomotives, heavy machinery, and scientific instruments which most countries of Asia and Africa are obliged to import from abroad.

The available information on the Chinese economy is thus ambiguous enough for each observer persuasively to place the emphasis on the successes or on the failures, on the results or on their cost, as he chooses. The attitudes of Asian and African governments in this respect vary exceedingly, depending not

---

* See in particular William W. Hollister, 'Capital Formation in Communist China', *China Quarterly*, No. 17, 1964, pp. 49, 52.

† See Yüan-li Wu, *The Steel Industry in Communist China* (New York: F. A. Praeger, 1965), who gives (p. 285) figures of 8·6 million tons of acceptable-quality steel (out of a total of 13·35 million tons) for 1959, and 12 million tons of acceptable quality (out of a total of 18·45 million tons) for 1960. The remainder represents the output of the 'backyard furnaces' subsequently abandoned because their output was useless for modern industry. Following the 'depression' of 1961–2, output is reported to have returned to a level of 10 to 12 million tons a year.

only on their political tendencies, but on what they hope to gain from Peking's policies of cultivating good relations through cultural exchange and technical and economic assistance. On the whole, China's influence, though significant, is clearly far less than it was a year or two ago. It is certainly less than might have been expected, given the tactless behaviour of the Soviets in the past, in their dealings both with the Chinese and with other Asian and African peoples, and China's natural advantages as the leader and champion of the former colonial peoples. The reasons for this are multiple. China's very limited capacity to offer economic assistance, as compared with that of the Soviet Union, is a factor, but one which should not be overrated. Gifts and loans are valuable, but China's own experience in dealing with economic and social problems similar to those of other developing countries should logically give her a privileged position in supplying technical assistance. If Peking has not exploited this fact more effectively, it is not merely because the Chinese image has been weakened by economic difficulties. The psychological and cultural differences which exist within the Afro-Asian world, as well as between Asia and Europe, are probably even more important. The solemn and puritanical attitude of the Chinese experts has rendered them almost as alien and incomprehensible to the populations among whom they work as the condescension of the Soviets or the conspicuous consumption of the Americans. Moreover, Mao's attitude of unremitting hostility not only toward the imperialist enemy, but toward those whom he regards as creatures of the imperialists, is a source of anxiety to many Asian and African leaders. During the latter half of 1965, the Chinese themselves contributed markedly to the doubts about their intentions both by their bellicose stance during the Indo-Pakistani conflict, and by their apparent involvement in the abortive *coup d'état* in Indonesia.

The Indonesian adventure remains an incomprehensible blunder; but Chinese policy during the border war between India and Pakistan was, in fact, perfectly logical in terms of Mao's goals, both revolutionary and diplomatic. On the one hand, his object, as in 1962, appears to have been to create political disorder in India and to discredit the Indian example of non-alignment by showing that in the last analysis the country was dependent on aid from the 'imperialists'. But at the same time, the Chinese intervention in the autumn of 1965, in an attempt to prolong the conflict, was intended to serve the larger interest of weakening the international order as a whole, which Mao regards as inherently reactionary because it is based on the joint defence of the *status quo* by the United States and the Soviet Union.

Although this attitude is perfectly logical in the double context of Mao's guerrilla experience and the sharp conflict which exists between China and the United States, it is, as already suggested, ill-calculated to win friends in Asia and Africa. Very few among the leaders of the newly independent countries are prepared to 'struggle against imperialism' otherwise than verbally, at the risk of becoming involved in armed combat. The Soviet line toward the Indo-Pakistani conflict, culminating in the Taskhent Agreements, therefore brought Moscow substantial gains in its rivalry with Peking for influence among the non-European countries.

Soviet action in defence of the *status quo* naturally drew from Peking loud cries of 'treason' and 'collusion with American imperialism'. But this does not mean that the Chinese themselves are as consistently and implacably opposed to all compromises with principle as their revolutionary rhetoric would lead one to believe.

Certainly to the extent that Asia and Africa are 'the centre of world contradictions' and 'the storm centres of world revolution', Mao and his comrades are persuaded that in the long run they can overthrow the imperialist system, with the aid of the revolutionaries from these continents. But for the moment, China is in no position to affront American power directly on a world scale. This fact is clearly recognized by the leaders of the Chinese army and serves as a basis for their strategic thinking.* It is also attested by China's extreme prudence over direct involvement in the

---

* This is confirmed most explicitly by the articles contained in the issues of a secret bulletin of the People's Army, dating from the winter of 1961, which found their way into the hands of the United States government and were released by the State Department in 1963. For an analysis of these materials, which are of undoubted authenticity, see the articles of John Wilson Lewis, Alice Langley Hsieh, and John Gittings in *China Quarterly*, No. 18, 1964, pp. 67–117. A complete English translation has now been published: *The Politics of the Chinese Red Army*, edited by J. Chester Cheng (Stanford, Hoover Institution Publications, 1966).

Vietnam war.* It is therefore necessary to take advantage of the 'contradictions among imperialists', and to gather together, in the broadest possible united front, all the countries situated in the 'intermediate zone' between the socialist camp and 'the biggest imperialism, United States imperialism'. And since the Soviet Union, in Mao's view, no longer participates in this struggle, but on the contrary seeks to come to an agreement with the United States to dominate the world, China must take upon herself the primary responsibility for the diplomatic strategy as well as for the revolutionary strategy designed to advance the interests of the peoples of Asia, Africa, and Latin America.

Side by side with a clear revolutionary line of support for wars of national liberation, there is thus a diplomacy which addresses itself not only to the non-aligned powers of Asia and Africa, but to such 'imperialist' governments as that of France, whenever they appear resolved to maintain their national independence.

If Mao no doubt sincerely identifies himself with the anti-imperialist struggles of other peoples, his primary concern remains, as it has always been, the fate of China. At the same time, it must be added that in his eyes China's internal evolution has now taken on decisive international importance. For to the extent that he sees China as the only genuinely socialist great power – the Soviet Union having definitively taken the road of revisionism and the restoration of capitalism – the ideological purity and firmness of will of the Chinese revolutionaries is henceforth the principal guarantee of ultimate victory on a world scale.

It is therefore of the utmost importance that China, in Mao's phrase, should not 'change colour' – i.e., alter her political character. In order to guard against this danger, the hard lessons of past struggles must be brought home to the young people who have grown to maturity since the victory of 1949.

This preoccupation with training succeeding generations of revolutionaries, in order that China may continue to play her role as the vanguard of the world revolution, lies at the heart of the

---

* In his interview of January 1965 with Edgar Snow, Mao expressed doubt on the likelihood of United States intervention in North Vietnam, but declared that in any case, China would not send her armies beyond her own borders, and would fight only if directly attacked by the United States. *The New Republic*, 27 February 1965, pp. 17, 22.

'Great Proletarian Cultural Revolution' that has swept across China during the past year. But in order to understand what Mao is about, we must first go back and pick up certain strands in Chinese internal policy since 1961.

The failure of the Great Leap Forward of 1958–9 led, as described in the previous chapter, to the adoption of a much more prudent and rational economic policy. This was accompanied, for a year or two, by a substantial relaxation of pressure in the ideological sphere. It was recognized that political zeal was no substitute for technical competence, the emphasis on indoctrination was reduced, and a more tolerant attitude prevailed in the domain of art and literature. It was obvious at the time that this whole trend must be repugnant to Mao, since it involved not only the abandonment of a political style peculiarly his own, but the humiliating recognition that his leadership had brought the country to the brink of disaster. On the one hand, he sought consolation in the field of foreign affairs, and especially in the quarrel with the Soviet Union. But after a pause of a year or two he also began once more to foster a radical political climate within the country. At the Ninth Plenum of the Central Committee of the Chinese Communist Party, in January 1961, a warning was issued against an attempted comeback by 'a very small number of unregenerate landlord and bourgeois elements'. The Tenth Plenum, in September 1962, declared that to combat such tendencies a prolonged and acute class struggle would be necessary, and henceforth there was a constantly rising crescendo of exhortations both to study Mao's thought, in order to arm oneself for the revolutionary struggle, and to be vigilant against the class enemy.

This new wave of ideological militancy was not accompanied by any corresponding change in economic policies, which continued on the whole to follow the realistic course laid down in 1960. A gaping contrast thus emerged between the style and the substance of Chinese internal policy. The contradiction was so striking that it was widely suggested that Mao must have reached some compromise with those in the top leadership oriented toward planned economic development, whereby they might pursue such concrete policies as they liked, provided that he was allowed to hide the retreat from the 'Great Leap' behind a suitable ideological smoke screen.

This view (to which I was personally inclined to subscribe down to the beginning of 1966) has now been proved to be altogether false. On the one hand, Mao is not a man to accept defeat or humiliation easily and gracefully, and it is now clear that he never forgave those who in 1959 pointed out the consequences of his 'guerrilla methods' for economic development, and of the rupture with the Soviets for Chinese armaments, and demanded a change to a more moderate approach in internal affairs and a more conciliatory attitude toward Moscow. Only P'eng Te-huai, a military man not known for his subtlety, who had blurted out his opposition too openly, had been eliminated in 1959, but Mao had not forgotten the others. On the other hand, to leave open the contradiction between the style and the substance of internal policy would make it very probable that, following Mao's disappearance, the prudent substance and its partisans would triumph and lead the country in a direction which Mao could only regard as 'bureaucratic' and 'revisionist'.

It would be pointless to speculate here about the way in which Mao went about strengthening his hand and preparing the ground for a new radical upsurge on the whole front of internal and external policy. We do not have as yet any solid information about what went on in the top leadership group in Peking in the years following 1960 though such information may soon be forthcoming. What is certain is that, beginning in 1964, Mao's efforts began to show definite results. It was in the course of this year that the call first went out to train successors in order to ensure that the Chinese revolution would remain on the course laid down by Mao.

The problem of 'revolutionizing' young people, in order to make of them revolutionaries for ever, both at home and abroad, was the central theme of the Ninth Chinese Communist Youth League Congress in June 1964.[6] It figured extensively in the last and most remarkable of the nine Chinese replies to the Soviet blast of 14 July 1963, entitled 'On Khrushchev's Phoney Communism and Its Historical Lessons for the World', issued on the first anniversary of the Soviet article. In this text Mao is credited with the view that a very long period of time is necessary to decide the issue of the struggle between capitalism and socialism. 'Several decades won't do it; success requires anywhere from one to several centuries.' During this period, the proletarian dictatorship must be maintained and strengthened. In a passage attributed to Mao in the form of a direct quotation, it is stated:

Class struggle, the struggle for production, and scientific experiment are the three great revolutionary movements for building a mighty socialist country. These movements are a sure guarantee that Communists will be free from bureaucracy and immune against revisionism and dogmatism, and will for ever remain invincible. They are a reliable guarantee that the proletariat will be able to unite with the broad working masses and realize a democratic dictatorship. If, in the absence of these movements, the landlords, rich peasants, counter-revolutionaries, bad elements and ogres of all kinds were allowed to crawl out, while our cadres were to shut their eyes to all this and in many cases fail even to differentiate between the enemy and ourselves but were to collaborate with the enemy and become corrupted and demoralized . . . then it would not take long, perhaps only several years or a decade, or several decades at most, before a counter-revolutionary restoration on a national scale inevitably occurred, the Marxist-Leninist party would undoubtedly become a revisionist party or a fascist party, and the whole of China would change its colour.*

According to Mao, the decisive factor in avoiding such a 'change of colour' is 'to train and bring up millions of successors who will carry on the cause of proletarian revolution'.[7] But the only true successors, in Mao's eyes, are those who are inspired by his thought and his example.

It is ironic that this call to train successors should have been put forward at the congress of the Chinese Communist Youth League, for at the very same time, during the first half of 1964, events were taking shape which would lead to the jettisoning of the Youth League in favour of a new and infinitely more militant organization: the Red Guards.

The most important factor in these developments was the growing role of the army. Early in 1964, a campaign was launched to 'learn from the People's Liberation Army'. The army was held up as a model of political loyalty and political consciousness, and

* This passage has now been identified as an extract from Mao's remarks, dated 9 May 1963, on documents from Chekiang Province regarding the participation of cadres in productive labour. See *Mao Chu-hsi Yü-lu* (*Quotations from Chairman Mao*), Peking, Political Department of the People's Liberation Army, 1966, pp. 36-7.

'political departments' similar to those in the army were set up in organizations responsible for administering economic enterprises.

The 'Learn from the Army' campaign was entirely in harmony with Mao's mentality and with the place of the army in the Chinese revolution. We have seen that, in Mao's road to power, the Red Army, if it did not actually replace the Communist Party as the incarnation of the 'will of the proletariat', was far more than a simple instrument of the party. To fit the army for its role in the national and social revolution, Mao developed methods of indoctrination which were applied to the troops before they were extended to the country at large. In the history of Chinese Communism, the army was thus the first school for the ideological transformation of the masses. Moreover, Mao regards the army as the natural repository of the ethos of struggle and sacrifice which is for him the hallmark of every true revolutionary movement. The army also tends naturally toward the combination of discipline and initiative which is Mao Tse-tung's constant preoccupation. It is thus not surprising that the heroes recommended as models to Chinese youth in the last few years have been soldiers.

The campaign launched in 1964 did not appear to involve the modification of the Chinese political system by the transfer of political authority to the army. It was, however, a portent of such developments in the future. How much so has only recently been revealed, as the Red Guards' bible, *Quotations from Chairman Mao* (cited in the note on p. 324, and of which I shall have more to say below), has become available outside China. For the first edition of this book, we now learn, was published in May, 1964, on the eve of the Chinese Communist Youth League Congress, and thereafter it was distributed widely as a reward for the meritorious study of Chairman Mao's works. And this volume, which was thus to play a key role in the ideological training of cadres of the party and other organizations, was published by the Political Department of the Army.*

It is clear today that the Army was also involved in two other

* This fact, and the preface containing the information about the first edition (not to be confused with Lin Piao's Foreword to the edition published in December 1966) have been omitted from the English and other foreign-language editions recently published in Peking.

trends which emerged during 1964, and which are central to the current 'Great Proletarian Cultural Revolution': the attack on tradition, and the increasingly extravagant cult of Mao Tse-tung and his thought. The anti-traditionalist orientation found expression in a campaign to reform the Peking opera. For the moment, this did not involve the elimination of all the traditional operas, but merely of some regarded as particularly reactionary, such as those about ghosts. But it also pointed the way to today's all-out onslaught on tradition.

There was also a new upsurge of the cult devoted to Mao and his thought. We have seen that such a cult has flourished since 1945, but in recent years it has attained levels previously unknown. It seemed at the time of the 'Great Leap' that extremes had been reached which would not easily be surpassed. The following is a typical example:

What I call 'belief' means believing in Mao Tse-tung's thought; moreover, this belief must be steadfast and immovable. In the course of China's revolutionary struggles and socialist construction, vast practical experience has demonstrated that Mao Tse-tung's thought is the only correct thought. It is the incarnation of Marxism-Leninism in China; it is the symbol of truth. Therefore, if a person at any time whatever, in any place whatever, regarding any question whatever, manifests wavering in his attitude toward Mao Tse-tung's thought, then, no matter if this wavering is only momentary and slight, it means in reality that the waverer departs from Marxist-Leninist truth and will lose his bearings and commit political errors. So we must follow Chairman Mao steadfastly and eternally! Forward, following a hundred per cent and without the slightest reservation the way of Mao Tse-tung![8]

It was hardly possible to go further than this in proclaiming the absolute and exclusive validity of Mao's thought. Developments since 1964, and especially in the course of 1966, have none the less lifted the Mao cult to a completely new level as regards its intensity and all-pervasiveness, and have also brought striking changes in the nature of that cult. To understand the significance and function of these tendencies, it is necessary to put them in the context of the current political situation as a whole.

In 1965 attention shifted to the countryside with the launching of a campaign for the ideological and economic 'clean-up' of the communes and the sending of large numbers of students to participate in this movement. Then, in November, came the preliminary stage in the 'cultural revolution' that was to begin in earnest in the spring of 1966. The first attacks were directed against Wu Han, vice-mayor of Peking, for his historical play *Hai Jui Dismissed from Office*, written during the ideological thaw of 1959–61, which was interpreted as a glorification of 'feudal' morality and a 'veiled criticism of contemporary people'. When the offensive against Wu Han was taken up again in April, its terms were infinitely more precise and more damning. His play, it was said, was 'directed precisely against the Lu Shan meeting [of July 1959] and against the Central Committee of the party headed by Comrade Mao Tse-tung, with a view to reversing the decisions of that meeting', and of bringing the 'right opportunists' [i.e., P'eng Te-huai] back to office, to administer their 'revisionist programme'.[10]

The new wave of criticism against Wu Han was rapidly extended to the whole of the party organization of the Peking area; P'eng Chen, the mayor of Peking, was reportedly imprisoned late in April, and his dismissal was officially announced at the beginning of June. What had appeared at the outset to be merely another 'rectification' campaign like that of 1957, which would have severely tested the party apparatus but left it largely intact, had turned into a vast unpheaval which was not only to lead to the purge of an unprecedented number of important figures in the party, but would shake the whole political system of the country to its foundations.

Before reviewing the extraordinary events in China since the spring of 1966, it will be well to pause and ask ourselves who launched this movement, and why. There is no doubt that it corresponds to Mao's temperament and political style, and that he fully supports it and gives his approval to all major decisions. Moreover, as already pointed out, the events of 1966 represent the climax of a five-year effort on Mao's part to see his policies of the Great Leap Forward vindicated once more. On the other hand, despite the constant efforts of the Chinese press, beginning with Mao's swim in the Yangtze in July 1966, to demonstrate that he is bursting with health and vigour (or perhaps precisely because of these efforts) there is reason to doubt that he is capable of taking direct charge of events himself.

If the current Great Proletarian Cultural Revolution is to a considerable extent stage-managed for Mao by someone else, a large part of the responsibility obviously rests on Lin Piao. Since he took charge of the People's Liberation Army following the elimination of P'eng Te-huai in 1959, Lin has played an increasingly important role in the affairs of state, but until the latter part of 1965 he lived discreetly out of the public eye. He was generally thought to be in bad health, and hence out of the running as a contender for Mao's succession. The publication of his article 'Long Live the Victory of the People's War', in September 1965, marked the beginning of an exceedingly rapid rise to prominence. In January 1966, Lin Piao's 'five-point principle' for the guidance of the army during the coming year, which made of Chairman Mao's works 'the highest instructions on all aspects of the work of the army', further emphasized his role in the campaign to enforce political orthodoxy just then getting under way.

Although it is clear that the army is not entirely united behind Lin Piao, he does speak, of course, for the group now in control of the military establishment. Thus, it is surely no accident that Lin's ascension into public view should have begun immediately after the issuance of the current edition of *Quotations from Chairman Mao*, the preface of which is dated 1 August 1965 – 1 August, the anniversary of the Nanchang Uprising, being Army Day in China.

Another key figure in the events of 1966, who also emerged from semi-retirement to play a leading role in the Great Proletarian Cultural Revolution, is none other than Mao's wife, today known not under her stage name of Lan-p'ing, but as Comrade Chiang Ch'ing. Since her marriage to Mao in Yenan she had played no open political role whatever, though according to some reports she persistently endeavoured to intervene in cultural affairs. Her appearance to welcome Mme Sukarno in 1962 (see picture following p. 176) was one of the very few occasions on which she had been seen in public in recent years. In the summer of 1966 she became a member of the Central Committee, as well as deputy head of the group responsible for running the 'cultural revolution', and figured prominently in the mass meetings held thereafter; on several occasions she sallied forth alone to resolve conflicts regarding the 'cultural revolution' at Peking University.

Chiang Ch'ing's rise to eminence found its culmination in her appointment as adviser on cultural work to the People's Liberation Army, which was announced on 28 November 1966, at a meeting celebrating the mass induction into the army of the Peking opera troupe and several other musical and theatrical organizations. In her speech on that occasion – which was greeted by a 'thunderous ovation' – Chiang Ch'ing revealed that her 'fairly systematic contact with certain sections of literature and art' had begun 'a few years' previously. We may assume that one of the first episodes in her intervention in this field was precisely the reform of the Peking opera beginning in 1964. As regards the substance of cultural policy, she affirmed flatly that the 'critical assimilation' of the Chinese heritage was 'impossible', thus completely reversing the position of her husband, who in the past had come out repeatedly in favour of the selective assimilation of all that was precious in China's past. She also displayed her discriminating knowledge of Western culture by lumping together 'rock-and-roll, jazz, strip-tease, impressionism, symbolism, abstractionism, fauvism, modernism' as things 'intended to poison and paralyse the minds of the people'.*

Assuming that leadership in the Great Proletarian Cultural Revolution belongs largely to the trio Mao Tse-tung – Lin Piao – Chiang Ch'ing, why did they decide to launch this movement? Fairly obviously, it was in order to deal with opposition within the Chinese Communist Party toward the radical policies they favour. It is not unlikely that Mao himself originally thought he could impose his will by a conventional type of rectification campaign, such as that which had been started in late 1965. When the opposition proved to be more widespread and powerful than he had imagined, he had recourse to stronger measures.

What was the intraparty debate about? We may assume that the central issues were those which have remained unresolved in China since 1958: mass action versus bureaucracy, political zeal

* *Peking Review*, No. 50, 9 December 1966, pp. 6–9. The speech as a whole, and particularly Chiang Ch'ing's remarks on the Peking opera troupe and its leading members, is an obvious and depressing example of a rancourous and spiteful person enjoying her hour of triumph over the party authorities who have so long slighted her and frustrated her ambitions, and the distinguished performers of the traditional theatre whose success she envies.

versus technical competence, men versus weapons, and men versus machines. If I am correct in assuming that for the past five years Mao has been waiting until the time was ripe to impose a new leap forward, economic policy must have been a major issue. This time Mao was resolved to eliminate opposition *before* launching a new leap, and his suspicion undoubtedly fell on all those who had shown a lack of enthusiasm for his policies in 1958–9, of whom Liu Shao-ch'i was evidently one. These sceptics perhaps also ventured to think that 'Mao Tse-tung's thought' placed too heavy an accent on the omnipotence of the human will, as compared to the rational elements in Marxism, and was better adapted to inspiring guerrilla fighters than to building a modern economy.*

Undoubtedly the war in Vietnam and the possibility of an American attack against China herself were also subjects of discussion. Some observers of the Chinese scene have made of this the central point and have suggested that the Great Proletarian Cultural Revolution as a whole should be viewed primarily as an attempt to prepare for a war with the United States, which Mao regards as henceforth inevitable. I cannot subscribe to this view. The events of the past year appear to me to be above all an attempt to reshape China and the Chinese people. But it is very likely that the anxiety inspired in Peking by events in Southeast Asia helped Mao and Lin Piao to impose their radical and uncompromising line on the Central Committee.

---

* A quarter of a century ago, in a letter written shortly before the beginning of the rectification campaign of 1942 which marked the first stage in the cult of Mao Tse-tung's thought, Liu Shao-ch'i declared flatly that the theoretical level of the Chinese Communist Party remained exceedingly low. Very few party members, Liu pointed out, were capable of reading Marx and Lenin in the original, and even these (of whom Mao was obviously not one) had rarely read the Marxist classics thoroughly. As a result, little headway had been made in the difficult task of the Sinification of Marxism. (Liu Shao-ch'i, 'Letter to Comrade Sung Liang', in *Lun Tang*, Dairen, Ta-chung Shu-tien, 1946, pp. 343–4.) As we saw earlier, in April 1945 at the Seventh Congress of the Chinese Communist Party, Liu Shao-ch'i hailed Mao's thought as an admirable example of the Sinification or nationalization of Marxism. But there is clearly reason to ask oneself whether he was perfectly sincere. The problem of the relations, both intellectual and personal, between Mao and Liu, which is highlighted by the current open conflict between the two men, has never been adequately studied.

Whatever the issues in the debate, it is clear from the events since the spring of 1966 that Mao's position by no means won universal acceptance throughout the party apparatus. For if it had, Mao would hardly have embarked on the extraordinary and perilous adventure of creating an entirely new organization, the 'Red Guards', which is beyond the control of the party officials except Mao and his henchmen. This enterprise is, of course, entirely without precedent in the forty-nine-year history of Communist régimes, which have always taken as their most fundamental axiom the predominance of the party over all other forms of political and social organization. It is also in contradiction to Mao's own principle, laid down in 1938: 'The party commands the gun; the gun must never be allowed to command the party.' For the Red Guards, although they harness the enthusiasm of adolescents delighted to occupy the centre of the stage, were created and guided by the army, and continue to take the army as their model and inspiration.

It would be oversimplifying a complex problem to suggest that Lin Piao has simply taken over control in Peking as the head of the army. Because the Chinese Communists came to power after decades of warfare, there has always existed in their ranks an interpenetration of the political and military élites, with the army playing a political role in the base areas and virtually all of the top party leaders actually exercising command in the field. Mao has most likely come to rely on Lin Piao and certain other military men, not because they are solidiers, but simply because they appear to be the 'reddest' Communists of all – that is to say, those whose political style corresponds most closely to Mao's own guerrilla mentality. But since 1949, the civil and military functions have been more sharply differentiated. It is therefore highly important that the army as such has played a key role not only in organizing the Red Guards in the summer of 1966, but in the preparatory stages leading to these developments. We have seen that what was to become the bible of the Red Guards was published by the Political Department of the army in May 1964, and republished in revised form in August 1965. And beginning with the first attacks on Wu Han in November 1965, the army newspaper has been in the forefront of the 'cultural revolution'.

In any case, the radical and guerrilla-minded faction of which Mao and Lin Piao are the leaders, and which undoubtedly cuts across both the party and the army, is embarked on an enterprise that is in total contradiction to the ideal of discipline characterizing most armies, namely the promotion of 'revolt'. A 'big-character poster' put up by the Red Guards of the Middle School attached to Tsinghua University in Peking in July, before the movement emerged into public view, and later reproduced with approval in the theoretical journal *Red Flag*, declares:

All present-day reactionaries and those of antiquity, in China and in other countries, say: Exploitation is justified, oppression is justified, aggression is justified, and revisionist rule is justified; but it is unjustifiable for the proletariat to rebel. It is Chairman Mao, our most respected and beloved leader and the greatest revolutionary teacher, who turned this pigheaded theory right side up. Chairman Mao has said: 'In the last analysis, all the truths of Marxism can be summed up in one sentence: To rebel (*tsao fan*) is justified.' The current great proletarian cultural revolution is a great revolutionary rebellion. We will stage a great rebellion against whoever is revisionist and opposed to Mao Tse-tung's thought.

To rebel is in the tradition of us proletarian revolutionaries, the tradition which the Red Guards must carry on and develop. We rebelled in the past, rebel now and will rebel in the future! We will rebel as long as there are classes and class struggle! We will rebel as long as there are contradictions! The revolutionary rebel spirit is needed for a hundred years, a thousand years, ten thousand years, and 100 million years to come!*

At first glance it appears exceedingly singular that Mao should encourage young people to revolt in a country which has been under Communist rule for seventeen years, especially as this revolt is directed against 'persons within the party who have been

---

* *Peking Review*, No. 38 (16 September 1966); Chinese text in *Hung Ch'i*, No. 11, p. 29. This poster was written on 27 July; it was published only on 21 August, after the first mass meeting attended by the Red Guards on 18 August. (The Red Guards unit at the Middle School attached to Tsinghua University was the first experimental unit, created by officers of the People's Liberation Army in late June.) The citation from Mao Tse-tung, which was published in lengthier form in the Peking press at the same time as the poster, is from Mao's speech at Yenan on the occasion of Stalin's sixtieth birthday in 1939. Full text in *Hsin-hua Yüeh-pao*, Vol. I, No. 3, pp. 581–2.

in authority, and have taken the capitalist road'.[11] To be sure, these persons are said to be only a handful, but in fact the resistance of the party apparatus is obviously much greater than these optimistic official statements would imply, and Mao's aim is not merely to eliminate a few individuals. He is bent on nothing less than smashing the entire party organization as it now exists, and building it up again from the bottom – no doubt incorporating into it in the process a great many revolutionary cadres and militants drawn from the Red Guards and others who have come to the fore in the course of the Great Proletarian Cultural Revolution. In order to attain this end, he has not shrunk back from the possible consequences of a period of disorganization. As the Red Guards of the Middle School attached to Tsinghua University wrote in their first poster, the aim is to 'turn the old world upside down, smash it to pieces, pulverize it, and create chaos – the greater the confusion the better!'[12]

What does Mao want to bring out of this chaos? His ambition is apparently to create a party organization of a new type, with built-in safeguards against 'bureaucracy'. In particular, the 'Cultural Revolution Groups' which emerged during the spring and summer of 1966 are to be made permanent.[13] They are to be chosen by a system of general elections modelled on that of the Paris Commune, and their members are to be subject to recall. Thus it will not be possible for the cadres to detach themselves from the masses and to become 'bourgeois authorities'.

In a country where two such virulent bureaucratic traditions as that of the Chinese mandarin and the Communist *apparatchik* make their influence felt simultaneously, the aim of fostering as an antidote direct participation by the citizen in public affairs is in itself a laudable one. It is also, as we have seen, characteristic of Mao's political style as it has developed over the past half century. Stalin's way was to rule through the party bureaucracy; and when he lost confidence in the party organization in the 1930s, he ruthlessly purged it with the aid of another organization – the secret police. Mao, on the other hand, has always insisted on action through the masses. Even today, we are told that the 'conscious action of hundreds of millions of people' is the essence of the 'proletarian dictatorship' as it is exercised in China as an instrument for rooting out those in the party leadership who are 'taking the capitalist road'.* In fact, we have seen that the shock force in Mao's assault on the party apparatus is discreetly but effectively guided by the army. And though the actions of the Red Guards have a greater appearance of spontaneity than those of the N.K.V.D., there is room for doubt as to whether a combination of juvenile fanaticism and mob violence constitutes a valid alternative to bureaucracy as a technique for governing a modern nation.

None the less, Mao's current policies in the organizational sphere may be regarded as to some extent the culmination (and the *reductio ad absurdum*) of his previous line. At the same time, they also mark in certain respects a sharp departure from his past behaviour. This is particularly the case as regards the rupture of the unity of the top leadership, which has been one of the most remarkable phenomena distinguishing the history of China under Mao from that of the Soviet Union under Stalin. It is possible that such was not Mao's original intention. Although resentment at past criticism of his 'guerrilla methods' has certainly not been absent from his motives during the past year, he may at first have been more concerned to struggle *for* certain policies and methods of work, than *against* those opposed to his policies. But from the beginning, the 'cultural revolution' was also aimed at asserting Mao's own authority, and that of Lin Piao as his successor. Moreover, the divergence regarding policy between Mao and Lin on the one hand, and Liu Shao-ch'i, Teng Hsiao-p'ing and the party hierarchy on the other, was so radical that the struggle to impose a policy rapidly became a struggle for political survival – and perhaps, ultimately, even for personal survival. Moreover, although Mao's ultimate aim was the creation of a Communist party of a new type, he was soon obliged, in the face of the resistance he encountered, to abandon temporarily the effort to reform the party from within in favour of an all-out assault on the party with the support of the army. Thus he united against himself men in the top leadership of the party who had not, in the past, necessarily shared the same views,

* See the editorial 'The Dictatorship of the Proletariat and the Great Proletarian Cultural Revolution', translated in *Peking Review*, No. 52, 1966, p. 20.

still less constituted a single faction, but who had in common their attachment to the organization which they had created and led.

The situation as regards the ideological content of the Great Proletarian Cultural Revolution is analogous to that in the domain of organization. Here, too, Mao's current policies, though in some respects they grow out of Mao's past, also constitute the negation and even the betrayal of the past.

We have seen that, as early as 1918, Mao regarded the 'three bonds' of Confucian morality as one of the 'evil demons' which must be destroyed if China was to progress. Ever since then he has remained persuaded that China could be regenerated only through a profound cultural revolution which would change the thinking and habits of the Chinese and fit them to survive in the modern world. For a brief moment he thought that this could be achieved by Ch'en Tu-hsiu's 'Mr Democracy' and 'Mr Science'. Then he was converted to Leninism; but if, in terms of political methods, this meant that he had abandoned individualism for collectivism, in the domain of culture Mao remained committed to certain Westernizing values. Chief among these were the replacement of the ideal of harmony with nature, which had always characterized Chinese society, with a Promethean urge to transform nature, the destruction of superstition, and the abandonment of traditional attitudes of submission to superiors such as fathers, husbands, clan elders, and bureaucrats.

In 1938, Mao put forward the slogan 'Sinification of Marxism' and continued thereafter to promote this idea. But if 'Sinification' meant dressing up Marxism in Chinese metaphors, illustrating it with Chinese examples, and also, to a certain extent, mixing it up with selected ideas and values judged 'progressive' but drawn from Chinese historical and philosophical taditions, the core of Mao's 'Sinified' Marxism remained modernizing and iconoclastic in relation to Chinese culture as a whole.

Though Mao's amalgam of Marxism with Chinese elements was easily assimilated by his compatriots, there was also the danger that they might retain primarily what was Chinese and familiar, and fail to grasp the new and specifically Marxist ideas indispensable to social revolution and economic development. And if the valorization of the past strengthened Chinese national pride, the language of traditional metaphors was not a suitable vehicle for affirming the universality of the Chinese revolution. It was therefore logical that a point should ultimately be reached at which the preservation of national values would appear as an encumbrance rather than a source of strength.

It seemed, when the Great Proletarian Cultural Revolution began in the spring of 1966, that this was precisely what was happening. Mao's slogan was no longer the 'Sinification of Marxism', but a 'proletarian culture', and instead of the emphasis on the great unity of 95 per cent of the Chinese people, there was constant talk of class struggle and proletarian dictatorship.

To be sure, no one doubted that 'proletarian dictatorship' meant rule by Mao, and that 'class struggle' meant primarily struggle against elements in the party opposed to Mao. But it did appear that even the sectarian and primitive version of Marxism propagated under the label of 'Mao Tse-tung's Thought' would serve to inculcate attitudes necessary to economic progress. A careful reading of the innumerable 'philosophical' articles by workers and peasants published in the Chinese press revealed that what their authors had learned from the study of Mao's thought was to be resourceful, to look at all sides of a problem, to test their ideas by experiment, and to work hard for the sake of the common good.

This rational kernel in the Great Proletarian Cultural Revolution, while it has not entirely disappeared from view, has been largely swallowed up in a mass movement which has attained levels – or at least forms – of irrationality previously unknown even in Stalin's Russia. In essence, this trend, which emerged in the middle of August, combines a cult of Mao's person of an entirely new type with the transformation of the 'Thought of Mao Tse-tung' from an ideology into a kind of Marxist Koran endowed with magical virtues.

We have seen frequently in the course of this book how thoroughly Mao's works, especially those dating from the 1920s and 1930s, were cut and rewritten for the current canon to make it appear that Mao's position had been more consistent than in fact it was. Nevertheless, the *Selected Works*, consisting of relatively long and relatively complete texts presented in chronological order, gave the reader some kind of picture of the development of Mao's thought, as he struggled over the years to adapt himself to changing circumstances.

In the *Quotations from Chairman Mao*, which became widely known in the summer of 1966 and were reproduced one by one in the press, any such historical perspective has disappeared. These short extracts of a few lines each are chosen in such a manner that they appear, not as fragments of analysis of concrete historical situations, but as universally valid precepts to be learned by heart. To be sure, the Red Guards and others who read these texts are supposed to 'apply' them. But what is to be applied is not Mao's method of analysis, but his conclusions, which amount essentially to the affirmation of the omnipotence of the human will and the exhortation to struggle relentlessly and accept no compromise whatsoever, either with 'U.S. imperialism' and 'Soviet revisionism' or with the objective difficulties encountered in the path of economic development.

Thus, of the two basic strands in Marxism, the critical analysis of historical circumstances is almost totally subordinated to the revolutionary will. This in itself, though it represents a simplification of the content of 'Mao Tse-tung's Thought', is in a sense a logical continuation of the trend toward intellectual impoverishment which has for several years past characterized Mao's China, as it characterized Stalin's Russia. But since August the Chinese have gone a step further in a direction which bears no relation whatever to Marxism and has no precedent even in the Soviet Union of the early 1950s. This involves, as I have already suggested, the conferring of magical virtues not only on Mao's thought but on the physical object – the little red plastic-bound volume of *Quotations from Chairman Mao* – which contains it.

This development is intimately linked with the growth and transformation of the Mao cult, which has attained in the past few months a level which leaves that of Stalin completely in the shade. This is true, first of all, in simple quantitative terms. Mao's photo and Mao's name are far more ubiquitously and insistently present in the Chinese press than were Stalin's in the

Soviet press fifteen years ago.* But qualitatively the difference is even more striking.

Until very recently, although Mao and his thought were the object of the highest respect, his physical presence as such did not play any great role in his leadership style. With the exception of the banquet and parade on the occasion of the Chinese national day, he seldom appeared in public. Though he was not obliged, like Stalin, to avoid crowds for the sake of security, he preferred to make known his views either through written statements or through speeches before closed groups of the party or state apparatus, and leave the mass meetings to others. A certain element of mystery and withdrawal was apparently thought desirable to enhance his prestige.

During the winter of 1965–6 Mao disappeared completely from public view, to such an extent that he was widely thought to be moribund or, in any case, permanently incapable of playing an active role. (I was strongly inclined to believe this myself at the time.) It is still not clear whether this absence was caused primarily by his health (which is certainly not good), by pressure from those opposed to his policies, or by a desire on his own part to prepare the dramatic events which began in the spring. It is now claimed that he was in Shanghai preparing the 'cultural revolution' and the Red Guard movement. In any case, Mao was photographed in early May with the Albanian delegation then visiting China, and reappeared to public view when he swam in the Yangtze on 16 July. He 'met the masses' on 10 August in a hall near the headquarters of the Chinese Communist Party in Peking. But it was the rally of 18 August in T'ien An Men Square, the first of several such gatherings, that marked the veritable starting point for the singular developments we are now witnessing in China.

It was on this occasion that the Red Guards made their first official appearance, though they had been seen on the streets of

* For example, in an issue of *People's Daily* chosen at random – that of 4 September 1966 – Mao's name appears 280 times in six pages. In the issue of *Pravda* published on 7 November 1952, the last Soviet national day during Stalin's lifetime, the Soviet dictator's name was mentioned only 99 times in six pages. Mao's photo occupies the entire front page of the 1 October 1966 issue of *People's Daily*; Stalin was generally satisfied with a quarter of a page.

Peking for several days previously. In the course of the afternoon's proceedings, a girl student placed a Red Guard armband on Mao's arm, thus symbolizing the personal union between the 'great teacher, great leader, great supreme commander, and great helmsman' (as Mao is henceforth called) and the young activists who are his instrument in carrying out the 'cultural revolution'. The Red Guards waved in the air their red-bound volumes of *Quotations from Chairman Mao*, thus producing a characteristic effect which has been repeated and amplified on each subsequent occasion. Two passages from official accounts of mass meetings in T'ien An Men Square project more vividly and concisely than any analysis the level attained today by the Mao cult and the way in which the cult of his thought is linked to the cult of his person.

First, from the account of the mass rally attended by one and a half million people to mark the national day on 1 October:

When the morning sun shed its shimmering rays over the city, throngs extending over dozens of *li* had already been converging on T'ien An Men Square and the boulevard east of it. Basking in the early sunshine, the crowds recited quotations from Chairman Mao's works and read the paean dedicated to him: 'The red sun rises before us. Its splendour reddens the great earth. Our great leader, beloved Chairman Mao, may you be with us for ever.' . . .

Red balloons which trailed big streamers with slogans slowly floated in the red sunshine and hovered above the Red Guards and Young Pioneers massed in the square. Then big characters 'Long live Chairman Mao!' formed by bouquets of flowers in the hands of more than a hundred thousand people appeared on the south side of the square. The square glowed with thousands upon thousands of hands waving their *Quotations from Chairman Mao*. The people thronged T'ien An Men Square, which is 400,000 square meters in area, as well as the wide street east of it, and the square became a roaring ocean of red. The shouting of slogans mingling with cheers sounded like spring thunders, unceasing and deafening.

At this moment many jotted down these words in the flyleaf of their *Quotations from Chairman Mao* to commemorate this moment of great joy: '10 a.m. exactly, October 1, 1966.'[14]

The rally on 18 October for the Red Guards is described in similar terms:

The sky was azure and Peking basked in the golden sunshine. When the sun and its myriad of resplendent rays appeared over the horizon, the great mass of Red Guards and revolutionary teachers and students, militant and alert and with red flags and portraits of Chairman Mao held high, had already been converging from all directions on T'ien An Men Square and the wide streets running into it. These young fighters, . . . each carrying a copy of the bright red-covered *Quotations from Chairman Mao*, made up contingents extending for over 50 *li*. They formed a magnificent stream of red. They recited over and over again passages from Chairman Mao's writings. . . .

At ten minutes to one the majestic strains of 'The East Is Red' were struck up, and the happiest moment which people had been looking forward to day and night had arrived.

Chairman Mao, our most, most respected and beloved leader, and his close comrade-in-arms Comrade Lin Piao, together with other leading comrades of the party centre, rode in nine open cars. As they neared the great mass of Red Guards and revolutionary teachers and students, loud bursts of joy roared from the square and the wide streets. Countless hands waved dazzling copies of *Quotations from Chairman Mao* and countless pairs of eyes turned towards the direction of the reddest red sun. Shouts and cheers of 'Long live Chairman Mao! Long live Chairman Mao!' thundered forth. . . .

When Chairman Mao drove past the ranks of the revolutionary teachers and students . . . , many students quickly opened their copies of *Quotations from Chairman Mao* and wrote the same words on the flyleaf 'At 1:10 p.m. on October 18, the most, most happy and the most, most unforgettable moment in my life, I saw Chairman Mao, the never-setting red sun.'[15]

Apart from the sheer intensity of adulation expressed in these texts – and in the events they describe – the growing personalization of the Mao cult is also strikingly conveyed. In the past, Mao was admired and even revered, but in a more abstract and impersonal way, either for his 'thought' or for his contributions as a revolutionary leader. Today, his very presence, and even more so the opportunity to shake his hand, causes such emotion that those who see or meet him are frequently described as weeping uncontrollably afterward.[16] This fetishism of the leader goes hand in hand with the fetishism of his works; the Red Guards carry their little red volumes with them wherever they go, and sleep with them by their side.

It is not easy to pass judgement on a phenomenon of this magnitude. Clearly more is involved than an artificially created mass hysteria. Although there is undoubtedly deep and widespread dissent both within the party and outside it, Mao probably still enjoys a degree of popular adhesion substantially greater than that in the Soviet Union under Stalin, who ruled by sheer police terror. At the same time, there is reason to wonder whether Mao's popularity has not already been gravely undermined by the massive use of violence in recent months. During the wave of terror unleashed by the Red Guards in August and September 1966, the number of people savagely beaten was probably several tens of thousands, of whom several thousand were actually beaten to death. And back of the Red Guards stands, as everyone knows, Lin Piao with his army. This situation is hardly calculated to encourage the public expression of dissent, but neither is it likely to strengthen the citizen's feeling of identification with his government.

In any case, there is probably more interference with the life of the individual than there was in Russia even during the purges of the thirties, though the consequences for the victims are not invariably death or the forced labour camps. Stalin ruthlessly liquidated political dissidence, real or imagined; the Red Guards enforce not only political orthodoxy, but a 'revolutionary' puritan morality as regards dress and behaviour. Moreover, Mao asks not merely passive acceptance of his political line, but active participation by every citizen.

Are certain social categories more inclined than others to support Mao's current policies? While all such generalizations can be only tentative and approximate, it is hard to escape the impression that the 'cultural revolution' is to a large extent a movement of the countryside against the cities, and of the peasants against the workers. This is not merely confirmed by a certain amount of direct testimony; it is entirely in keeping with the organizational and ideological content of the power struggle between Mao Tse-tung and Lin Piao on the one hand, and the leaders of the party on the other.

From the organizational standpoint, the army is largely a peasant army, and Mao has never forgotten that the urban workers did not lift a finger on behalf of the Communist victory of 1949, but waited passively for the army to occupy the cities. The party, on the other hand, although it was built up in the course of guerrilla struggles in the countryside, is naturally prone to attach the greatest importance to the urban proletariat, and we have seen that such has long been Liu Shao-ch'i's tendency.

On a deeper level, the mentality of the industrial workers is largely foreign to the China Mao is trying to create today. Having assimilated certain skills, they tend to regard technical knowledge as important, thus going against the grain of the current orthodoxy which regards the reciting of phrases from Mao's works, learned by rote, as a panacea for developing production. Moreover, because they are, in a still largely agricultural China, an élite in terms both of the modern knowledge they possess and the favoured economic situation they enjoy, they are out of tune with the extreme equalitarianism which is also one of the hallmarks of a policy which proposes to replace professional writers by collectives of workers, peasants, and soldiers. And they are probably less prone than the peasantry to participate in the extreme and primitive manifestations of xenophobia which have become more and more frequent since the middle of 1966.

On the other hand, it is by no means certain that a majority of the peasantry supports the 'cultural revolution' either. The similar but less radical policies of 1958–60 led to open revolt in certain rural areas; it would be surprising if this time the reaction were one of unbridled enthusiasm. Moreover, according to some reports the party organization has maintained its existence and its control better in the countryside than in the capital.

Mao's recent efforts to extend the 'cultural revolution' into the factories and communes would tend to indicate that he is not unduly confident of the support of either workers or peasants.

Understandably the most enthusiastic support comes from youth. The great majority of the Red Guards were born after 1949, and all of them have been taught during the whole of their conscious lives to regard Chairman Mao as the saviour of China and a kind and solicitous father figure. Moreover, they have not been steeped like their elders in the culture of the past, and this, joined to youthful exuberance, makes them the natural and enthusiastic instruments of the smashing of statues, burning of books, and defacing of pictures which occurred in Peking and other cities in August and September of 1966. Quantitatively, this vandalism probably has been less than in France at the time of the revolution, or in England at the time of the dissolution of the monasteries by Henry VIII. But given the profound respect for the heritage of the past which undoubtedly still exists among many older Chinese, the psychological shock may be even greater. The numerous suicides among the élite of China's writers and artists may well be the result not merely of the harassment to which they have been subjected by the Red Guards, but of despair at the wanton destruction of elements in China's literary and artistic heritage which only primitive-minded fanatics can regard as reactionary.

The forms taken by the Mao cult today appear even stranger against the background of this iconoclasm. Without assuming that they represent simply a new metamorphosis of the imperial tradition, it is clear that they owe a great deal more to certain patterns from the Chinese past than to Marxism. Ironically, the Great Proletarian Cultural Revolution, which presents itself as an attack on the 'bourgeois' and 'feudal' values of the past in the name of universal proletarian truth, is accompanied by developments which contradict both the universalist and the rationalist elements in Marxism.

In a way, these events are a logical culmination of Mao's life. A central theme of this biography has been Mao's participation, over half a century, in the effort of the Chinese to modernize and develop their country, and at the same time to remain themselves. While recognizing that Western ideas were needed, as his teacher Yang Ch'ang-chi wrote, to prod China into movement again, Mao has placed the accent from beginning to end on the need for the Chinese to do things in their own way, rather than merely copy the West. As we saw earlier, he attracted attention as early as the winter of 1918–19, at the meetings of Li Ta-chao's Marxist Study Group, by his efforts to combine socialism with ideas from China's past. The current Great Proletarian Cultural Revolution can be viewed as a deliberate and nostalgic attempt on Mao's part to relive the cultural revolution of the 4 May period, and to recapture his own youth by projecting another generation of adolescents on to the political stage as his own generation suddenly came to the fore fifty years ago, as the promoters of a new attempt to renovate China.

This time, of course, the whole process takes place in Marxist terms; it is not an encounter among diverse schools of thought, but reflects a debate within the élite about the correct interpretation of Marxism-Leninism today. This in itself is perhaps less important than the shift, behind a façade of Marxist verbiage, away from the participation in the world intellectual community which was the essence of the 'new culture' movement of Mao's youth. In the 4 May movement, the Confucian tradition was attacked in the name of Western scientific ideas and Western philosophies, from Dewey to Marx. Today, both the Chinese past and Western influences (including those emanating from a 'revisionist' and 'decadent' Soviet Union) are rejected in the name of 'Mao Tse-tung's Thought'. This new dogma is not the synthesis of Marxism and the Chinese heritage which has long been Mao's professed aim, but a kind of revolutionary litany to the glory of Mao and of the Chinese people. In this context, moreover, the leaders in Peking are no longer content to describe Africa and Asia in general as the storm-centre of the revolutionary tempests. In recent months, it has been declared most emphatically that China is the centre and the leader of the world revolution.

Given the high value which the Chinese have always placed on their own civilization, it is not surprising – quite apart from Mao's own vanity – that the thought of their leader should be presented as a contribution to revolutionary theory in some ways more important than that of Marx himself. But it is a tragedy for the people of China that the current orthodoxy should so completely lose sight of the element of criticism and protest inherent not only in Marxism, but in Mao's own earlier writings and actions. Mao has struggled all his life against the supersitions of the old society; he is now engaged in implanting new superstitions. In one of the three brief 'standing articles' – so called because they are recommended to the Chinese people today for constant study – Mao wrote in 1944: 'If we have shortcomings, we are not afraid to have them pointed out and criticized, because we serve the people.'[17] Those who venture to criticize him openly today are liable to be beaten to death by the Red Guards.

It would assuredly be wrong to interpret current developments as simply the effort of an ageing despot to cling to power and feed his own vanity. Mao is no doubt sincerely convinced that his leadership and his teachings are indispensable to the salvation of

China. But it is impossible not to see in the Great Proletarian Cultural Revolution an attempt by Mao to erect in his own lifetime a monument to himself more lasting than the pyramids: a China which will apply his thought and revere his name for centuries to come. In January 1965 Mao told Edgar Snow that future generations in China would assess the work of the revolution in accordance with their own values and their own experience. A thousand years hence, he said, even Marx and Lenin might appear a little ridiculous.[18] In abandoning this detached attitude and endeavouring to fix his own place in history, he has not increased his ultimate stature.

Agnes Smedley said of Mao thirty years ago that 'his spirit dwelt within itself, isolating him'.* Such isolation easily leads, in the long run, to loss of contact with reality and impatience with every form of opposition, even from one's oldest and most faithful comrades. But the violence of Mao's reaction to the implicit and explicit criticism of his policies in recent years is no doubt further increased by an uneasy feeling that perhaps his experience may no longer be relevant to China's problems today.

If this is indeed the case, it is not so much a reflection on his abilities as the consequence of an objective situation. The gulf between the traditionalist China in which Mao grew to manhood and the intellectual and organizational needs of a society which is endeavouring to assimilate the most advanced forms of science and technology in the second half of the twentieth century is too vast to be bridged by a single human being. Certain of Mao's comrades and rivals in the Chinese Communist Party were unquestionably better prepared than he, by education and turn of mind, to absorb the political and economic lessons which a modernizing China would have to learn from the Soviet Union and the West. But by this very token they were less deeply rooted in Chinese tradition and out of touch with the real needs and aspirations of Chinese society in the 1920s and 1930s. Mao was supremely well attuned to those needs, and thus he was able to play the role he did. He is today not willing to recognize that the long period during which his ideas and methods were in harmony with China's needs has come to an end. Hence the constant and increasingly shrill campaigns directed at showing that every

* See above, p. 210.

word in his writings is still valid and that every lesson of his guerrilla experience is still applicable today.

In the end, it may well be that Mao's 'Sinified' Marxism will undergo an evolution within the Chinese and world context not unlike that of Marxism within the European context. Marx based the whole dynamism of his system on the revolutionary force of a class, the proletariat, which he regarded as totally alienated from the existing society and totally deprived of the fruits of the economic development of that society. If this doctine has come to appear increasingly irrelevant in the advanced industrial nations of Europe and America, this has been in direct proportion to the growing participation of the workers in the benefits of society, and their growing sense of identification with existing society. Moreover, by drawing attention to injustice, Marxism in fact contributed to its own obsolescence. Mao, as we have seen, today largely substitutes an 'external proletariat' – the peoples of Asia, Africa, and Latin America – for the urban working class which was seen by Marx as the bearer of redemption for society as a whole. Just as it was the integration of the workers into American and European society which partly invalidated the Marxist protest against existing society, so the integration of China into the world community can ultimately be expected to diminish the virulence of the Maoist protest against the 'domination of imperialism'. And just as the most effective answer to the Marxist revolutionary movement within Western society proved to be not repression, but the reform of injustice, so the most effective answer to the Maoist challenge is not mere military containment, but a positive effort to draw China into the family of nations and aid her people to enjoy their fair share of the world's resources.

It is certain that the hysterical and xenophobic atmosphere which prevails in Peking at the present time does not contribute to such hopeful developments. But Americans cannot view the situation with any complacency. Although, as I said earlier, I am convinced that the primary impetus of the current upheaval grows out of Mao's implacable determination to transform the face of China and place his mark on it for ever, there is little doubt that the external situation has played a significant role in debates within the Chinese Communist Party, and also in relations between the party and the army. The Red Guards have been justified in official pronouncements not only as the vanguard of the cultural revolution, but also as the reserve force of the People's Liberation Army in the coming war for the defence of China against American attack. Even though we may be persuaded that no government of the United States would be stupid and irresponsible enough deliberately to engage the country in a war against China, we can hardly be surprised that events in Vietnam have inspired such fears in Peking. The continuance of this war on China's doorstep can hardly be expected to encourage liberalism and tolerance in that country. Nor can the daily spectacle of a mighty industrial nation raining tons of bombs on a small Asiatic people be expected to attenuate Mao's vision of the world today as primarily a struggle between the 'revolutionary countryside' of Asia, Africa, and Latin America on the one side, and the 'imperialist' cities of Europe and North America on the other. Above all, the situation in Vietnam contributes to the survival of Mao's guerrilla myth, according to which the only valid and genuinely revolutionary answer to any problem whatsoever lies in armed struggle.

As the events of the past year have unfolded in Peking, there have been two opposite reactions in the United States: anxiety lest these radical tendencies ultimately lead to foreign adventures, and satisfaction at the prospect that China would be too much occupied with her internal problems to intervene in Vietnam or elsewhere. Although the first possibility cannot be ruled out, the second seems much more likely. In the coming years, we may well see a China turning on herself in bitterness and frustration, away from a world in which the ostensibly socialist countries such as the U.S.S.R. are, in Chinese eyes, abandoning their principles to worship the golden calf of individual well-being, and the peoples of Asia, Africa, and Latin America have, on the whole, shown themselves unable or unwilling to carry on effective revolutionary struggles inspired by the Chinese example. Although this seems less disturbing in the short run than the perspective of military expansion in south-east Asia, it is clear that in the long run no one could benefit from a situation in which a nation of 700 or 800 million was thus cut off from the rest of humanity. To avoid such a divorce between China and the world is one of the few fixed and clear objectives that can be formulated as we enter a period in which American policy-makers may find themselves faced with the most unpredictable problems and choices.

Indeed, the only thing that can be said with certainty about the future of Mao's China is that almost anything is possible. In the struggle at the top, the odds seem slightly in favour of a victory by Mao and his faction. But even if, as the gentle and retiring Chiang Ch'ing is said to desire, the whole group of those opposed to Mao, beginning with Liu Shao-ch'i, were eliminated once and for all by summary execution, many obstacles would remain to the establishment of a viable political system based on the principles and methods of the Great Proletarian Cultural Revolution. Elements in the party organization at the middle and lower levels might oppose Mao's attempt to rule the country through the army and the Red Guards. It is even possible that party and/or military leaders in the provinces, some of whom have successfully resisted the penetration of the Red Guard movement into their territory, might assert their autonomy – either to await the outcome of events, or to take up the standard of revolt in the name of the party against Mao's personal rule.

In the short run, it is quite possible that Mao's prestige and the overwhelming superiority of the armed forces at his disposal may enable him to crush all opposition and maintain control of the whole country. But even in that case it is hard to see how a policy which proposes to make China great and strong through a combination of mass hysteria and Maoist incantations can be accepted indefinitely by the scientists, technicians, managers, and skilled workers of the modern sector of the economy.

Thus, although it is impossible to predict the trials and convulsions through which the Chinese people may pass in the near future, it seems inevitable that ultimately Mao's heaven-storming policies will be replaced by others better adapted to the dull but efficient rationality of modern industrial society. This does not mean that China must necessarily follow the example of the Soviet Union, still less that those who oppose Mao today are, as Mao himself would have it, 'Khrushchevite revisionists'. The Soviet variant of Communism exercises no great intellectual fascination on anyone nowadays, and no leadership group could conceivably survive in Peking that did not take as its first objective the maintenance of political and ideological independence – though under Mao's successors relations with Moscow need not be as bad as they are now.

Whatever the precise nature of the policies adopted in China in the coming years, it is hard to see how the Mao cult in its present extravagant form can long survive his disappearance. Indeed, his successors will be virtually obliged to denounce it in order to divest themselves of the responsibility both for the terror and bloodshed engendered by the 'cultural revolution', and for the economic disasters to which these developments may be expected to lead.

When and if this occurs, Mao's revolutionary achievements will not necessarily be repudiated or lose their significance. They will simply be put in their proper historical perspective as the experience of a particular phase which cannot necessarily serve as a model for the future. It is clear that this would never satisfy Mao. He would wish to go down in history as the man who not only laid the foundations of the new order in his country, but set its course for all time. But centuries hence, though Mao, like Marx, may be out of date, he will not be forgotten. This should be enough for the ambition of any man.

Notes

1. Schram, *Political Thought of Mao*, p. 181.
2. *Peking Review*, No. 43, 1963, p. 6. Translation modified on the basis of the Chinese text in *Hung Ch'i*, No. 20, 22 October 1963. (The Chinese terms *tsao ti hen* and *hao ti hen*, which I translated, in the Hunan report, 'an awful mess' and 'very good indeed', are the same as those rendered 'terrible' and 'fine' in *Peking Review*.)
3. *Peking Review*, No. 35, 1965.
4. *Peking Review*, No. 36, 1965, p. 24.
5. Quotations from Liu Shao-ch'i's speech at the Seventh Party Congress in 1945; Carrère d'Encausse and Schram, *Le Marxisme et l'Asie*, p. 363.
6. *Peking Review*, No. 28, 1964, pp. 6–22.
7. *Peking Review*, No. 29, 1964, pp. 24–6.
8. Liu Tzu-chiu, 'Mao Tse-tung szu-hsiang shih wo-men sheng-li ti ch'i-chih', *Cheng-chih Hsüeh-hsi*, No. 19, 1959, pp. 3–4.
9. On the 'Learn from the Army' campaign, see the article of John Gittings, *The China Quarterly*, No. 18 (April-June 1964), pp. 153–9.
10. *Peking Review*, No. 22 (27 May), 1966, p. 6.
11. This has been stated repeatedly; see, for example, Lin Piao's speech of 1 October, *Peking Review*, No. 41 (7 October), 1966, p. 10.
12. *Peking Review*, No. 37 (9 September), 1966, p. 21. (Translation slightly modified on the basis of the Chinese text in *Hung Ch'i*, No. 11, p. 27.)
13. Decision of the Central Committee of the Chinese Communist Party adopted on 8 August 1966, *Peking Review*, No. 33 (12 August), 1966, pp. 9–10.
14. *Peking Review*, No. 41 (7 October), 1966, p. 4.
15. *Peking Review*, No. 43 (21 October), 1966, pp. 5, 8.
16. *Peking Revtew*, No. 42 (14 October), 1966, p. 23.
17. *Selected Works*, Vol. III, p. 227.
18. *The New Republic*, 27 February 1965, p. 23.

Index

China (*continued*)
authority of the father in, 20, 269–70; territorial losses, 23; beginnings of reformist ideas, 23–4; problems of military strength, 26; uncertain state of education, 35; ideas for its complete Westernization, 38–9; weight of authority and tradition, 39; acceleration of social and intellectual change, 44; turns towards Russia, 47; changes in political scene, 49; introduction of a vernacular language, 54–5; idea of provincial autonomy, 63; and foreign atrocities, 83; its cultural and social liberation, 93–4; role in world revolution, 148, 234, 254, 308, 321, 344; nature of its rural society, 167; need for economic and military aid from U.S., 224; full-scale civil war, 240, 241–4; economic and cultural backwardness, 257; changes in, under People's Republic, 261; climate of terror, 268; plight of twentieth-century intellectuals, 269, 337, 343; evolution of present-day policy, 277, 322–3; attitude to romanticism, 293; sees the whole world through its own experience, 313; influence on other governments, 318–19; limited capacity to give economic aid, 319; relations with America, 320–21; leadership succession problem 323–4, 349; rupture of top leadership, 334, 345; xenophobia in, 342, 346; need for integration in world community, 346–8
America, 320–21

China, Central, establishment of revolutionary government, 97; tactics of New Fourth Army against Japan, 206; friction in, 218

China, North, 219

China, North-Eastern Region, 284

Chinese Communist Party, 29n, 38, 44; leaders of, 47; preparations for its creation, 58; its beginnings, 60–63, 64n; effect of Soviet advice on, 60–61; attitude towards Marxism, 62, 64; date of creation, 63n, 65n, 66; its six small groups (1921), 64; and collaboration with 'bourgeois nationalists', 67, 69; membership, 68, 75; collaboration with Kuomintang, 70, 72, 79, 83, 92, 106, 200–202; supports Wu P'ei-fu, 72n; Mao's position in, 76, 78, 92, 93, 109, 134n, 181, 192, 216, 232; and the peasantry, 78, 79, 85, 90, 91, 92; and Sun Yat-sen, 80; organizes demonstrations of 30 May, 82n; capitalists' fear of, 84; growth in dimension, 96; management of their own affairs, 103; Chiang Kai-shek's break with, 106; break with the Kuomintang, 106–7, 113, 115, 129; attacks on, 112–13; and peasant demonstrations, 114; alleged agreement of Mao with central committee, 135; its dogmas, 135, 135n; and Red Army principles of attack, 146; and Japanese aggression, 165, 169, 172–3, 192ff, 203–4; belief in imminent victory of the revolution, 169; fellow-feeling with Russian Communists, 192; 'three-thirds' system of government, 205; anti-Japan student demonstrations, 206, 206n; its part in the revolution, 215; enormous increase in power, 216–17; end of real

*Some other books published by Penguins
are described on the following pages.*

The Political Thought of Mao Tse-tung

Stuart R. Schram

Today 'Mao Tse-tung's thought' reigns supreme in China, and its power is felt throughout the world. But do most of us know what he really does think, and has thought, in the course of a revolutionary career which now spans half a century?

In this Pelican book Professor Schram looks behind the current concept of Mao Tse-tung's thought as a series of maxims enshrined in the little red book to examine the historical genesis of Mao's adaptation of Marxism to Chinese conditions. Covering Mao's concepts of ideology, social theory, guerrilla warfare and foreign relations, he analyses the multiple roots of his thinking: Marxism-Leninism, of course, but also the Chinese classics, Western liberalism and Chinese nationalism, each modified by his own remarkable personality.

This revised and updated edition of what has become a standard work on the subject, containing a fuller introduction and over thirty new texts, carries the discussion of Mao's theories and policies down to the end of 1968. While showing how, in various contexts, Mao has adapted himself to circumstances, it also emphasizes the basic continuity of his approach to revolution, from Yenan to the conquest of power and from the 'Great Leap Forward' to the 'Great Proletarian Cultural Revolution'. Throughout all these developments the question remains posed: How far is communism an end for Mao, and how far a means to China's re-emergence as a great nation?

Not for sale in the U.S.A. or Canada

The Birth of Communist China

C. P. Fitzgerald

This Pelican, which is a fully revised edition of the author's *Revolution in China*, sets out to assess the significance of the Chinese Revolution.

After sketching in the background of China's long history and social structure C. P. Fitzgerald, who is now Professor of Far Eastern History at Canberra, opens his main account at the fall of the Manchu Emperors in 1911 and traces the origins of revolution through the early republic of Sun Yat-sen and the Nationalist dictatorship of Chiang Kai-shek to the military campaigns of Mao Tse-tung. He assesses the varying influences of Confucianism and Christianity, of East and West, and of the Japanese and Russians on this massive movement, and makes it abundantly clear that the China of today is not an inexplicable freak but a logical development of its immensely long past.

Professor Fitzgerald has a gift for fluent narrative and a long experience of China, and his interpretation of one of the central political events of this century is as readable as it is reliable.

The Immortals

I lost my proud poplar, and you your willow,
Poplar and willow soar lightly to the heaven of heavens.
Wu Kang, asked what he has to offer,
Presents them respectfully with cassia wine.

The lonely goddess in the moon spreads her ample sleeves
To dance for these faithful souls in the endless sky.
Of a sudden comes word of the tiger's defeat on earth,
And they break into tears of torrential rain.

In the spring of 1957, Li Shu-i wrote a poem in memory of her husband, Liu Chih-hsün, killed in battle in 1933, and sent it to Mao. In reply, Mao wrote this poem commemorating both Liu and his own wife, Yang K'ai-hui, executed by the Kuomintang in 1930. In 1923–7, Liu Chih-hsün had been a member of the Hunan Provincial Committee of the C.C.P. and Secretary of the Hunan Provincial Peasant Association. He had taken part in the Nanchang uprising of 1 August 1927 and fought subsequently in the Red Army until his death. For her part, Li Shu-i had been a good friend of Yang K'ai-hui.

Yang K'ai-hui's surname means 'poplar', and Liu Chih-hsün's surname 'willow'. The two characters Yang-liu taken together also mean willow; hence the second line evokes both the ascent of the souls of the two fallen comrades and the image of willow catkins floating lightly in the breeze. According to an ancient legend Wu Kang, who had committed certain crimes in his search for immortality, was condemned to cut down a cassia tree on the moon. Each time he raises his axe the tree becomes whole again, and he thus has to go on felling it for all eternity. The tiger in the seventh line refers to the Kuomintang régime, and the last couplet describes the emotion of Mao's lost companion at the final triumph of the revolution. It is not surprising that the official interpretation should find in this poem 'a large element of revolutionary romanticism'.*

* Tsang K'e-chia, 'On Mao Tse-tung's Poems', in Mao Tse-tung, *Nineteen Poems* (Peking, Foreign Languages Press, 1958), p. 60. The above translation is adapted from that given in this volume, p. 30. On 'revolutionary romanticism' as the hallmark of Mao's writings, see p. 293.